Fundamental Ideas of

ANALYSIS

Fundamental Ideas of ANALYSIS

MICHAEL C. REED

Duke University

John Wiley & Sons, Inc.

New York   Chichester   Weinheim   Brisbane   Toronto   Singapore

Cover art: **Matisse, Henri.**
Interior with a Violin Case. Nice, (winter 1918-19).
Oil on canvas, 28 3/4 x 23 5/8"(73 x 60 cm).
The Museum of Modern Art, New York. Lillie P. Bliss Collection.
Photograph ©1998 The Museum of Modern Art, New York.

©1998 Succession Henri Matisse, Paris/Artists Rights Society
(ARS), New York

MATHEMATICS EDITOR	Barbara Holland
MARKETING MANAGER	Leslie Hines
PRODUCTION EDITOR	Ken Santor
DESIGNER	Maddy Lesure
ILLUSTRATION AND COMPOSITION	John Davies

This book was typeset in Palatino by John Davies with the LATEX Documentation System and printed and bound by the Hamilton Printing Company. The cover was printed by Phoenix Color Corporation.

Library of Congress Cataloging in Publication Data
Reed, Michael C.
        Fundamental ideas of analysis / Michael C. Reed.
             p.   cm.
        Includes bibliographical references (p. 403 – 405) and index.
        ISBN 0-471-15996-4 (cloth : alk. paper)
        1. Mathematical analysis. I. Title.
QA300.R44   1998
515–dc21   97-20683
                                                                          CIP

Printed in the United States of America
10 9 8 7 6 5 4 3

for Rhonda

I prove a theorem and the house expands:
the windows jerk free to hover near the ceiling,
the ceiling floats away with a sigh.

– from the poem "Geometry", by Rita Dove
Poet Laureate of the United States, 1993-1995.

Preface

The ideas and methods of mathematics, long central to the physical sciences, now play an increasingly important role in a wide variety of disciplines. This success, in fields as diverse as biology, economics, operations research, robotics, cryptology, and finance, raises difficult questions about the undergraduate curriculum. Many mathematics majors now plan careers in other fields, and non-majors form a significant part of the student population in advanced undergraduate courses. Because of the interests of these students, and the (appropriate) use of machine computation, some courses which used to be highly theoretical (e.g., ordinary differential equations) now emphasize methods and computational experiments. There is a bewildering array of undergraduate courses, both pure and applied. What should we teach these students? Where will they learn the power and beauty of pure mathematics? How can we make clear the coherence and continuity of the central ideas of mathematics?

Because of all of these issues, the undergraduate analysis course is more important than ever. For in this course students learn that mathematics is more than just methods that work. Analysis provides theorems that *prove* that results are true and provides techniques to estimate the errors in approximate calculations. In this course, students are asked for the first time to construct long proofs and it is from the proofs that they obtain deep understanding of the ideas. Finally, the ideas and methods of analysis play a fundamental role in ordinary differential equations, probability theory, differential geometry, numerical analysis, complex analysis, and partial differential equations, as well as in most areas of applied mathematics. For all these reasons, analysis should be a central, if not *the* central, course in the undergraduate curriculum.

This analysis text makes the connections to the rest of the curriculum visible. The standard topics for a one term undergraduate real analysis course are covered in the unstarred sections of chapters 1 through 6.

But, in addition, numerous examples are given which show the ways in which real analysis is used in ordinary and partial differential equations, probability theory, numerical analysis, number theory, and so forth. In this way the importance of the technical questions becomes clear to the students and the entire course is more lively and interesting. These "connections" are developed in the starred sections of chapters 1 through 6 and in chapters 7 – 10. Throughout the text the connections are also emphasized in examples, problems, and projects. The development of the standard topics does not depend on the material in the starred sections, so the instructor can choose which starred sections to include and which to omit. Furthermore, with one or two exceptions, the starred sections and chapters are independent of each other. My intention is to provide materials which make it easy for the individual instructor to construct a lively and interesting course that is appropriate, both in content and in level of difficulty, for his or her own students. Assistance is available in a separate *Instructor's Manual* in which the starred sections and chapters, the projects, and the harder problems are thoroughly discussed.

By teaching most of the starred sections and chapters, an instructor can use this text for a two semester course. There *are* too many subjects in the undergraduate curriculum that a student "must" know, at least if each is taught as a separate undergraduate course. The solution is, I believe, to decide which subjects in analysis are central and to include them as topics in a required year-long course. My choices are evident from the table of contents: Students should know the existence theory for ordinary differential equations; they should know Cauchy's theorem and Fourier series; they should understand why probability theory is a branch of analysis; and, along the way, they should see a partial differential equation and some numerical analysis. Of course, I hope that this text will prove useful for individuals and departments whose choices would not be identical to mine.

Because of its philosophy, this text differs in a variety of ways from other texts in real analysis. The properties of the real numbers are assumed, and set theory and point set topology play only a minor role. The development of the analytical tools is lean. The point is not to prove the best possible theorem, but to show why a theorem is necessary, to prove it, and then to use it. No strong distinction is made between classical and modern analysis. Rather, the modern ideas are introduced when they arise naturally to clarify and unify explicit calculations that the students have already made. These differences are changes of emphasis rather than radical departures from the traditional texts.

Chapters 1 and 2 proceed quite slowly since many students encounter long proofs here for the first time. The level of difficulty is a little higher in the unstarred sections chapters 3 through 6 and still higher in starred sections and in chapters 7 through 10. Note, however, that some sections in the later chapters, for example Sections 7.3 and 10.2, are easy and can be taught without teaching the other material in the chapter.

Each section of the book has a large selection of problems designed to help the students understand definitions and learn to construct proofs. Each chapter ends with a few projects which can be assigned to students individually or in groups. Some topics, for example the wave equation, the Laplace transform, the zeta function, and Chebyshev's inequality, are introduced in the problems and projects. Throughout the text, and in the problems and projects, it is assumed that the students have available and can use graphing hand calculators.

The illustrations in the text were created by John Davies, Senior Systems Programmer in the Mathematics Department at Duke, who also contributed many ideas to the design of the book. He has my deep gratitude. Thanks are also due to my editor at Wiley, Barbara Holland, whose suggestions and ideas helped to shape the presentation of the material.

Many others have made valuable suggestions: Lester Coyle, Greg Lawler, Harold Layton, Laura McKinney, Lang Moore, William Pardon, Isaac Reed, Kirsten Travers. I am grateful to the reviewers: Morton Brown at the University of Michigan, Robert Cole at the Evergreen State College, Steven London at the University of Houston-Downtown, Brad Osgood at Stanford University, Daniel Robinson at the Georgia Institute of Technology, Eric Schecter at Vanderbilt University, Joe Thrash at the University of Southern Mississippi-Gulf Park, and Robert Underwood at the Colorado School of Mines. They greatly influenced the choice of material in the book. Particularly important for me was the feedback from colleagues who taught out of a preliminary edition last year: Stephen Abbott at Middlebury College, Jose Barrionuevo at University of South Alabama, Danielle Carr at Bryn Mawr College, James Donaldson at Howard University, Edward Effros at UCLA, Tony Phillips at SUNY Stonybrook, and David Schaeffer at Duke. I thank them all.

Finally, I am particularly grateful to my wife, Rhonda Hughes, whose idea it really was.

Michael Reed

Contents

CHAPTER 1

Preliminaries

This chapter has two purposes. First, we review the properties of real numbers and establish our notation for sets and functions. Second, we provide simple examples of mathematical proofs which can serve as models for students with no previous experience in proving theorems.

1.1 The Real Numbers

The reader is surely familiar with the basic arithmetic properties of addition, $x + y$, and multiplication, xy, of real numbers.

(P1) $x + y = y + x$ for all x and y.
(P2) $(x + y) + z = x + (y + z)$ for all x, y, and z.
(P3) $x + 0 = x$ for all x.
(P4) For every x there is number $-x$ so that $x + (-x) = 0$.
(P5) $xy = yx$ for all x and y.
(P6) $(xy)z = x(yz)$ for all x, y, and z.
(P7) $x \cdot 1 = x$ for all x.
(P8) For every $x \neq 0$ there is number x^{-1} so that $xx^{-1} = 1$.
(P9) $x(y + z) = xy + xz$ for all x, y, and z.

Addition and multiplication are commutative (P1 and P5) and associative (P2 and P6). Property P9 is called the distributive property. We occasionally write $x \cdot y$ instead of xy to emphasize that we are multiplying x and y. A set with operations of addition and multiplication which satisfy these nine properties is called a **field**. Thus, the set of real numbers, which we denote by $\mathbb{R}$, is a field. There are other fields. Any set with two elements can be made into a field by defining addition and multiplication appropriately (see problem 7). The complex numbers, $\mathbb{C}$, are a field that contains $\mathbb{R}$. We define $\mathbb{C}$ and study some of its properties in Section

6.5. It is convenient to think of the real numbers as the points on a line:

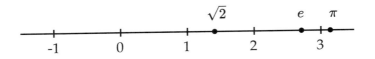

Figure 1.1.1

The real numbers have an order relation $\leq$, "less than or equal to", which has the following properties:

(O1) For all x and y, either $x \leq y$ or $y \leq x$.
(O2) If $x \leq y$ and $y \leq x$, then $x = y$.
(O3) If $x \leq y$ and $y \leq z$, then $x \leq z$.
(O4) If $x \leq y$, then for any z we have $x + z \leq y + z$.
(O5) If $x \leq y$, then for any $0 \leq z$, we have $xz \leq yz$.

We can define $\geq$ and $<$ in terms of $\leq$. We say that $x \geq y$ if and only if $y \leq x$, and we say that $x < y$ if and only if $x \leq y$ and $x \neq y$. In terms of the line in Figure 1.1.1, $x < y$ means that x is to the left of y.

All the rules for the manipulation of elementary algebraic expressions, for example, removing parentheses or the laws of exponents, can be proven from (P1) through (P9). The usual rules for the manipulation of inequalities can be derived from (P1)–(P9) and (O1)–(O5). We illustrate how this can be done in the following proposition. The reader is asked to construct other such proofs in problems 1 through 6.

Proposition 1.1.1

(a) For all real numbers x, y, and z, if $x + z = y + z$, then $x = y$.

(b) Additive inverses are unique.

(c) For all real numbers x and y, $(-x)(-y) = xy$.

(d) For all real numbers x and y, if $x \leq y$, then $-x \geq -y$.

Proof. To prove (a), let $-z$ be a real number such that $z + (-z) = 0$. We know that such a number exists by P4. Since $x + z = y + z$, we know that

$$(x + z) + (-z) \;=\; (y + z) + (-z)$$

$$\begin{aligned}
x + (z + (-z)) &= y + (z + (-z)) &&\text{(by P2)} \\
x + 0 &= y + 0 &&\text{(by P4)} \\
x &= y. &&\text{(by P3)}
\end{aligned}$$

This proves (a).

To prove (b), suppose that $z + x = 0$ and $z + y = 0$. Then $x + z = z + x = 0$ and $y + z = z + y = 0$ by P1. Thus, $x + z = 0 = y + z$, so by part (a), we conclude that $x = y$.

To prove (c), we use the fact that $z \cdot 0 = 0 = 0 \cdot z$ for all real numbers z (problem 4) and the distributive rule (P9) to conclude that

$$\begin{aligned}
xy + x(-y) &= x(y + (-y)) &&= x \cdot 0 = 0 &&&(1) \\
xy + (-x)y &= (x + (-x))y &&= 0 \cdot y = 0. &&&(2)
\end{aligned}$$

It follows from (1) and part (b) that $x(-y) = -(xy)$. Since this equality holds for all real numbers x and y, it holds if we replace x by $-x$. Thus $(-x)(-y) = -((-x)y)$. But, by (2) and (P4), $(-x)y = -(xy)$, so we can use the result of problem 3 to conclude that

$$(-x)(-y) = -((-x)y) = -(-(xy)) = xy.$$

To prove (d) we suppose that $x \le y$. Let $z = (-x) + (-y)$. Then, by (O4), $x + (-x) + (-y) \le y + (-x) + (-y)$. By the commutative and associative properties, (P1), (P2), (P5), and (P6),

$$x + (-x) + (-y) = (x + (-x)) + (-y) = 0 + (-y) = (-y) + 0 = -y.$$

Similarly,

$$y + (-x) + (-y) = (-x) + y + (-y) = (-x) + 0 = -x.$$

Thus $-y \le -x$, which is the same as $-x \ge -y$ by the definition of $\ge$. $\square$

We will normally use $\square$ to denote the end of a formal proof or the beginning of a theorem. Using $\le$, we can now define the **absolute value** of a real number:

$$|x| \equiv \begin{cases} x & \text{if } x \ge 0 \\ -x & \text{if } x < 0 \end{cases}$$

We use $\equiv$ instead of $=$ when the left-hand side is *defined* in terms of the right-hand side.

The following proposition states the most important properties of absolute value.

Proposition 1.1.2

(a) $|x| \geq 0$ for all x and $|x| = 0$ if and only if $x = 0$.

(b) $|xy| = |x||y|$ for all x and y.

(c) $|x + y| \leq |x| + |y|$ for all x and y (the triangle inequality).

Proof.   To prove (a), we first suppose that $x \geq 0$. Then, by definition, $|x| = x$ so $|x| \geq 0$. On the other hand, if $x < 0$, then $|x| = -x$, so again $|x| \geq 0$. Thus, in both cases, $|x| \geq 0$. If $x \neq 0$ then $-x \neq 0$, so the only number with absolute value zero is 0. This proves (a). The proof of (b) is left as an exercise (problem 1).

To prove (c) we argue as follows. For all x, we know that $x \leq |x|$ since $|x| = x$ if x is nonnegative and $x < 0 < -x = |x|$ if x is negative. Similarly, $y \leq |y|$. Thus, by using (O4) twice,

$$x + y \ \leq \ |x| + y \ \leq \ |x| + |y|.$$

Therefore, if $x + y \geq 0$, we have that $|x+y| = x+y \leq |x|+|y|$. If $x+y < 0$, we proceed as follows. We always know that $x \geq -|x|$ since this is clearly true if x is nonnegative and $x = -|x|$ if x is negative. Similarly, $y \geq -|y|$. Thus, again using (O4) twice,

$$x + y \ \geq \ -|x| + y \ \geq \ -|x| - |y| \ = \ -(|x| + |y|).$$

Using part (d) of Proposition 1.1.1, we see that $-(x + y) \leq |x| + |y|$. Therefore, since $|x+y| = -(x+y)$ if $x+y < 0$, we conclude that $|x+y| \leq |x| + |y|$, which completes the proof of (c).   ❏

There are two other important properties of the real numbers. The real numbers are **complete**, which means roughly that every sequence of real numbers that looks as if it is converging does indeed converge to a limit. This is discussed in Section 2.4. Second, the real numbers satisfy the **Archimedian property**, which says that if $a > 0$ and $b > 0$, then there is an integer n so that $na > b$. If we think about the real numbers, this seems obvious. However, there are ordered fields which do not have the Archimedian property (ordered fields are defined in problem 12). Throughout the discussion of completeness in Section 2.4 we assume that the Archimedian property holds for $\mathbb{R}$.

Problems

1. Prove that if x and y are real numbers, then $|xy| = |x||y|$. Hint: check all the cases.

2. Use the field properties of the real numbers to provide a careful proof of the elementary algebraic identity $(x + y)^2 = x^2 + 2xy + y^2$.

3. Prove that $-(-x) = x$ for all real numbers x. Hint: show that $(-x)+(x) = 0$ and then use part (b) of Proposition 1.1.1.

4. Prove that all real numbers z satisfy $z \cdot 0 = 0 = 0 \cdot z$. Hint: first prove that $0 + z \cdot 0 = z \cdot 0 + z \cdot 0$ and then use part (a) of Proposition 1.1.1.

5. Use part (d) of Proposition 1.1.1 to prove that if $x \leq y$ and $z \leq 0$, then $zy \leq zx$.

6. Suppose that $0 \leq x \leq y$.

    (a) Using the properties (O1)–(O5), prove that $0 \leq x^2 \leq y^2$. Is the conclusion true if we omit the hypothesis that $0 \leq x$?

    (b) Using mathematical induction, prove that $0 \leq x^n \leq y^n$ for all positive integers n. Induction is discussed in Section 1.4.

7. Let $\mathbb{Z}_2$ be a set consisting of two elements which we denote by 0 and 1. Define an operation of addition, $+$, and an operation of multiplication, $\cdot$, by the following rules:

$$0 + 0 = 0; \quad 1 + 1 = 0; \quad 1 + 0 = 1 = 0 + 1$$

$$0 \cdot 0 = 0; \quad 1 \cdot 1 = 1; \quad 1 \cdot 0 = 0 = 0 \cdot 1.$$

   Prove that $\mathbb{Z}_2$ is a field.

8. Prove that if x and y are real numbers, then $2xy \leq x^2 + y^2$.

9. Let x_1, x_2, and x_3 be real numbers. Prove that

$$|x_1 + x_2 + x_3| \leq |x_1| + |x_2| + |x_3|.$$

10. Prove that all real numbers x and y satisfy

$$\left| |x| - |y| \right| \leq |x - y|. \tag{3}$$

    Hint: apply the triangle inequality to $x = (x - y) + y$ and then reverse the roles of x and y.

11. Let a and b be real numbers with $a < b$. Prove that there are integers m and $n \neq 0$ so that

$$a < \frac{m}{n} < b.$$

    Hint: First use the Archimedian property to prove that an n exists so that $bn - an > 1$. Then argue from this that there is an integer m so that $an < m < bn$.

12. Let $\mathcal{F}$ be a field. Suppose that there is a set $P \subset \mathcal{F}$ which satisfies the following properties:

    (a) For each x in $\mathcal{F}$, exactly one of the following three statements holds: x is in P; $-x$ is in P; $x = 0$.

    (b) If x is in P and y is in P, then $x + y$ is in P and xy is in P.

If such a P exists, $\mathcal{F}$ is called an **ordered field**. For any set $P \subset \mathcal{F}$, define $x \leq y$ if and only if $(y - x)$ is in P or $x = y$. Prove that P satisfies the properties (a) and (b) if and only if $\leq$ satisfies the properties (O1)–(O5).

1.2 Sets and Functions

Sets are defined by describing their members or by the bracket notation $\{x \mid p(x)\}$, which indicates the collection of objects x so that the proposition $p(x)$ is true. Thus, if a and b are real numbers,

$$S \equiv \{x \mid a \leq x \leq b\}$$

defines the set S of real numbers that are greater than or equal to a and less than or equal to b. S is also denoted by $[a, b]$ and called a **closed** interval because it contains its endpoints.

$$T \equiv \{x \mid a < x < b\}$$

defines the set T of real numbers which are greater than a and less than b. T is also denoted by (a, b) and called an **open** interval. Occasionally, we will use **half-open** intervals $[a, b) \equiv \{x \mid a \leq x < b\}$. The half-open interval $(a, b]$ is defined analogously. Sometimes we specify a set by listing its members. For example, the **natural numbers** are the set $\mathbb{N}$, where

$$\mathbb{N} \equiv \{1, 2, 3, \ldots\}$$

and the **integers** are the set

$$\mathbb{Z} \equiv \{\ldots, -2, -1, 0, 1, 2, \ldots\}.$$

The set of real numbers that can be written in the form $\frac{m}{n}$ where m and n are integers and $n \neq 0$ are called the **rational numbers** and denoted by $\mathbb{Q}$. A real number which is not rational is called **irrational**. If x belongs to a set X, we write

$$x \; \epsilon \; X,$$

and, if it does not belong to X, we write

$$x \notin X.$$

We use the standard notation for the union, $S \cup T$, and intersection, $S \cap T$, of sets:

$$S \cup T \equiv \{x \mid x \in S \text{ or } x \in T\}$$
$$S \cap T \equiv \{x \mid x \in S \text{ and } x \in T\}.$$

We say that S is a **subset** of T if every element of S is also in T, in which case we write $S \subseteq T$. If $S \subseteq T$ and $S \neq T$, we say that S is **strictly contained** in T. We denote the set with no members by $\emptyset$ and note that it is a subset of every set X since every $x \in \emptyset$ is in X. We remark that when we talk about sets and set operations we are assuming that there is a universal set which contains all the sets we are talking about. For example, if we are talking about intervals of real numbers or rational numbers, the universal set is $\mathbb{R}$.

If S is a subset of X, the **complement** of S in X, denoted by S^c, is the set of elements of X which are not in S; that is,

$$S^c \equiv \{x \in X \mid x \notin S\}.$$

The definition of S^c depends on the set X that contains S. Thus, if $S \subseteq T \subseteq X$ and T is strictly contained in X, then the complement of S as a subset of T is different than the complement of S as a subset of X. In cases like this we will denote the complement of S in T by $T \setminus S$ and the complement of S in X by $X \setminus S$. More generally, if T and S are any subsets of X, we define $T \setminus S \equiv \{x \in X \mid x \in T \text{ and } x \notin S\}$. The reader who is unaccustomed to unions, intersections, and complements should work problems 1–4.

If S and T are sets, we define their **Cartesian product**, denoted $S \times T$, as the set of all ordered pairs where the first element of the pair comes from S and the second element of the pair comes from T:

$$S \times T \equiv \{(s, t) \mid s \in S \text{ and } t \in T\}.$$

Thus, if $S = [2, 3]$ and $T = [1, 4]$, then $[2, 3] \times [1, 4]$ is the set of ordered pairs (x, y) such that $2 \leq x \leq 3$ and $1 \leq y \leq 4$. That is, $[2, 3] \times [1, 4]$ is just the rectangle in the plane with vertices at the points $(2, 1), (3, 1), (2, 4)$, and $(3, 4)$. Since the Euclidean plane $\mathbb{R}^2$ is just the set of *all* ordered pairs (x, y) where $x \in \mathbb{R}$ and $y \in \mathbb{R}$, we see that $\mathbb{R}^2 = \mathbb{R} \times \mathbb{R}$. Note that there is a possibility of confusion because we are using the notation (a, b) to mean two different things: an open interval on the real line and a point in the plane. We will usually say "the open interval (a, b)" or "the point (a, b)" to distinguish between the two possibilities.

In pre-calculus, students are taught that a function (of one variable) is "a rule" which assigns to every real number x another real number $f(x)$ called the "value" of f at the "argument" x. Typically, such functions are given by formulas like $f(x) = 2x^2 - 1$ or $f(x) = x \sin x$. We will analyze such functions but we will also consider functions which are not given by simple explicit formulas. Further, we shall need functions whose arguments and values are in more general sets than the real numbers $\mathbb{R}$. Thus, it makes sense to be more careful about what a function is. The idea of Cartesian product allows us to say precisely what we mean by a function.

Definition. Let S and T be sets. A **function** from a set S to a set T is a subset, F, of $S \times T$ such that each $s \in S$ occurs in at most one ordered pair in F. For each pair, $(s, t) \in F$, we call t the **value** of the function at s, and if the name of the function is f we will write

$$s \xrightarrow{\ f\ } t$$

or

$$t = f(s).$$

Note the distinction between F and f. The set F is a subset of the set of ordered pairs, while $f(s)$ is the name of the second element of the ordered pair whose first element is s. The symbol f is the name of the "rule" which assigns $f(s)$ to s.

Definition. The set $\{s \mid (s, t) \in F\}$ is called the **domain** of f, and the set $\{t \mid (s, t) \in F\}$ is called the **range** of f. These sets are sometimes denoted by $Dom(f)$ and $Ran(f)$ for short.

Thus $Dom(f)$ is simply the subset of members of S which occur as the first element of an ordered pair in F. $Ran(f)$ is the subset of members of T which occur as the second element of an ordered pair in F. We say that f "is a function from S to T" to indicate that the domain of f is S and the range of f is a subset of T.

Example 1 Let f be the function from $\mathbb{R}$ into $\mathbb{R}$ that is given by the formula $f(s) = s^2 - 2$. Since $S = \mathbb{R}$ and $T = \mathbb{R}$, we know that $S \times T = \mathbb{R}^2$. The set F consists of all ordered pairs of real numbers of the form $(s, s^2 - 2)$. That is,

$$F = \{(s, s^2 - 2) \mid s \in \mathbb{R}\}.$$

Thus, the set F consists of the points in the plane that we normally refer to as the graph of f. See Figure 1.2.1(a). The requirement in the definition of function that each s occur as the first element of at most one ordered pair ensures that for each $s \in Dom(f)$ the function has exactly one value. That is, the vertical line passing through $(s, 0)$ intersects the graph of f in exactly one point. The domain of f is $\mathbb{R}$, and the range of f is the set $[-2, \infty)$. We use the symbol $[a, \infty)$ to denote the set $\{x \in \mathbb{R} \mid x \geq a\}$. Similarly, the symbol $(-\infty, b]$ denotes the set $\{x \in \mathbb{R} \mid x \leq b\}$, and (a, ∞) and $(-\infty, b)$ are defined analogously.

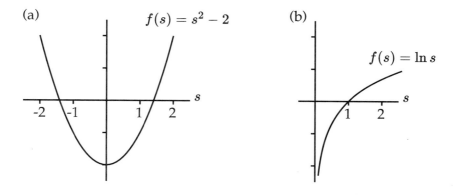

Figure 1.2.1

Example 2 Let f be the function from $\mathbb{R}$ into $\mathbb{R}$ that is given by the formula $f(s) = \ln s$ for $s > 0$. The reader is probably familiar with the natural logarithm function; it is defined formally in Example 2 of Section 4.3. Once again $S \times T = \mathbb{R}^2$. The set F, which is given by

$$F = \{(s, \ln s) \mid s \in \mathbb{R} \text{ and } s > 0\},$$

is the graph of the natural logarithm function; see Figure 1.2.1 (b). However, in this case the only members of $S = \mathbb{R}$ which occur as the first element of ordered pairs in F are positive numbers. Thus $Dom(f)$ is the interval $(0, \infty)$. On the other hand, any real number is the natural logarithm of some $s > 0$; so $Ran(f) = \mathbb{R}$.

Example 3   Let $\mathcal{M}^{(2)}$ denote the set of two-by-two matrices with real entries. Let f be the function from $\mathcal{M}^{(2)}$ to $\mathbb{R}$ that assigns to each matrix $\begin{pmatrix} a & b \\ c & d \end{pmatrix}$ its determinant $ad - bc$. Then,

$$F = \left\{ \left(\begin{pmatrix} a & b \\ c & d \end{pmatrix}, ad - bc \right) \middle| \begin{pmatrix} a & b \\ c & d \end{pmatrix} \epsilon \mathcal{M}^{(2)} \right\}$$

and

$$f\left(\begin{pmatrix} a & b \\ c & d \end{pmatrix} \right) = ad - bc.$$

In this case $Dom(f) = \mathcal{M}^{(2)}$ and $Ran(f) = \mathbb{R}$, since, given any real number q, it is easy to choose $a, b, c,$ and d so that $ad - bc = q$.

Example 4   For each point (a, b) in $\mathbb{R}^2$, let $f(a, b)$ be the two-by-two matrix given by

$$f(a, b) = \begin{pmatrix} a & b \\ b & a \end{pmatrix}.$$

Then f is a function from $\mathbb{R}^2$ to $\mathcal{M}^{(2)}$. The domain of f is all of $\mathbb{R}^2$, and the range of f is the subset of $\mathcal{M}^{(2)}$ consisting of all two-by-two matrices which are symmetric.

Example 5   We denote by $C[0, 1]$ the set of real-valued functions on the interval $[0, 1]$ that are continuous. We will give a technical definition of "continuous" in Chapter 3. For the moment, just assume that saying that a function g is continuous means that the graph of g doesn't have any jumps in it. We will see in Chapter 3 that one can integrate any continuous function on a finite interval. We define a function I on $C[0, 1]$ by

$$I(g) \equiv \int_0^1 g(x)\, dx.$$

I is a function from $C[0, 1]$ to $\mathbb{R}$. The set F is

$$F = \{(g, I(g)) \mid g \, \epsilon \, C[0, 1]\}.$$

The domain of I is $C[0, 1]$ and the range of I is $\mathbb{R}$ since, given any real number a, we can easily find a function g so that $I(g) = a$.

Notice that, by the formal definition, a function from S to T is a set of ordered pairs $F \subseteq (S \times T)$ such that each $s \in S$ occurs at most once. However, following common terminology, we shall use the word "function" for the expression which gives the second element of the ordered pair in terms of the first.

Definition.   Let f be a function from a set S to a set T. If $Ran(f) = T$ we say that the function is **onto**. If for each $t \in Ran(f)$ there is only one $s \in S$ so that $f(s) = t$, then we say that f is **one-to-one**. Other texts sometimes use "surjective" instead of "onto" and "injective" instead of "one-to-one".

The function $f(x) = x^2 - 2$ in Example 1 is not onto since its range is $[-2, \infty)$, and it is not one-to-one since for every $x \in \mathbb{R}$, $f(x) = f(-x)$. The function $f(x) = \ln x$ in Example 2 certainly looks like it is onto since it appears that for every real number t there is an $s > 0$ so that $\ln s = t$. Furthermore, it appears to be strictly increasing; that is, if $s_1 < s_2$, then $\ln s_1 < \ln s_2$. Thus, it should be one-to-one. Once we give the formal definition of the natural logarithm, these facts will be easy to prove. The function f in Example 3 is certainly onto since every real number can be the determinant of a two-by-two matrix. It is not one-to-one since two different matrices can have the same determinant. The function f in Example 4 is one-to-one because if the two pairs (a_1, b_1) and (a_2, b_2) are different, then the matrices $f(a_1, b_1)$ and $f(a_2, b_2)$ are different. However, f is not onto $\mathcal{M}^{(2)}$ since every point in $Ran(f)$ is a symmetric matrix, but not all two-by-two matrices are symmetric. Finally, the function I in Example 5 is onto since its range is $\mathbb{R}$, but it is not one-to-one since two different functions, g_1 and g_2, can have the same integral, $I(g_1) = I(g_2)$.

The significance of the one-to-one property is that it allows us to define inverse functions. Suppose that f is a one-to-one function from a set S to a set T. We define f^{-1}, called the **inverse function** of f, to be the function from T to S, with domain $Ran(f)$, so that for each $t \in Ran(f)$, the value of f^{-1} at t is the unique element $s \in S$ such that $f(s) = t$. See Figure 1.2.2. The functions f and f^{-1} are related as follows. For each $t \in Ran(f)$, we have

$$f(f^{-1}(t)) = t,$$

and for each $s \in S$, we have

$$f^{-1}(f(s)) \;=\; s.$$

It is clear that f^{-1} is one-to-one and that the inverse function of f^{-1} itself is the original function f. We will see later that the inverse function of the natural logarithm is the exponential function.

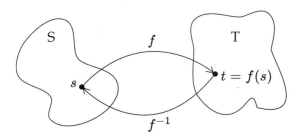

Figure 1.2.2

Many of the functions which we will consider in this book are functions from $\mathbb{R}$ to $\mathbb{R}$. In this case there are several natural ways to make new functions from two given functions f and g. We define the **sum** of the functions, $f + g$, by

$$(f + g)(x) \;\equiv\; f(x) + g(x)$$

on the domain

$$Dom(f + g) \;\equiv\; Dom(f) \cap Dom(g).$$

We define the **product** of the functions by

$$fg(x) \;\equiv\; f(x)g(x)$$

on the same domain. Finally, we define the **composition** of the two functions f and g, denoted by $f \circ g$, by

$$(f \circ g)(x) \;\equiv\; f(g(x))$$

on the domain

$$Dom(f \circ g) \;\equiv\; \{x \in \mathbb{R} \mid x \in Dom(g) \text{ and } g(x) \in Dom(f)\}.$$

Thus, to compute the value of $f \circ g$ at x, we first compute the number $g(x)$ and then compute $f(g(x))$. The reason for the complicated expression for the domain of $f \circ g$ is that x must be in the domain of g and $g(x)$ must be in the domain of f. Note that $f \circ g$ and $g \circ f$ are not normally the same function.

Example 6   Let f be the function $f(x) = \frac{1}{x}$ with domain $\{x \in \mathbb{R} \mid x \neq 0\}$ and let g be the function $g(x) = \sin x$. The domain of g is $\mathbb{R}$, so the domain of $f \circ g$ is the set of x so that $g(x) \neq 0$ since 0 is not in $Dom(f)$. Thus,

$$Dom(f \circ g) \;=\; \mathbb{R} \backslash \{0, \pm\pi, \pm 2\pi, \ldots\}$$

and

$$(f \circ g)(x) \;=\; \frac{1}{\sin x}.$$

On the other hand, $\sin x$ is defined everywhere, so

$$Dom(g \circ f) \;=\; \{x \in \mathbb{R} \mid x \neq 0\}$$

and

$$(g \circ f)(x) \;=\; \sin \frac{1}{x}.$$

Thus, $f \circ g$ and $g \circ f$ are not the same function.

Problems

1. Let S be the open interval $(1, 2)$ and let T be the closed interval $[-2, 2]$. Describe the following sets:

    (a) $S \cup T$.

    (b) $S \cap T$.

    (c) $\mathbb{R} \backslash S$.

    (d) $T \backslash S$.

    (e) $\mathbb{R} \backslash (T \backslash S)$.

2. Describe the following sets of real numbers:

    (a) $\{x \in \mathbb{R} \mid x^2 > 2\}$.

    (b) $\{x \in \mathbb{R} \mid |x| \leq 3\}$.

    (c) $\{x \in \mathbb{R} \mid |x - 2| \leq 3\}$.

    (d) $\{x \in \mathbb{Q} \mid |x| \leq 1\}$.

3. If A, B, and C are sets, prove that $A \cap (B \cup C) = (A \cap B) \cup (A \cap C)$.

4. If A, B, and C are sets, prove that $A \backslash (B \cup C) = (A \backslash B) \cap (A \backslash C)$.

5. In each case, describe the Cartesian product $S \times T$.

   (a) $S = [0, 1]$ and $T = \{x \mid x \geq 0\}$.

   (b) $S = [0, 1]$ and $T = [2, 4] \cup [5, 6]$.

   (c) $S = \mathbb{N}$ and $T = \mathbb{N}$.

   (d) $S = \mathbb{N}$ and $T = \mathbb{R}$.

6. Generate the graph of each of the following functions on $\mathbb{R}$ and use it to determine the range of the function and whether it is onto and one-to-one:

   (a) $f(x) = x^3$.

   (b) $f(x) = \sin x$.

   (c) $f(x) = e^x$.

   (d) $f(x) = \frac{1}{1+x^4}$.

7. Each of the following functions has domain $\{x \in \mathbb{R} \mid x \neq 0\}$. For each, generate its graph and use it to determine the range of the function and whether it is onto and one-to-one:

   (a) $f(x) = \frac{1}{x}$.

   (b) $f(x) = \frac{1}{x^2}$.

   (c) $f(x) = \sin \frac{1}{x}$.

   (d) $f(x) = \ln |x|$.

8. Let $\mathcal{P}$ be the set of polynomials of one real variable. If $p(x)$ is such a polynomial, define $I(p)$ to be the function whose value at x is

$$I(p)(x) \equiv \int_0^x p(t)\, dt.$$

   Explain why I is a function from $\mathcal{P}$ to $\mathcal{P}$ and determine whether it is one-to-one and onto.

9. In each case, determine the range of $f + g$ and fg and say whether the function is onto and whether it is one-to-one.

   (a) $f(x) = x$ and $g(x) = x$.

   (b) $f(x) = x$ and $g(x) = \frac{1}{2}\sin x$.

   (c) $f(x) = \sin^2 x$ and $g(x) = \cos^2 x$.

10. In each case, determine the domain and compute a formula for $f \circ g$ and $g \circ f$:

(a) $f(x) = 3x^2 + 2$ and $g(x) = e^x$.

(b) $f(x) = 4 - x^2$ and $g(x) = \ln x$.

(c) $f(x) = \frac{1}{x}$ and $g(x) = \ln x$.

1.3 Cardinality

Two sets, S and T, are said to have the same number of elements, or the same **cardinality**, if there is a one-to-one function, f, from S onto T, that is $Dom(f) = S$ and $Ran(f) = T$. This definition is reasonable because if S and T have the same cardinality, then the function f gives a way of matching up each element of S with a corresponding element of T. We say that f establishes a **one-to-one correspondence** between S and T.

Proposition 1.3.1 Let S, T, and U be sets. If S and T have the same cardinality and T and U have the same cardinality, then S and U have the same cardinality.

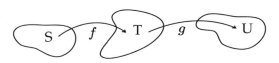

Figure 1.3.1

Proof.   Since S and T have the same cardinality, there is a one-to-one function, f, from S onto T. On the other hand, since T and U have the same cardinality, there is a one-to-one function, g, from T onto U. Then, $g \circ f$ is a function from S to T with domain S and range U. If $s_1 \neq s_2$, then $f(s_1) \neq f(s_2)$ since f is one-to-one. Thus, since g is one-to-one, $g(f(s_1)) \neq g(f(s_2))$. Therefore, $g \circ f$ is one-to-one, which proves, by definition, that S and U have the same cardinality.   ❑

A set is called **finite** if it is empty or if it has the cardinality of the set $\{1, 2, 3, \ldots, n\}$ for some n. Otherwise it is called **infinite**. This definition just states formally the way in which we usually determine the size of finite sets. We count the elements! That is, we make a one-to-one correspondence between the elements of the set and the integers, 1, 2, $\ldots$ until we run out of set elements. Of course it is clear what we mean by the size of a finite set. The purpose of the definition is to say what we mean when sets are infinite. One of the interesting aspects of the set theory that was developed in the 19th century was the discovery that there are infinite sets of different sizes. An infinite set is called **countable** if it has the cardinality of the natural numbers $\mathbb{N}$.

Example 1 Consider the set of integers $\mathbb{Z} = \{\ldots, -2, -1, 0, 1, 2, \ldots\}$. To show that $\mathbb{Z}$ is countable, we must construct a one-to-one function from $\mathbb{Z}$ to $\mathbb{N}$ with domain $\mathbb{Z}$ and range $\mathbb{N}$. It is easy to check that

$$f(n) = \begin{cases} 2n & \text{if } n \geq 1 \\ 1 - 2n & \text{if } n \leq 0 \end{cases}$$

has these properties. Notice that this means that, according to our definition of cardinality, $\mathbb{Z}$ and $\mathbb{N}$ have the same cardinality even though $\mathbb{N}$ is strictly contained in $\mathbb{Z}$.

Proposition 1.3.2   If S is an infinite subset of a countable set T, then S is countable.

Proof.   Since T is countable, there is a one-to-one correspondence, f, from $\mathbb{N}$ onto T. The elements of S are a subset of $T = \{f(n) \,|\, n \in \mathbb{N}\}$. Let n_1 be the smallest integer in $\mathbb{N}$ so that $f(n_1) \in S$. Then, let n_2 be the smallest integer bigger than n_1 such that $f(n_2) \in S$. Continuing in this manner we define n_k to be the smallest integer bigger than n_{k-1} such that $f(n_k) \in S$. It is easy to check that the function g defined by

$$k \xrightarrow{\;g\;} f(n_k)$$

is a one-to-one function from $\mathbb{N}$ onto S. Therefore, S is countable.   ❏

In some cardinality proofs it is convenient to use the fundamental theorem of arithmetic, which, for completeness, we shall state formally. Recall that the prime numbers are the set, $\mathbb{P}$, of positive integers that have no divisors besides themselves and the number 1.

❏ **Theorem 1.3.3 (The Fundamental Theorem of Arithmetic)** Every positive integer $N \geq 2$ can be written uniquely as a finite product of positive integral powers of primes:

$$N = p_1^{s_1} p_2^{s_2} \ldots p_n^{s_n}.$$

A discussion and proof of the fundamental theorem of arithmetic can be found in most undergraduate texts on number theory; see, for example,

[33]. We use it in the following proposition. For other uses, see Proposition 1.4.2 in the next section or Section 6.6.

Proposition 1.3.4   If S and T are countable sets, then $S \times T$ is countable.

Proof.   Since S and T are countable, there exist one-to-one functions, f and g, with domains equal to $\mathbb{N}$ and ranges equal to S and T, respectively. Each element of $S \times T$ is of the form $(f(n), g(m))$ for some $n \in \mathbb{N}$ and $m \in \mathbb{N}$. Let h be the function from $S \times T$ to $\mathbb{N}$ defined by

$$(f(n), g(m)) \xrightarrow{\ h\ } 2^n 3^m.$$

By the fundamental theorem of arithmetic, h gives a one-to-one correspondence between $S \times T$ and an infinite subset of $\mathbb{N}$. According to Proposition 1.3.2, this infinite subset is countable. Thus, by Proposition 1.3.1, we conclude that $S \times T$ is countable.   ❑

Example 2   The set of pairs of positive integers is countable since it is equal to $\mathbb{N} \times \mathbb{N}$.

We can combine these propositions to show that $\mathbb{Q}$ is countable.

❑ **Theorem 1.3.5**   The set of rational numbers, $\mathbb{Q}$, is countable.

Proof.   Every positive rational number can be written in the form $\frac{m}{n}$, where m and n are integers, $n \neq 0$, and the fraction $\frac{m}{n}$ is in lowest terms (that is, n and m have no common factor). The function f, defined by

$$\frac{m}{n} \xrightarrow{\ f\ } (m, n),$$

gives a one-to-one correspondence between $\mathbb{Q}_+$, the set of positive rationals, and an infinite subset of $\mathbb{N} \times \mathbb{N}$. Since $\mathbb{N} \times \mathbb{N}$ is countable, Propositions 1.3.1 and 1.3.2 guarantee that $\mathbb{Q}_+$ is countable. Since $\mathbb{Q}$ can be written as $\mathbb{Q}_- \cup \{0\} \cup \mathbb{Q}_+$, parts (b) and (c) of problem 3 guarantee that $\mathbb{Q}$ is countable.   ❑

So far, all the infinite sets that we have considered have been countable, and, therefore, the following theorem may come as a surprise.

❑ **Theorem 1.3.6** The set of real numbers between 0 and 1, that is $(0, 1)$, is not countable.

Proof. In the proof we will use several facts about the decimal expansion of real numbers. Decimal expansions are discussed in project 4 of Chapter 6. Every real number x between 0 and 1 has a decimal expansion $x = 0.x_1 x_2 x_3 \ldots$ where each x_i is an integer between 0 and 9. Each decimal expansion corresponds to a different number, except for expansions ending in 0's, which correspond to numbers that can also be represented by expansions ending in 9's. For example, $\frac{1}{2}$ can be written as $.5000\ldots$ or as $.4999\ldots$.

The proof is by contradiction. Suppose that $(0, 1)$ were countable. Then there would be a one-to-one function f from $\mathbb{N}$ to $(0, 1)$ with domain $\mathbb{N}$ and range $(0, 1)$. Denote by $x_j^{(n)}$ the jth integer in the decimal expansion of $f(n)$. That is,

$$
\begin{aligned}
f(1) &= 0.x_1^{(1)} x_2^{(1)} x_3^{(1)} \ldots x_j^{(1)} \ldots \\
f(2) &= 0.x_1^{(2)} x_2^{(2)} x_3^{(2)} \ldots x_j^{(2)} \ldots \\
f(3) &= 0.x_1^{(3)} x_2^{(3)} x_3^{(3)} \ldots x_j^{(3)} \ldots \\
&\vdots \\
f(n) &= 0.x_1^{(n)} x_2^{(n)} x_3^{(n)} \ldots x_j^{(n)} \ldots
\end{aligned}
$$

We now choose a sequence of integers $y_1, y_2, y_3, \ldots$ as follows. Choose y_1 to be any integer between 2 and 8 which is not equal to $x_1^{(1)}$. Choose y_2 to be any integer between 2 and 8 which is not equal to $x_2^{(2)}$. Continuing in this manner we choose y_n to be any integer between 2 and 8 which is not equal to $x_n^{(n)}$. The decimal expansion

$$y \equiv .y_1 y_2 y_3 \ldots y_n \ldots$$

corresponds to a unique real number between 0 and 1 since it doesn't end in 0's or 9's. However, y cannot equal $f(1)$ since the decimal expansion for y differs from the decimal expansion for $f(1)$ in the first place. On the other hand, y cannot equal $f(2)$ since the decimal expansion for y differs from the decimal expansion for $f(2)$ in the second place. Continuing, we see that y cannot equal $f(n)$ since the decimal expansion for y differs from the decimal expansion for $f(n)$ in the nth place. That is, y does not equal any of the numbers $f(n)$ for any n, so y is not in the range of f. But this is a contradiction since f was assumed to have range $(0, 1)$. Therefore, $(0, 1)$ is not countable. ❑

An infinite set that is not countable is called
uncountable. Since $\mathbb{R}$ contains the interval
$(0, 1)$, $\mathbb{R}$ is also uncountable. In fact, $\mathbb{R}$ can be
put into one-to-one correspondence with $(0, 1)$.
Consider the function $f(x) = \tan x$ on the in-
terval $(-\frac{\pi}{2}, \frac{\pi}{2})$. The graph of f, shown in Figure 1.3.2, shows that f is one-to-one and its
range is $\mathbb{R}$. Thus, f is a one-to-one correspon-
dence between $(-\frac{\pi}{2}, \frac{\pi}{2})$ and $\mathbb{R}$. Since the function $g(x) = \pi x - \frac{\pi}{2}$ gives a one-to-one correspondence between the interval $(0, 1)$ and the
interval $(-\frac{\pi}{2}, \frac{\pi}{2})$, the composition $f \circ g$ gives a
one-to-one correspondence between $(0, 1)$ and $\mathbb{R}$.

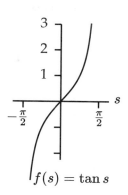

Figure 1.3.2

The reader is asked to show in problem 2 that the union of two countable sets is countable. Since $\mathbb{R}$ is the union of the rational numbers and
the irrational numbers, and the rational numbers are countable, the irrational numbers must be uncountable. In this sense, there are a lot more
irrational numbers than rational numbers.

Most of the ideas in this section were introduced by Georg Cantor
$(1845 - 1918)$.

Problems

1. Prove that the set $S \equiv \{5, 10, 15, 20, \ldots\}$ is countable by constructing a
    one-to-one function from S onto $\mathbb{N}$.

2. Prove that the set of real numbers of the form e^n, $n = 0, \pm 1, \pm 2, \ldots$ is
    countable.

3. Prove that:

    (a) The union of two finite sets is finite.

    (b) The union of a finite set and a countable set is countable.

    (c) The union of two countable sets is countable.

4. Using Propositions 1.3.2 and 1.3.4 we proved that $\mathbb{N} \times \mathbb{N}$ is countable (Ex-
    ample 2). Construct a function which gives the correspondence explicity.
    Hint: look at the pairs (n, m) in the plane and figure out how to count
    them!

5. Prove that the set of two-by-two matrices with rational entries is count-
    able.

6. Prove that the set of two-by-two matrices with real entries is uncountable.

7. Find an explicit one-to-one correspondence between the interval $(-1, 7)$ and the real numbers, $\mathbb{R}$.

8. Suppose that for each natural number n, the set A_n is countable. Denote the union of the sets A_n by $A \equiv \cup_n A_n$. Prove that A is countable.

9. For each of the following sets, say whether it is finite, countable, or uncountable:

   (a) The set of functions from a finite set to a finite set.

   (b) The set of functions from a finite set to a countable set.

   (c) The set of functions from a countable set to finite set with two or more elements.

   (d) The set of all finite subsets of the integers. Hint: prove that for each n the set of finite subsets of size n is countable.

1.4 Methods of Proof

Learning how to construct proofs is an important part of a first course in analysis. Since each problem and each proof is different, there is no "method" to follow that will yield a correct proof. The best way to begin is to read and reread the proofs in the text, following the logic line by line. Next, the easy problems should be attempted (usually the first few in a problem set). Begin by asking why the result is true. How is the result related to other results that you know? What information do you already have about the objects in question? Your instructor will give you feedback on the correctness of your logic and the style of your proofs. Often a result can be proven in different ways, so you will benefit by comparing your proofs with those of other students. There is no substitute for thought and hard work. But everyone, including a professional mathematician, gets stuck sometimes, so you should not hestitate to ask for help. Above all, you should not be discouraged. Constructing correct and elegant mathematical proofs is difficult. Learning how to do it is a major intellectual accomplishment.

A typical theorem statement consists of a set, P, of declarative sentences, called the "hypotheses" or the "assumptions," and a set, Q, of declarative sentences called the conclusions. The statement of the theorem is that if all the sentences in P are true, then it follows that the sentences in Q are true. That is, P implies Q. Propositions 1.3.1, 1.3.2, and 1.3.4 have exactly this form. Sometimes a theorem statement has only conclusions, Q. This is the case for Theorems 1.3.5 and 1.3.6. This is done for brevity when it is assumed that the reader knows what the assumptions, P, are. For example, for Theorem 1.3.5, the unwritten assumptions are: (1) "The real numbers satisfy the properties P1 through P9 of Section 1.1"; (2) "$\mathbb{Q}$ is the set of real numbers of the form $\frac{m}{n}$." The first sentence could be listed as an assumption for every theorem in this book, but that would be annoying, so the assumption is unwritten but implicit in the theorem statement. The second statement is just the definition of $\mathbb{Q}$. Another example of unwritten hypotheses is Proposition 1.1.2, which seems to have only conclusions. It is very important to remember that a theorem which asserts that P implies Q does not assert that either P or Q is true. It asserts only that *if* P is true, then Q is also true.

Just because P implies Q does *not* mean that Q implies P which is called the **converse** statement. For example, consider the set of students at a certain college. Let P be the statement "x is male" and Q be the statement "x is blond." Suppose that P implies Q (which might be true for that particular school). That does not mean necessarily that Q implies P. In the case where P implies Q *and* Q implies P, we say that P and Q are **equivalent** or that P is true **if and only if** Q is true. If P implies Q we often say that P is "sufficient" for Q because if P is true, then Q is true. If Q implies P then we say that P is "necessary" for Q because Q can't be true without P being true. Thus, saying that P is both necessary and sufficient for Q is the same as saying that P and Q are equivalent.

"Not P" is the statement that not all the sentences in P are true. The statements "P implies Q" and "not Q implies not P" are logically equivalent. Here is how one can see this. First, suppose that P implies Q. Suppose that not Q is true. If not P is false, then P is true, which would imply that Q is true. Therefore, not P must be true. Thus, not Q implies not P. On the other hand, suppose that not Q implies not P. Suppose P is true. Then Q must be true because otherwise not P would be true since not Q implies not P. The fact that "P implies Q" and "not Q implies not P" are equivalent is useful because it means that we have two ways of proving that P implies Q. We can take P as true and prove that the statements Q are true. This is called a **direct proof**. Or we can take the statement not Q as our hypothesis and try to prove that the statement not P is true. This is called a **contrapositive** proof. We give both a direct and a contrapositive proof for the following simple proposition.

Proposition 1.4.1   Suppose that $x^2 - x > 0$ and $x > 0$. Then $x > 1$.

Direct Proof. Suppose that $x^2 - x > 0$ and $x > 0$. Since $x(x - 1) \neq 0$, neither x nor $x - 1$ can be zero. Since their product is positive they are either both positive or both negative. Thus, since x is positive, $x - 1$ must be positive. That is, $x - 1 > 0$, which implies $x > 1$.   ❏

Contrapositive Proof. Suppose not Q; that is, $x \leq 1$. We want to show not P, that is that $x^2 - x > 0$ and $x > 0$ cannot both be true. If $x > 0$, then, multiplying $x \leq 1$ by x, we obtain $x^2 \leq x$, which means that $x^2 - x \leq 0$. On the other hand, suppose $x(x - 1) = x^2 - x > 0$; then, since $x - 1 \leq 0$, we must have $x < 0$ since the product is positive. Thus, $x^2 - x > 0$ and $x > 0$ cannot both be true.   ❏

Another common method of proof is **proof by contradiction**. To prove that P implies Q one assumes that P is true and that Q is not true and then shows that this leads to a contradiciton. The proof of Theorem 1.3.6 used proof by contradiction. Here is an example of this method, created by Euclid more than 2000 years ago. The statement is another example of omitted hypotheses. The unstated hypothses are, "The real numbers satisfy (P1)–(P9)" and "there exists a real number, called $\sqrt{2}$, whose square is 2".

Proposition 1.4.2 $\sqrt{2}$ is irrational.

Proof.   If $\sqrt{2}$ is rational then we can write $\sqrt{2}$ as a ratio of integers in lowest terms, $\sqrt{2} = \frac{m}{n}$. Lowest terms means that the numerator and denominator have no common factor. Thus,

$$\sqrt{2}n = m.$$

Squaring both sides yields

$$2n^2 = m^2 \tag{4}$$

which shows that 2 is a factor of m^2. By the fundamental theorem of arithmetic, the factors of m^2 are just the factors of m each taken twice. Thus 2 is a factor of m, which means that we can write $m = 2k$ for some integer k. Substituting in (4) yields

$$2n^2 = (2k)^2 = 4k^2$$

or

$$n^2 = 2k^2.$$

It follows that 2 is a factor of n^2 and therefore, by the same reasoning as above, 2 is a factor of n. But if 2 is a factor of both m and n, then the fraction $\frac{m}{n}$ is not in lowest terms, which is a contradiction. ❏

Another method of proof is **proof by induction**. Suppose that for each positive integer n, $Q(n)$ is a statement. Suppose that we know that $Q(1)$ is true. Then we show for each k that *if* $Q(k)$ is true, then $Q(k+1)$ is true. The statement that $Q(k)$ is true is called the induction hypothesis. The statement that if $Q(k)$ is true then $Q(k+1)$ is true is called the induction step. Since we know $Q(1)$ is true, we conclude from the induction step that $Q(2)$ is true. Since $Q(2)$ is true we conclude, again from the induction step, that $Q(3)$ is true. Continuing in this manner we see that we have proved that $Q(n)$ is true for each n. Thus, a proof by induction consists of verifying $Q(1)$ and proving the induction step. Here is a simple example from number theory.

Proposition 1.4.3 If n is a positive integer, then

$$1 + 2 + \cdots + n = \frac{n(n+1)}{2}. \tag{5}$$

Proof. $Q(n)$ is statement (5). $Q(1)$ is certainly true since $1 = (1) \cdot (2)/2$. Now, suppose that $Q(k)$ is true for some k. That is,

$$1 + 2 + \cdots + k = \frac{k(k+1)}{2}.$$

Then, by this induction hypothesis,

$$
\begin{aligned}
1 + 2 + \cdots + k + (k+1) &= \frac{k(k+1)}{2} + (k+1) \\
&= (k+1)(\tfrac{k}{2} + 1) \\
&= \frac{(k+1)(k+2)}{2} \\
&= \frac{(k+1)((k+1)+1)}{2}.
\end{aligned}
$$

We have shown that if $Q(k)$ is true, then $Q(k+1)$ is true. Therefore, by induction, $Q(n)$ is true for all n. ❏

The following example shows another simple use of induction.

Example 1   Suppose that we know that if f and g are differentiable functions, then $f + g$ is differentiable and $(f + g)' = f' + g'$. We would like to extend this to finite sums. Thus, $Q(n)$ is the following statement: If $f_1, f_2, \ldots, f_n$ are differentiable functions, then $f_1 + f_2 + \ldots + f_n$ is differentiable and $(f_1 + f_2 + \ldots + f_n)' = f_1' + f_2' + \ldots + f_n'$. We are assuming that we already know that this is true for $n = 2$. So, suppose that $Q(k)$ is true for some $k \geq 2$. We want to prove $Q(k + 1)$. Let $f_1, f_2, \ldots, f_{k+1}$ be differentiable functions and write

$$f_1 + f_2 + \ldots + f_{k+1} = (f_1 + f_2 + \ldots + f_k) + f_{k+1}.$$

The first term on the right is differentiable by the induction hypothesis and the second term is differentiable by assumption. Therefore, since $Q(2)$ is true, $(f_1 + f_2 + \ldots + f_{k+1})$ is differentiable and

$$\begin{aligned}
(f_1 + f_2 + \ldots + f_{k+1})' &= (f_1 + f_2 + \ldots + f_k)' + f_{k+1}' \\
&= f_1' + f_2' + \ldots + f_k' + f_{k+1}'
\end{aligned}$$

by the induction hypothesis.

Problems

1. Give five examples which show that P implies Q does not necessarily mean that Q implies P.

2. Proposition 1.1.2 seems to consist only of conclusions. What are the unwritten hypotheses?

3. Suppose that $a, b, c,$ and d are positive numbers such that $a/b < c/d$. Prove that
$$\frac{a}{b} < \frac{a + c}{b + d} < \frac{c}{d}.$$

4. Suppose that $0 < a < b$. Prove that:
    (a) $a < \sqrt{ab} < b$.
    (b) $\sqrt{ab} \leq \frac{1}{2}(a + b)$.

5. Suppose that x and y satisfy $\frac{x}{2} + \frac{y}{3} = 1$. Prove that $x^2 + y^2 > 1$. Hint: try a contrapositive proof.

6. Prove that if n is a positive integer, then $n^3 + 5n$ is divisible by 6.

7. Prove that for any n real numbers, $x_1, x_2, \ldots, x_n$,
$$|x_1 + x_2 + \cdots + x_n| \leq |x_1| + |x_2| + \cdots + |x_n|.$$

8. Prove that for all positive integers n,

$$1^3 + 2^3 + \cdots + n^3 = (1 + 2 + 3 + \cdots + n)^2.$$

9. Let $x > -1$ and n be a positive integer. Prove Bernoulli's inequality:

$$(1 + x)^n \geq 1 + nx.$$

10. In order to disprove the implication that P implies Q, one often provides an example in which P is true but Q is not. Such an example is called a **counterexample** to the statement that P implies Q. For each of the following *incorrect* statements, identify P, identify Q, and provide a counterexample.

    (a) If an integer is divisible by 2, then it is divisible by 4.

    (b) All quadratic polynomials have two real roots.

    (c) If a function f from $\mathbb{R}$ to $\mathbb{R}$ is one-to-one, then the function f^2 is one-to-one.

    (d) If a function f from $\mathbb{R}$ to $\mathbb{R}$ is one-to-one and bounded, then f^{-1} is one-to-one and bounded.

11. Suppose that $c < d$.

    (a) Prove that there is a $q \in \mathbb{Q}$ so that $|q - \sqrt{2}| < d - c$. Hint: use problem 11 of Section 1.1.

    (b) Prove that $q - \sqrt{2}$ is irrational.

    (c) Prove that there is an irrational number between c and d.

12. Prove that e is irrational by supposing that $e = \frac{m}{n}$ and deriving a contradiction. Use the fact that $e = \sum_{j=0}^{\infty} \frac{1}{j!}$. Let s_k be the partial sum $s_k = \sum_{j=0}^{k} \frac{1}{j!}$.

    (a) Prove that

$$e - s_k < \frac{1}{(k+1)!} \left\{ 1 + \frac{1}{k+1} + \left(\frac{1}{k+1} \right)^2 + \cdots \right\}.$$

    (b) Prove that $e - s_k < \frac{1}{k(k!)}$ for all positive integers k.

    (c) If $e = \frac{m}{n}$, prove that $n!e$ and $n!s_n$ are integers.

    (d) If $e = \frac{m}{n}$, prove that $n!(e - s_n)$ is an integer between 0 and 1, which is absurd.

CHAPTER 2

Sequences

This chapter is devoted to studying sequences of real numbers and their limits. Most of the important concepts of analysis are defined by limiting operations. For example, the derivative is the limit of the difference quotient and the integral is the limit of Riemann sums. Furthermore, many proofs in analysis and applied mathematics involve "taking the limit" of a sequence or series of elementary functions that get close to the function one is trying to find. These limiting operations are simplest to study in the case of sequences of real numbers, so we begin there. Furthermore, there are many beautiful questions about sequences which make them interesting to study in their own right. Finally, sequences occur frequently in other branches of mathematics and in applications; see Section 2.3, Section 2.7, the problems, and the projects.

2.1 Convergence

Intuitively, a sequence is simply an infinite list of numbers, containing a first number, a second, and so forth:

$$2, \ 4, \ 6, \ 8, \ 10, \ 12, \ldots$$
$$1, \ -1, \ 1, \ -1, \ 1, \ -1, \ldots$$
$$1, \ 1\tfrac{1}{2}, \ 1\tfrac{3}{4}, \ 1\tfrac{7}{8}, \ 1\tfrac{15}{16}, \ldots$$

More formally, a **sequence** is a function from the natural numbers, $\mathbb{N}$, to the real numbers, $\mathbb{R}$. We usually give names to the values of the function, for example, by letting a_1 denote the value of the function at 1, a_2 denote the value of the function at 2, and so forth. We represent the entire sequence by $\{a_n\}_{n=1}^{\infty}$, where n is the index which runs from 1 to ∞, indicating that the successive terms of the sequence are $a_1, a_2, a_3, \ldots$.

Often, a sequence is specified by giving an explicit formula for the terms that shows how they depend on n. For example, the first sequence above is given by the formula $a_n = 2n$, the second sequence by the formula $a_n = (-1)^{n+1}$, and the third by $a_n = 2 - 2^{-n+1}$. However, sometimes a sequence is specified by giving an algorithm for computing the terms. For example, if

$$a_1 = 2$$

and we are told that

$$a_{n+1} = 6a_n$$

for all n, then it is easy to see that $a_2 = (6)(2)$, $a_3 = (6)^2(2)$, and so forth. Thus, the formula for the nth term of the sequence is $a_n = 2(6^{n-1})$. When we specify a sequence, the name of the index doesn't matter, so $\{a_n\}_{n=1}^{\infty}$ and $\{a_k\}_{k=1}^{\infty}$ specify the same sequence. Sometimes, a sequence is given where the index n starts at some integer other than 1. In such a case, one can define a new index for the same sequence that starts at 1. For example, if the sequence $\{a_n\}_{n=2}^{\infty}$ is given, we can set $k = n - 1$ and observe that $\{a_{k+1}\}_{k=1}^{\infty}$ specifies the same sequence. When the index runs from 1 to ∞, we will often drop the explicit statement of the range of the index and write simply $\{a_n\}$ instead of $\{a_n\}_{n=1}^{\infty}$.

The idea of a "limit" of a sequence is very simple. We say that a sequence $\{a_n\}$ converges to a limit a if the terms in the sequence, a_n, get closer and closer to a as n gets larger. So, the first sequence above does not converge to a limit because each term is two bigger than the previous term, so the sequence can't get closer and closer to any finite number. The second sequence doesn't converge because it keeps hopping back and forth between -1 and 1. The third squence converges to 2 because, as n gets larger, the terms get closer and closer to 2.

These ideas seem so simple and clear that it is reasonable to ask why we need a technical definition of convergence. The answer is that we will encounter lots of sequences whose convergence and limits are not obvious. Even when the sequence is given by an explicit formula, it may be difficult to see immediately whether it converges and, if so, what the limit is. This is the case with the following two sequences,

$$a_n = n \sin \frac{1}{n}$$

and

$$a_n = (1 + \frac{1}{n})^n,$$

which are often introduced in calculus. At other times, the sequence may be specified by giving a_0 and a recursive formula for the rest of the terms, for example,

$$a_{n+1} = \frac{1}{2}a_n + 2$$

or

$$a_{n+1} = 2a_n(1 - a_n).$$

Again, it is not easy to see immediately whether either sequence converges and, if so, what the limit is. Given $a_0 = .75$, for example, you can easily compute the first 10 terms of each of these sequences. They look like they are converging, and you'll even be able to make a reasonable guess for the limit. However, to *prove* that they converge we need a technical criterion for "limit" and a lot of practice in verifying the criterion in simple cases.

Definition.   We say that a sequence $\{a_n\}$ **converges** to a **limit** a if, for every $\varepsilon > 0$, there is an integer $N(\varepsilon)$ so that

$$|a_n - a| \leq \varepsilon, \qquad \text{for all } n \geq N. \qquad (1)$$

If $\{a_n\}$ converges to a, then we write

$$\lim_{n \to \infty} a_n = a \qquad \text{or} \qquad a_n \to a.$$

Here is what this definition means geometrically. Let a small number $\varepsilon > 0$ be given. Consider the interval of length 2ε, which has a at its center; see Figure 2.1.1.

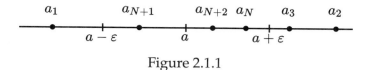

Figure 2.1.1

There must exist an N so that all the terms a_n in the sequence for $n \geq N$ lie in the interval. Note that for convergence to hold this must be true for *each* ε. That is, for each ε there must exist such an N. The size of N will normally depend on ε, smaller ε requiring larger N, which is why we wrote $N(\varepsilon)$ in the definition.

Example 1   Consider the sequence $\{a_n\}$, whose n^{th} term is given by $a_n = 1 + \frac{1}{\sqrt{n}}$. As n gets larger, $\sqrt{n}$ gets larger, so a_n should converge to 1.

Let's prove this by checking the definition of convergence. Let $\varepsilon > 0$ be given. We want to show that we can choose $N(\varepsilon)$ so that (1) holds, that is, so that

$$|a_n - 1| = \frac{1}{\sqrt{n}} \le \varepsilon$$

for all sufficiently large n. The inequality on the right is equivalent to

$$1 \le \varepsilon\sqrt{n},$$

which is equivalent to

$$\frac{1}{\varepsilon^2} \le n.$$

Choose N to be any integer greater than or equal to $\frac{1}{\varepsilon^2}$. Then, if $n \ge N$, we have

$$n \ge N \ge \frac{1}{\varepsilon^2}.$$

Therefore,

$$\frac{1}{\sqrt{n}} \le \frac{1}{\sqrt{N}} \le \varepsilon.$$

This shows that (1) holds for $n \ge N$. Since we have shown how to choose $N(\varepsilon)$ for every $\varepsilon > 0$, we have proven that $a_n \to 1$. Throughout the proof we used properties of the order relation among real numbers. For example, we used the fact that if $0 < x \le y$, then $0 < y^{-1} \le x^{-1}$.

Example 2 Suppose that $\{a_n\}$ is the sequence given by

$$a_n = \sqrt{2 + \frac{3}{n}}.$$

Since $\frac{3}{n}$ gets smaller and smaller as n gets larger, the expression under the square root sign gets closer and closer to 2. So, it is intuitively reasonable that this sequence has a limit and the limit is $\sqrt{2}$. In order to prove it, let $\varepsilon > 0$ be given. We want to show that

$$\left|\sqrt{2 + \frac{3}{n}} - \sqrt{2}\right| \le \varepsilon \tag{2}$$

for all sufficiently large n. Multiplying and dividing by $\sqrt{2 + \frac{3}{n}} + \sqrt{2}$, we find

$$\left|\sqrt{2 + \frac{3}{n}} - \sqrt{2}\right| = \left|\frac{(\sqrt{2 + \frac{3}{n}} - \sqrt{2})(\sqrt{2 + \frac{3}{n}} + \sqrt{2})}{(\sqrt{2 + \frac{3}{n}} + \sqrt{2})}\right|$$

$$= \frac{3/n}{\sqrt{2 + \frac{3}{n}} + \sqrt{2}}$$

$$\leq \frac{3/n}{2\sqrt{2}}.$$

In the last step we replaced the denominator $\sqrt{2 + \frac{3}{n}} + \sqrt{2}$ by something smaller, namely $2\sqrt{2}$, so the fraction got larger. Thus, we want

$$\frac{3/n}{2\sqrt{2}} \leq \varepsilon,$$

which is equivalent to

$$n \geq \frac{3}{2\varepsilon\sqrt{2}}.$$

Choose N to be any integer $\geq 3/2\varepsilon\sqrt{2}$. Then, if $n \geq N$, we have $n \geq 3/2\varepsilon\sqrt{2}$, which is equivalent to $3/2n\sqrt{2} \leq \varepsilon$. Therefore $n \geq N$ implies that (2) holds. We have verified the criterion (1) in the definition of limit, so we conclude that $\sqrt{2 + \frac{3}{n}} \to \sqrt{2}$.

Example 3 Suppose that $a_1 = 5$ and the rest of the members of the sequence $\{a_n\}$ are given by the recursion relation

$$a_n = ra_{n-1}$$

where $|r| < 1$. Then, $a_2 = 5r$; $a_3 = 5r^2$; and, in general, $a_n = 5r^{n-1}$. Since we are assuming $|r| < 1$, it seems reasonable that $a_n \to 0$. Let's prove it. Let $\varepsilon > 0$ be given. We want to show that for n large enough

$$|a_n - 0| = 5|r|^{n-1} \leq \varepsilon, \tag{3}$$

or equivalently,

$$\left(\frac{1}{|r|}\right)^{n-1} \geq \frac{5}{\varepsilon}.$$

If we take the natural logarithm of both sides we see that

$$(n-1)\ln\frac{1}{|r|} \geq \ln\frac{5}{\varepsilon}$$

or

$$n \geq \frac{\ln\frac{5}{\varepsilon}}{\ln\frac{1}{|r|}} + 1. \tag{4}$$

Note that $\ln \frac{1}{|r|} > 0$ since $|r| < 1$. Choose N to be any integer such that

$$N \geq \frac{\ln \frac{5}{\varepsilon}}{\ln \frac{1}{|r|}} + 1.$$

Then $n \geq N$ implies that (3) holds, since (4) and (3) are equivalent. Therefore, by the definition of limit, we conclude that $a_n \to 0$.

If a sequence $\{a_n\}$ does not converge, we say that it **diverges**. There are different ways in which a sequence can diverge. The sequence $a_n = (-1)^n$ diverges because it keeps hopping back and forth between 1 and -1. On the other hand, the sequence $a_n = n^2$ diverges because the terms get large. It is convenient to have some special terminology for this second case.

Definition. We say that a sequence $\{a_n\}$ **diverges to** ∞ if, given any number M, there is an N so that $n \geq N$ implies that $a_n \geq M$. If $\{a_n\}$ diverges to ∞, we write $a_n \to \infty$ or $\lim_{n\to\infty} a_n = \infty$. Similarly, we say that a sequence $\{a_n\}$ **diverges to** $-\infty$ if, given any number M, there is an N so that $n \geq N$ implies that $a_n \leq M$. If $\{a_n\}$ diverges to $-\infty$, we write $a_n \to -\infty$ or $\lim_{n\to\infty} a_n = -\infty$.

In terms of the number line, saying that $a_n \to \infty$ means that, given any number M, there is an N so that $n \geq N$ implies that a_n is to the right of M.

Example 4 Let's consider the sequence defined recursively in Example 3 but now with the assumption $|r| \geq 1$. We consider three cases. If $r = 1$, then $a_n = 5$ for all n, so the sequence converges to 5. If $r > 1$, then, since $a_n = 5r^{n-1}$, it is intuitively clear that a_n should diverge to $+\infty$. Let's check that the criterion in the definition is satisfied. Let M be given; we want to show that

$$5r^{n-1} \geq M \tag{5}$$

for n large enough. If $M \leq 0$, the inequality holds for all n so there is nothing more to prove. If $M > 0$, the inequality (5) is equivalent to

$$(n-1)\ln r + \ln 5 \geq \ln M$$

since the natural logarithm is an increasing function. Rearranging this inequality, we find

$$n \geq \frac{\ln M - \ln 5}{\ln r} + 1. \tag{6}$$

Thus, if we choose N to be any integer greater than the right-hand side of (6), we see that $n \geq N$ implies (5). Therefore, $5r^{n-1} \to \infty$ as $n \to \infty$ if $r > 1$.

If $r \leq -1$, then the values of $5r^{n-1}$ jump back and forth between positive numbers ≥ 1 and negative numbers ≤ -1, so the sequence does not converge, nor does it diverge to ∞ or to $-\infty$.

Example 5 Consider the sequence $a_n = 2 - \frac{1}{n}$. Each term is larger than the previous one, but the sequence does not diverge to ∞. In fact, it converges to 2. Thus, just because a sequence is increasing does not mean that it diverges to ∞.

Problems

1. Compute enough terms of the following sequences to guess what their limits are:

    (a) $a_n = n \sin \frac{1}{n}$.

    (b) $a_n = (1 + \frac{1}{n})^n$.

    (c) $a_{n+1} = \frac{1}{2} a_n + 2$, $a_1 = .5$.

    (d) $a_{n+1} = 2.5 \, a_n (1 - a_n)$, $a_1 = .3$.

2. Prove directly that each of the following sequences converges by letting $\varepsilon > 0$ be given and finding $N(\varepsilon)$ so that (1) holds.

    (a) $a_n = 1 + \frac{10}{\sqrt[2]{n}}$.

    (b) $a_n = 1 + \frac{1}{\sqrt[3]{n}}$.

    (c) $a_n = 3 + 2^{-n}$.

    (d) $a_n = \sqrt{\frac{n}{n+1}}$

3. Prove directly that each of the following sequences converges by letting $\varepsilon > 0$ be given and finding $N(\varepsilon)$ so that (1) holds:

    (a) $a_n = 5 - \frac{2}{\ln n}$ for $n \geq 2$.

(b) $a_n = \frac{3n+1}{n+2}$. Hint: to determine the limit, divide numerator and denominator by n.

(c) $a_n = \frac{n^2+6}{2n^2-2}$ for $n \geq 2$.

(d) $a_n = \frac{2^n}{n!}$. Hint: look at the ratio of successive terms.

4. Prove directly that each of the following sequences diverges to ∞ or diverges to $-\infty$:

   (a) $a_n = 2^n$.

   (b) $a_n = -n^2$.

   (c) $a_n = \sqrt{\ln n}$.

   (d) $a_n = \frac{n!}{2^n}$.

5. Prove that a sequence $\{a_n\}$ can have at most one limit.

6. Suppose that $a_n \to a$ and let b be any number strictly less than a. Prove that $a_n > b$ for all but finitely many n.

7. Suppose that $a_n \to a$ and that $a_n \geq b$ for each n. Prove that $a \geq b$.

8. Suppose that $a_n \to 0$ and $b_n \to \infty$. Show that we cannot draw any definite conclusions about the sequence $c_n = a_n b_n$ by giving examples which satisfy these hypotheses and

   (a) $c_n \to 0$.

   (b) $c_n \to \infty$.

   (c) $c_n \to 1$.

   (d) $\{c_n\}$ does not converge, nor does it diverge to $+\infty$ or $-\infty$.

9. Before we gave the formal definition of convergence, we said intuitively that $\{a_n\}$ converges to a if the a_n get "closer and closer" to a. Did we mean that each successive term was closer to a than the previous one? No. We used "closer and closer" in an imprecise way. This illustrates why one needs technical definitions which say exactly what one means.

   (a) Find a sequence $\{a_n\}$ and a real number a so that

   $$|a_{n+1} - a| < |a_n - a|$$

   for each n, but $\{a_n\}$ does not converge to a. Thus a sequence can get "closer and closer" to a without converging to a.

   (b) Find a sequence $\{a_n\}$ and a real number a so that $a_n \to a$ but so that the above inequality is violated for infinitely many n. Thus a sequence can converge without getting "closer and closer."

2.2 Limit Theorems

As we have seen in the last section, proving that limits exist by using the definition of "limit" can be hard work. Instead, one can sometimes use known limits and general limit theorems of the kind we will prove in this section. For example, let $\{a_n\}$ and $\{b_n\}$ be sequences and suppose that $a_n \to a$ and $b_n \to b$. Since a_n gets close to a and b_n gets close to b, it is intuitively clear that $a_n b_n$ should get close to ab. That is, if we define a new sequence, $\{c_n\}$, by setting $c_n = a_n b_n$, it should be true that $\{c_n\}$ has a limit and the limit should be ab. As a warm-up to proving the limit theorems, we prove two propositions. The first shows that convergent sequences are automatically bounded.

Proposition 2.2.1   Suppose that $b_n \to b$. Then the sequence $\{b_n\}$ is **bounded**; that is, there is a number, M, so that

$$|b_n| \ \leq \ M, \quad \text{for all } n. \tag{7}$$

Proof. By the definition of convergence, we can choose N so that $n \geq N$ implies $|b_n - b| \leq 1$. Then, using the triangle inequality, we see that

$$
\begin{aligned}
|b_n| &= |b_n - b + b| \\
&\leq |b_n - b| + |b| \\
&\leq 1 + |b|
\end{aligned}
$$

for $n \geq N$. Define $M \equiv \max\{|b_1|, |b_2|, ..., |b_{N-1}|, 1 + |b|\}$. Then, by the way that M is defined, we see that $|b_n| \leq M$ for $1 \leq n \leq N$. In addition, $|b_n| \leq 1 + |b| \leq M$ for $n \geq N$. Thus, the inequality in (7) holds for all n. ❏

The second proposition shows that if a sequence is sandwiched betweeen two sequences that converge to the *same* limit, then the sequence converges to that limit too.

Proposition 2.2.2 Let $\{a_n\}$ and $\{b_n\}$ be sequences such that $a_n \to L$ and $b_n \to L$. Let $\{c_n\}$ be a sequence such that $a_n \leq c_n \leq b_n$ for all n. Then $c_n \to L$.

Proof. Let $\varepsilon > 0$ be given. Since $a_n \to L$, we can choose an N_1 such that $n \geq N_1$ implies $|a_n - L| \leq \varepsilon$. Since $b_n \to L$, we can choose an N_2 such that $n \geq N_2$ implies $|b_n - L| \leq \varepsilon$. If we choose $N = max\{N_1, N_2\}$, then

$n \geq N$ implies that both $n \geq N_1$ and $n \geq N_2$. Suppose $n \geq N$. Then either $c_n \leq L$ or $c_n > L$. In the first case,

$$a_n \ \leq \ c_n \ \leq \ L$$

so

$$|L - c_n| \ \leq \ |L - a_n| \ \leq \ \varepsilon.$$

In the second case,

$$L \ < \ c_n \ \leq \ b_n$$

so

$$|L - c_n| \ \leq \ c_n - L \ \leq \ |b_n - L| \ \leq \ \varepsilon.$$

In either case, $n \geq N$ implies

$$|L - c_n| \ \leq \ \varepsilon,$$

so we have proven that $c_n \to L$.   ❏

Example 1 Let $c_n = 1 + \frac{\cos n \sin^2 n}{n}$. The sequence $\{c_n\}$ looks complicated, but notice that $-1 \leq \cos n \sin^2 n \leq 1$ for each n. Thus,

$$1 - \frac{1}{n} \ \leq \ c_n \ \leq \ 1 + \frac{1}{n}.$$

Since $\{1 - \frac{1}{n}\}$ and $\{1 + \frac{1}{n}\}$ both converge to 1, the sequence $\{c_n\}$ must converge to 1, too, according to Proposition 2.2.2.

❏ **Theorem 2.2.3** Let $\{a_n\}$ and $\{b_n\}$ be sequences and suppose that $a_n \to a$ and $b_n \to b$. Then,

$$\lim_{n \to \infty} (a_n + b_n) \ = \ a + b.$$

Proof. Let $\varepsilon > 0$ be given. Since $a_n \to a$, we can choose N_1 so that $n \geq N_1$ implies $|a_n - a| \leq \frac{\varepsilon}{2}$. (Why?) Since $b_n \to b$, we can choose N_2 so that $n \geq N_2$ implies that $|b_n - b| \leq \frac{\varepsilon}{2}$. Define $N = max\{N_1, N_2\}$. Then, if $n \geq N$,

$$
\begin{aligned}
|(a_n + b_n) - (a + b)| \ &= \ |(a_n - a) + (b_n - b)| \\
&\leq \ |(a_n - a)| \ + \ |(b_n - b)| \\
&\leq \ \frac{\varepsilon}{2} \ + \ \frac{\varepsilon}{2} \\
&= \ \varepsilon.
\end{aligned}
$$

This proves, by the definition of limit, that $\lim_{n \to \infty}(a_n + b_n) = a + b$. In the second line we used the triangle inequality. ❏

❏ **Theorem 2.2.4** Let $\{a_n\}$ be a sequence and suppose that $a_n \to a$. Then, for any constant κ,

$$\lim_{n \to \infty}(\kappa a_n) = \kappa \lim_{n \to \infty}(a_n) = \kappa a.$$

❏ **Theorem 2.2.5** Let $\{a_n\}$ and $\{b_n\}$ be sequences and suppose that $a_n \to a$ and $b_n \to b$. Then,

$$\lim_{n \to \infty}(a_n b_n) = ab.$$

Proof. Let $\varepsilon > 0$ be given. To see how to choose N, we write

$$
\begin{aligned}
|a_n b_n - ab| &= |a_n b_n - ab_n + ab_n - ab| \\
&\leq |(a_n - a)b_n| + |a(b_n - b)| \\
&= |a_n - a||b_n| + |a||b_n - b|.
\end{aligned}
$$

We want to show that the sum on the right-hand side is less than ε if we choose n large enough. We'll work on the second term first. If $a = 0$, then the second term is $\leq \frac{\varepsilon}{2}$ for all n. If $a \neq 0$, then, since $b_n \to b$, we can choose an N_1 so that $n \geq N_1$ implies

$$|b_n - b| \leq \frac{\varepsilon}{2|a|}.$$

Thus for such n the second term will be less than $\frac{\varepsilon}{2}$.

Since $b_n \to b$, Proposition 2.2.1 allows us to choose an M so that (7) holds. Also, $a_n \to a$, so we can choose N_2 so that $n \geq N_2$ implies

$$|a_n - a| \leq \frac{\varepsilon}{2M}.$$

Now choose $N = \max\{N_1, N_2\}$. Then,

$$
\begin{aligned}
|a_n b_n - ab| &\leq |a_n - a||b_n| + |a||b_n - b| \\
&\leq \frac{\varepsilon}{2M}M + |a|\frac{\varepsilon}{2|a|} \\
&= \varepsilon
\end{aligned}
$$

if $n \geq N$. Therefore, $a_n b_n \to ab$. ❏

❏ **Theorem 2.2.6**   Let $\{a_n\}$ and $\{b_n\}$ be sequences and suppose that $a_n \to a$ and $b_n \to b$. Suppose that $b \neq 0$ and $b_n \neq 0$ for any n. Then,

$$\lim_{n \to \infty} \left(\frac{a_n}{b_n}\right) = \frac{a}{b}.$$

The proofs of Theorem 2.2.4 and Theorem 2.2.6 are left to the student (problems 3 and 5).

Example 2   Let $a_n = (1 + \frac{1}{\sqrt{n}})^2 + (2 - 2^{-n})$. Since we proved in Section 2.1 that $1 + \frac{1}{\sqrt{n}} \to 1$, Theorem 2.2.5 implies that

$$\lim_{n \to \infty} \left(1 + \frac{1}{\sqrt{n}}\right)^2 = \left(\lim_{n \to \infty} \left(1 + \frac{1}{\sqrt{n}}\right)\right)^2 = 1^2 = 1.$$

Also, we know that

$$\lim_{n \to \infty} (2 - 2^{-n}) = 2,$$

so Theorem 2.2.3 guarantees that $\lim_{n \to \infty} a_n$ exists and

$$\lim_{n \to \infty} a_n = \lim_{n \to \infty} \left(1 + \frac{1}{\sqrt{n}}\right)^2 + \lim_{n \to \infty} (2 - 2^{-n}) = 3.$$

Example 3   In problem 3b of Section 2.1, you were asked to show directly that the sequence $a_n = \frac{3n+1}{n+2}$ has a limit. We write

$$\frac{3n + 1}{n + 2} = \frac{3 + \frac{1}{n}}{1 + \frac{2}{n}}.$$

Since the numerator on the right-hand side has limit 3 and the denominator has limit 1, Theorem 2.2.6 assures us that $\{a_n\}$ has a limit and that the limit is 3.

Notice how much easier the proof in Example 3 was than the direct proof in problem 3b of Section 2.1. There is an important point here about the role of theory in mathematics. Suppose that we'd been given six problems just like problem 3b except that the coefficients in the numerator and denominator were different. For each problem the direct proof of convergence would have been long, and after a while we would have noticed that the pattern of proof was the same. We would be doing lots of hard work (basically the *same* hard work) each time to show that the limit of the ratio is the ratio of the limits. That's when one sees the need for proving Theorem 2.2.6. The proof of Theorem 2.2.6 contains all the hard work which we were doing over and over in the separate problems. Once Theorem 2.2.6 is proved, it can be used to do each of the six problems easily.

Problems

1. Prove that each of the following limits exists by using Theorem 2.2.3 – Theorem 2.2.6:

   (a) $a_n = 5(1 + \frac{1}{\sqrt[3]{n}})^2$.

   (b) $a_n = \frac{3n+1}{n+2}$.

   (c) $a_n = \frac{n^2+6}{3n^2-2}$.

   (d) $a_n = \frac{5+(\frac{2}{3n})^2}{2+\frac{2n+5}{3n-2}}$.

2. Find the limits of the following sequences:

   (a) $a_n = e^{-n} \sin n$.

   (b) $a_n = (\sin n) \sin \frac{1}{n}$.

   (c) $a_n = (\cos n)(\ln n)^{-1}$ for $n \geq 2$.

3. Prove Theorem 2.2.4.

4. Let $\{b_n\}$ be a bounded sequence and suppose that $a_n \to 0$. Prove that $a_n b_n \to 0$.

5. Prove Theorem 2.2.6. Hint: first show that there are numbers, M_1 and M_2, so that $0 < M_1 \leq |b_n| \leq M_2$ for all n.

6. Let $p(x)$ be any polynomial and suppose that $a_n \to a$. Prove that

$$\lim_{n \to \infty} p(a_n) = p(a).$$

7. Let $\{a_n\}$ and $\{b_n\}$ be sequences and suppose that $a_n \leq b_n$ for all n and that $a_n \to \infty$. Prove that $b_n \to \infty$.

8. Let $a_n = \sqrt{n+1} - \sqrt{n}$. Prove that $a_n \to 0$.

9. (a) Let $\{a_n\}$ be the sequence in problem 1(c) of Section 2.1. Prove that $\{a_n\} \to 4$. Hint: subtract 4 from both sides.

   (b) Consider the sequence defined by the recursive relation $a_{n+1} = \alpha a_n + 2$. Show that if $|\alpha| < 1$ the sequence has a limit independent of a_1.

10. The Euclidean plane $\mathbb{R}^2 \equiv \mathbb{R} \times \mathbb{R}$ is the set of ordered pairs (x, y), where $x \in \mathbb{R}$ and $y \in \mathbb{R}$. Note that we can add pairs, $(x_1, y_1) + (x_2, y_2) = (x_1 + x_2, y_1 + y_2)$, and multiply pairs by real numbers, $\alpha(x, y) = (\alpha x, \alpha y)$. These operations correspond to vector addition and scalar multiplication. We define a notion of "size" for points in $\mathbb{R}^2$ by

$$\|(x, y)\| \equiv \sqrt{x^2 + y^2}.$$

That is, the size of (x, y) is just the Euclidean distance from the point to the origin.

(a) Let x_1, x_2, y_1, y_2 be real numbers. Using the inequality in problem 8 of Section 1.1, prove that

$$|x_1 x_2 + y_1 y_2| \leq \sqrt{x_1^2 + y_1^2} \sqrt{x_2^2 + y_2^2}.$$

(b) Prove that for any two points (x_1, y_1) and (x_2, y_2),

$$\|(x_1, y_1) + (x_2, y_2)\| \leq \|(x_1, y_1)\| + \|(x_2, y_2)\|.$$

(c) Let $p_n = (x_n, y_n)$ be a sequence of points in the plane and let $p = (x, y)$. We say that $p_n \to p$ if $\|p_n - p\| \to 0$. Prove that $p_n \to p$ if and only if $x_n \to x$ and $y_n \to y$.

2.3 Two-state Markov Chains

Consider the following problem. We check a phone every minute to see whether it is busy. Let p_n denote the probability that it is free on the nth check and q_n denote the probability that it is busy on the nth check. We now make some simple assumptions that allow us to compute p_{n+1} and q_{n+1} from p_n and q_n. Suppose that the phone is free when we check. Then we assume that with probability q it will be busy on the next check. Since it must either be free or busy, it will remain free with probability $1 - q$. Similarly, we assume that if it is busy when we check, then the probability that it will be free when we next check will be p, and therefore the probability that it will remain busy is $1 - p$. Then p_{n+1} and q_{n+1} can be given in terms of p_n and q_n by the formulas

$$p_{n+1} = (1 - q)p_n + pq_n \qquad (8)$$
$$q_{n+1} = qp_n + (1 - p)q_n. \qquad (9)$$

The first formula says simply that the probability that the phone will be free on the $n+1$ check is the probability that it was free on the nth check times the probability that if it's free it remains free plus the probability that it was busy on the nth check times the probability that if it's busy it will switch to free. Similar reasoning gives the second formula. If we are given the probability that the phone starts free, p_0, and the probability that it starts busy, q_0, then we can use these recursive formulas to compute p_n and q_n for each n. Although it would be complicated to write out the formulas explicitly, we can see what is happening if we reformulate the recursion using vectors and matrices. Think of the pair (p_n, q_n) as a two-dimensional column vector. Then (8) and (9) can be rewritten as

$$
\begin{pmatrix} p_{n+1} \\ q_{n+1} \end{pmatrix} = \begin{pmatrix} 1-q & p \\ q & 1-p \end{pmatrix} \begin{pmatrix} p_n \\ q_n \end{pmatrix}. \tag{10}
$$

So, if we let A denote the matrix

$$
A = \begin{pmatrix} 1-q & p \\ q & 1-p \end{pmatrix} \tag{11}
$$

and $v_n = (p_n, q_n)$, we can write

$$
v_{n+1} = A v_n
$$

so

$$
v_n = A^n v_0.
$$

This formula shows that we can obtain the probabilities at the nth check by applying the nth power of the matrix A to the vector of initial probabilities $v_0 = (p_0, q_0)$. Note that we must have $p_0 + q_0 = 1$ since the phone is either free or busy initially. By adding equations (8) and (9) we see that $p_{n+1} + q_{n+1} = p_n + q_n$. Thus $p_n + q_n = 1$ for all n.

We have created two sequences of real numbers, $\{p_n\}$ and $\{q_n\}$. We would like to know how they behave for large n. If we wait a long time, what are the probabilities that the phone will be free or busy? How do these probabilities depend on the initial probabilities p_0, q_0? First we assume that $\{p_n\}$ and $\{q_n\}$ converge to limits $\bar{p}$ and $\bar{q}$, respectively, and ask what those limits might be. Using equation (8) and Theorems 2.2.3 and 2.2.4 from Section 2.2, we see that

$$
\bar{p} = \lim_n p_n
$$

$$
= \lim_n p_{n+1}
$$

$$= \lim_n ((1-q)p_n + pq_n)$$

$$= (1-q)\lim_n p_n + p\lim_n q_n$$

$$= (1-q)\bar{p} + p\bar{q}.$$

In similar fashion,

$$\bar{p} + \bar{q} = \lim_{n\to\infty} (p_n + q_n) = 1.$$

This gives us two linear equations that must be satisfied by any limits $\bar{p}$ and $\bar{q}$. The equations are independent and thus have a unique solution, which is easily calculated to be:

$$\bar{p} = \frac{p}{p+q}, \qquad \bar{q} = \frac{q}{p+q}.$$

We shall show that $p_n \to \bar{p}$ and $q_n \to \bar{q}$. Using the equations (8), (9), and the equations above satisfied by $\bar{p}$ and $\bar{q}$, we calculate

$$(p_{n+1} - \bar{p})^2 + (q_{n+1} - \bar{q})^2$$
$$= ((1-q)p_n + pq_n - \bar{p})^2 + (qp_n + (1-p)q_n - \bar{q})^2$$
$$= ((1-q)(p_n - \bar{p}) + p(q_n - \bar{q}))^2 + (q(p_n - \bar{p}) + (1-p)(q_n - \bar{q}))^2.$$

To make the algebra easier, define $\alpha = p_n - \bar{p}$. Since $p_n + q_n = 1$ and $\bar{p} + \bar{q} = 1$, notice that $-\alpha = q_n - \bar{q}$. Thus,

$$(p_{n+1} - \bar{p})^2 + (q_{n+1} - \bar{q})^2 = ((1-q)\alpha - p\alpha)^2 + (q\alpha - (1-p)\alpha)^2$$
$$= \alpha^2\{(1-q)^2 - 2p(1-q) + p^2\}$$
$$\qquad\qquad + \alpha^2\{q^2 - 2q(1-p) + (1-p))^2\}$$
$$= 2\alpha^2(1 - (p+q))^2$$
$$= (1 - (p+q))^2\{(p_n - \bar{p})^2 + (q_n - \bar{q})^2.\}$$

If we iterate this equality, starting with the case $n = 1$, we obtain

$$(p_{n+1} - \bar{p})^2 + (q_{n+1} - \bar{q})^2 = (1 - (p+q))^{2n}\{(p_1 - \bar{p})^2 + (q_1 - \bar{q})^2\}. \tag{12}$$

From now on we assume that $0 < p < 1$ and $0 < q < 1$. It follows that $0 < p+q < 2$, which implies that $|1 - (p+q)| < 1$. Therefore,

$$\lim_{n\to\infty} (1 - (p+q))^n = 0$$

and so it follows from Theorem 2.2.4 that the right-hand side of (12) converges to zero. Thus, the left-hand side of (12) converges to zero too. Since

$$(p_{n+1} - \bar{p})^2 \leq (p_{n+1} - \bar{p})^2 + (q_{n+1} - \bar{q})^2,$$

it follows from Proposition 2.2.2 that $\lim_{n \to \infty} (p_{n+1} - \bar{p})^2 = 0$. On the other hand, if a sequence of positive numbers converges to zero, then the sequence of square roots converges to zero too (problem 1). Thus, $\lim_{n \to \infty} (p_n - \bar{p}) = 0$, so we have shown that $p_n \to \bar{p}$. A similar sequence of steps shows that $q_n \to \bar{q}$.

What this convergence means is that if we wait a long time, then the probability that the phone will be free will be approximately $\bar{p}$ and the probability that it will be busy will be approximately $\bar{q}$. Notice that this is true no matter what the initial probabilities p_0 and q_0 are.

Probabilistic systems such this are called stochastic processes. The one that we have described is called a Markov chain because time is discrete and at each time the probabilities are determined from the probabilities at the previous time. What we have described is a two-state Markov chain because at each time the system (that is, the phone) has only two possible states. The probabilities $\bar{p}$ and $\bar{q}$ are called invariant probabilities because $(\bar{p}, \bar{q}) = A(\bar{p}, \bar{q})$. If we start with the probabilities $\bar{p}$ and $\bar{q}$, then every time we check the probabilities will be $\bar{p}$ and $\bar{q}$. In the language of probability theory, we have proven that a two-state Markov chain has unique invariant probabilities to which the system converges as $n \to \infty$. A somewhat easier proof of this result can be given by using the tools of linear algebra. This is outlined in Project 2.

Problems

1. Suppose that $a_n \geq 0$ and that $a_n \to 0$. Prove that $\sqrt{a_n} \to 0$.

2. Let $p = .7$ and $q = .5$ and suppose that $p_0 = .5$ and $q_0 = .5$. Compute as many iterates of equations (8) and (9) as you need to make good guesses for $\lim_{n \to \infty} p_n$ and $\lim_{n \to \infty} q_n$.

3. Let p and q have the same values as in problem 2. Let $p_0 = .2$ and $q_0 = .8$ and compute as many iterates as you need to verify that p_n and q_n converge to the same limits as in problem 2, even though the initial probabilities are different.

4. Let $p = .8$ and $q = .9$ and suppose that $p_0 = .5$ and $q_0 = .5$. What are $\lim_{n \to \infty} p_n$ and $\lim_{n \to \infty} q_n$ in this case? Is the convergence to these limits more or less rapid than the convergence in problem 2? Why?

5. Suppose that $p = .2$ and $q = .5$. How large must we choose n to be sure that p_n and q_n are within 10^{-3} of the limiting values $\bar{p}$ and $\bar{q}$ no matter what p_0 and q_0 are?

6. In a certain city, taxis are sometimes on the north side and sometimes on the south side, but they must be in one place or the other. Let p_n and q_n denote the probabilities that a particular taxi is on the north side and the south side, respectively, on the nth time that we check. If a taxi is on the north side when we check, it has a probability of .5 of having switched to the south side when we next check. If it is on the south side, it has a probability of .3 of having switched north when we check again. If there are 100 taxis, about how many will be on the north and south sides if we wait a long time?

7. Suppose that taxis can be on the north side or the south side, and let p_n and q_n be the probabilities of finding them there on the nth check. Suppose that between any two times that we check, each taxi has a finite probability r of going home to the taxi barn (neither north or south) from which it doesn't emerge. Explain why p_n and q_n satisfy a recursion relation of the form

$$
\begin{aligned}
p_{n+1} &= b_{11}p_n + b_{12}q_n \\
q_{n+1} &= b_{21}p_n + b_{22}q_n,
\end{aligned}
$$

where the coefficients b_{ij} all satisfy $0 < b_{ij} < 1$ and in addition $b_{11} + b_{21} < 1$ and $b_{12} + b_{22} < 1$. By summing the two equations, prove that $p_n \to 0$ and $q_n \to 0$ as $n \to \infty$. What does this mean?

8. Suppose that the taxi in problem 7 can reemerge from the taxi barn and go to the north side or the south side with certain probabilities. Formulate (but do not solve) a three-state Markov chain for the probabilities p_n, q_n, and r_n that the taxi will be on the north side, the south side, or in the taxi barn, respectively.

2.4 Cauchy Sequences

In the first three sections of this chapter we proved that many different sequences converge to limits. In all cases, the idea of the proof was the same. First we guessed the limit, a. Then we subtracted the purported limit from the terms of the sequence, a_n, making a new sequence $\{a_n - a\}$. Then we proved that the sequence of differences converges to zero; that is, $(a_n - a) \to 0$. Since this is equivalent to $a_n \to a$, we concluded that the sequence $\{a_n\}$ converges and the limit is, indeed, a. But what if we can't guess the limit a? Is there a method by which we can tell if a sequence converges just by looking at the terms of the sequence itself? Just such a criterion was introduced by the mathematician Augustin-Louis Cauchy (1789–1857).

Definition. A sequence $\{a_n\}$ is called a **Cauchy sequence** if, given any $\varepsilon > 0$, there exists N so that

$$|a_n - a_m| \leq \varepsilon \quad \text{if} \quad n \geq N \quad \text{and} \quad m \geq N.$$

That is, a sequence is a Cauchy sequence if, given any $\varepsilon > 0$, all the terms of the sequence are within ε of each other from some index N on. We emphasize that for *each* ε there must exist an N. In general, if ε is smaller, one will have to choose N larger. Notice that there is no mention of a limit a.

Example 1 We will prove that the sequence $a_n = \frac{3n+1}{n+2}$ is a Cauchy sequence. Let $\varepsilon > 0$ be given. To see how we should choose N we first estimate:

$$\left| \frac{3n+1}{n+2} - \frac{3m+1}{m+2} \right| = \left| \frac{5(n-m)}{(n+2)(m+2)} \right|$$

$$\leq \frac{5n}{(n+2)(m+2)} + \frac{5m}{(n+2)(m+2)}$$

$$\leq \frac{5}{m+2} + \frac{5}{n+2}.$$

In the last step we used the fact that $\frac{k}{k+2} < 1$ for each positive integer k. Choose N so that $N \geq 10/\varepsilon$, or equivalently $5/N \leq \varepsilon/2$; the reason for this choice will be clear shortly. If $n \geq N$ and $m \geq N$, then

$$\frac{5}{m+2} + \frac{5}{n+2} \leq \frac{5}{m} + \frac{5}{n} \leq \frac{5}{N} + \frac{5}{N} \leq \varepsilon/2 + \varepsilon/2.$$

Since we have been able to choose an appropriate N for each ε, we have proven that $\{a_n\}$ is a Cauchy sequence.

The sequence in the example converges (problem 1(b) of Section 2.2). In fact, every convergent sequence is automatically a Cauchy sequence, as the following proposition shows.

Proposition 2.4.1   If $\{a_n\}$ converges to a finite limit, then $\{a_n\}$ is a Cauchy sequence.

Proof.   Let $a = \lim a_n$ and let $\varepsilon \geq 0$ be given. Since $a_n \to a$, we can choose N so that $n \geq N$ implies that $|a_n - a| \leq \varepsilon/2$. Therefore if $n \geq N$ and $m \geq N$, we find

$$
\begin{aligned}
|a_n - a_m| &= |(a_n - a) + (a - a_m)| \\
&\leq |(a_n - a)| + |(a - a_m)| \\
&\leq \varepsilon/2 + \varepsilon/2.
\end{aligned}
$$

Thus, $\{a_n\}$ is a Cauchy sequence.   ❑

The more difficult question is whether the converse to Proposition 2.4.1 is true. That is, if a sequence is a Cauchy sequence, does it necessarily converge to a limit? Before taking up this question, recall that the **rational numbers**, $\mathbb{Q}$, are the real numbers that can be expressed as quotients, m/n, where m and n are integers and $n \neq 0$. The **irrational numbers** are the real numbers that are not rational. The reader is certainly familiar with decimal expansions of real numbers; we discuss decimal expansions rigorously in Project 4 of Chapter 6. In terms of decimals, the rational numbers are just the real numbers whose decimal expansions are repeating after some point. The following irrational numbers,

$$
\begin{aligned}
\pi &= 3.14159\ldots \\
e &= 2.71828\ldots \\
\sqrt{2} &= 1.41421\ldots
\end{aligned}
$$

occur frequently. The number π is the ratio of the circumference to the diameter of any circle; e can be defined as the unique number so that the derivative of e^x is again e^x, that is, so that the slope of the tangent line to the graph of e^x at the point (x, e^x) is e^x; $\sqrt{2}$ can be defined as the length of the hypotenuse of a right triangle whose other two sides both have length 1. Now one must *prove* that the numbers so defined are irrational and devise ways of computing as much of their decimal expansions as one needs. We gave Euclid's proof that $\sqrt{2}$ is irrational in Proposition 1.4.2 and outlined a proof that e is irrational in problem 12 of Section 1.4. We proved in Section 1.3 that the set of rational numbers is countable and the set of irrational numbers is uncountable.

The rational numbers have the very important property that they are **dense** in the real numbers. This means that, given any real number a, there is a sequence of rational numbers, $\{r_n\}$, so that $r_n \to a$. If a is rational we just pick every member of the sequence to be equal to a. If a is irrational, we choose r_n to be the rational numbers given by taking the first n terms of the decimal expansion of a followed by zeros. So, in the case of π we would have

$$
\begin{aligned}
r_1 &= 3.0 \\
r_2 &= 3.1 \\
r_3 &= 3.14 \\
r_4 &= 3.141 \\
r_5 &= 3.1415,
\end{aligned}
$$

and so forth. Notice that if a number r has a decimal expansion that agrees with the decimal expansion of a through the mth decimal place, then the number differs from a by less than 10^{-m}. Thus it is clear that the sequence of rationals $\{r_n\}$ converges to a.

We now return to the question of whether Cauchy sequences of real numbers necessarily have limits. A set of numbers which has this property is called **complete**. Let us suppose that we didn't know about irrational numbers, so that our number system consisted of only the rational numbers, $\mathbb{Q}$. Suppose that we had a Cauchy sequence $\{r_n\}$ of rational numbers. Could we be sure that it converged to a limit? Clearly not, as the example of the sequence $\{r_n\}$ above shows. Since the sequence $\{r_n\}$ cannot converge to any other real number (problem 5 of Section 2.1), and since π is not in $\mathbb{Q}$, the sequence $\{r_n\}$ does not converge to any limit in $\mathbb{Q}$, even though it is a Cauchy sequence. Therefore, $\mathbb{Q}$ is *not* complete because not all Cauchy sequences in $\mathbb{Q}$ have limits in $\mathbb{Q}$. If we want to enlarge $\mathbb{Q}$ to make a complete set of numbers, we will have to include at least all the irrational numbers, for each of these is a limit of a Cauchy sequence of rationals. Indeed, this idea of characterizing all real numbers as the set of "limits" of Cauchy sequences of rationals is one of the methods of constructing the real numbers from the rationals. Even if we include the irrational numbers, the question of completeness isn't settled. Would a Cauchy sequence of irrationals necessarily have a limit which is an irrational or rational number? The answer, though not obvious, is "yes". This is proven as part of the construction of the set of real numbers from the rationals. Since we are not giving this construction here, we will merely assume the result, that is, that the real numbers are complete.

Axiom of Completeness. Every Cauchy sequence of real numbers converges to a finite real number.

Combining the Axiom of Completeness and Proposition 2.4.1, we have:

❑ **Theorem 2.4.2**   A sequence of real numbers is a Cauchy sequence if and only if it converges to a finite real number.

A sequence $\{a_n\}$ is said to be **monotone increasing** if $a_n \leq a_{n+1}$ for all n, and it is said to be **monotone decreasing** if $a_n \geq a_{n+1}$. The proof of the following theorem depends in the last step on the completeness of $\mathbb{R}$. The theorem itself is extremely useful since it is often easy to prove that a sequence is bounded. If it is also monotone, then we know that the sequence must converge.

❑ **Theorem 2.4.3**   Every bounded monotone sequence converges.

Proof.   Let $\{a_n\}$ be a bounded monotone increasing sequence; the proof for the decreasing case is similar. We will construct a sequence of intervals, $[b_n, c_n]$, of decreasing length so that each interval contains the tail of the sequence $\{a_n\}$. Since $\{a_n\}$ is bounded there is an M so that $a_n \leq M$ for all n. Let $b_1 = a_1$ and $c_1 = M$. Since $\{a_n\}$ is monotone increasing, every term of the sequence is in the interval $[b_1, c_1]$.

Figure 2.4.1

Note that $c_1 - b_1 = M - a_1$. Now, let m_1 be the midpoint of the interval $[b_1, c_1]$. If there is one term of the sequence in the interval $[m_1, c_1]$, then all the rest of the terms of the sequence must be in $[m_1, c_1]$ since the sequence is monotone increasing. In this case, let N_2 be an integer so that a_{N_2} is in $[m_1, c_1]$, and define our second interval by $b_2 = m_1$ and $c_2 = c_1$, i.e. the right half of $[b_1, c_1]$. See case 1 in Figure 2.4.1. If no points of the sequence are in the interval $[m_1, c_1]$, we define our second interval to be the left half of $[b_1, c_1]$; that is, $b_2 = b_1$ and $c_2 = m_1$; see case 2 in Figure 2.4.1. In either case, we know that

$$a_n \in [b_2, c_2], \qquad \text{for } n \geq N_2$$

and

$$c_2 - b_2 = \frac{1}{2}(M - a_1).$$

We now continue in this manner, successively dividing our intervals in half and choosing one of the two subintervals. Suppose that the interval $[b_j, c_j]$ and N_j have been defined so that

$$a_n \in [b_j, c_j], \qquad \text{for } n \geq N_j$$

and

$$c_j - b_j = \left(\frac{1}{2}\right)^{j-1}(M - a_1).$$

Let m_j be the midpoint of $[b_j, c_j]$. If there is an integer N_{j+1} so that $a_{N_{j+1}} \in [m_j, c_j]$, then we choose the right subinterval, that is, $b_{j+1} = m_j$ and $c_{j+1} = c_j$. If none of the a_n are in $[m_j, c_j]$, then we choose the left subinterval, i.e. $b_{j+1} = b_j$ and $c_{j+1} = m_j$. In either case,

$$a_n \in [b_{j+1}, c_{j+1}], \qquad \text{for } n \geq N_{j+1}$$

and

$$c_{j+1} - b_{j+1} = \left(\frac{1}{2}\right)^{j}(M - a_1).$$

Now, let $\varepsilon > 0$ be given. Choose j so that $2^{-j+1}(M - a_1) \leq \varepsilon$. Then, if $n \geq N_j$ and $m \geq N_j$, both a_n and a_m are in the interval $[b_j, c_j]$, so

$$|a_n - a_m| \leq 2^{-j+1}(M - a_1) \leq \varepsilon.$$

Thus, $\{a_n\}$ is a Cauchy sequence. By Theorem 2.4.2, it therefore converges to a finite limit. ❏

Example 1 It is easy to check that the sequence $a_n = \frac{n}{n+2}$ is monotone increasing, and it is clearly bounded by 1. Therefore, by Theorem 2.4.3, the sequence converges. Of course, this can also be proved directly.

Example 2 Consider the sequence $a_n = \sqrt[n]{n}$. Let f be the function

$$f(x) \equiv e^{\frac{\ln x}{x}}$$

defined for $x > 0$. Notice that for all positive integers n,

$$f(n) = e^{\frac{\ln n}{n}} = (e^{\ln n})^{\frac{1}{n}} = \sqrt[n]{n},$$

so the sequence $\{a_n\}$ consists of the values of f at the positive integers. Differentiating f, we find

$$f'(x) = \frac{1 - \ln x}{x^2} e^{\frac{\ln x}{x}},$$

which is negative if $x \geq e$ since $\ln x$ is an increasing function and $\ln e = 1$. Thus the numbers $a_n = f(n)$ are monotone decreasing for $n \geq 3$. Since they are bounded below by 1, Theorem 2.4.2 guarantees that the sequence $\sqrt[n]{n}$ converges. Problem 11 outlines the proof that the limit is 1.

Notice that if $\{a_n\}$ is monotone increasing and is *not* bounded, then for each M there is an element of the sequence, say a_N, so that $a_N \geq M$. Because the sequence is monotone increasing, we have $a_n \geq M$ for *all* $n \geq N$. Since we can choose such an N for each M, $\{a_n\}$ diverges to $+\infty$ according to the definition in Section 2.1. Thus every increasing sequence converges to a finite limit or diverges to $+\infty$. Similarly, every decreasing sequence converges to a finite limit or diverges to $-\infty$.

Problems

1. Prove directly, by verifying the definition, that $a_n = 1 + \frac{1}{\sqrt{n}}$ is a Cauchy sequence.

2. Prove that the rational numbers are dense in the real numbers without using decimal expansions. Hint: use problem 11 of Section 1.1.

3. Suppose that $\{a_n\}$ and $\{b_n\}$ are both Cauchy sequences and that $a_n \leq b_n$ for each n. Prove that $\lim_{n \to \infty} a_n \leq \lim_{n \to \infty} b_n$.

4. Suppose that $\{x_n\}$ and $\{y_n\}$ are both Cauchy sequences. Prove that $\lim_{n \to \infty} (x_n - y_n)$ exists.

5. Suppose that $\{a_n\}$ is a Cauchy sequence. Prove that $\{a_n^2\}$ is a Cauchy sequence. Is the converse true?

6. Let $\{b_n\}$ be a sequence of positive numbers such that $b_n \to 0$. Suppose that the terms of a sequence $\{a_n\}$ satisfy

$$|a_m - a_n| \leq b_n \qquad \text{for all } m \geq n.$$

Prove that $\{a_n\}$ is a Cauchy sequence.

7. Suppose that the terms $\{a_n\}$ satisfy $|a_{n+1} - a_n| \leq 2^{-n}$ for all n. Prove that $\{a_n\}$ is a Cauchy sequence.

8. A sequence of points in $\mathbb{R}^2$, $\{p_n\}$, is called a Cauchy sequence if, given $\varepsilon > 0$, there is an N so that $\|p_n - p_m\| \leq \varepsilon$ for all $n \geq N$ and $m \geq N$. The norm $\|\cdot\|$ is defined in problem 10 of Section 2.2. Prove that every Cauchy sequence in $\mathbb{R}^2$ has a limit in $\mathbb{R}^2$, that is, $\mathbb{R}^2$ is complete.

9. Suppose that $a_1 > 0$ and $a_{n+1} = a_n + \frac{1}{a_n}$. Prove that $\{a_n\}$ diverges to ∞. Hint: if $\{a_n\}$ converges to a nonzero limit, find an equation that would be satisfied by the limit.

10. Let $a_1 = \sqrt{2}$, and let a_n for $n \geq 2$ be defined recursively by the formula

$$a_{n+1} = \sqrt{2 + \sqrt{a_n}}.$$

   (a) Prove by induction that $\sqrt{2} \leq a_n \leq 2$ for all n.

   (b) Prove that $\{a_n\}$ is a Cauchy sequence and conclude that $\{a_n\}$ converges.

11.   (a) Compute enough terms of the sequence $a_n = \sqrt[n]{n}$ to guess the limit.

   (b) Let $a_n = \sqrt[n]{n} - 1$. Use the binomial theorem to prove that

$$n = (1 + a_n)^n \geq \frac{n(n-1)}{2} a_n^2.$$

   (c) Rearrange the inequality and prove that $\sqrt[n]{n} \to 1$.

12. Let $a_n = (1 + \frac{1}{n})^n$.

   (a) Prove that $f(x) = (1 + \frac{1}{x})^x$ is increasing for $x > 1$. Hint: show that $f'(1) > 0$ and $f''(x) > 0$ for $x \geq 1$.

   (b) Use the binomial theorem to show that a_n is bounded above by $1 + \sum_{k=0}^{n} \frac{1}{k!}$.

   (c) Prove that $\lim_{n \to \infty} a_n$ exists.

13. Let $\{a_n\}$ be monotone and bounded and define $\{b_n\}$ by

$$b_n = \frac{a_1 + a_2 + \ldots + a_n}{n}.$$

Prove that $\{b_n\}$ is monotone and bounded and therefore has a limit.

14. Let $\{a_n\}$ be a sequence and suppose that $a_n \to a$. Define the sequence $\{b_n\}$ as above and prove that $b_n \to a$.

2.5 Supremum and Infimum

Let S be a set of real numbers. We say that $m \in S$ is a **maximal** element if $m \geq x$ for all $x \in S$. Some sets have a maximal elements, and some sets don't. The number 1 is the maximal element of the closed interval $[0, 1]$, but the open interval $(0, 1)$ has no maximal element. Nevertheless, we want to characterize the relationship of 1 to the interval $(0, 1)$. Notice that 1 is an upper bound for the interval. In fact, of all the upper bounds it is the *least*. To generalize this idea to more complicated sets, we make a formal definition.

Definition. A number b is an **upper bound** for a set S if $x \leq b$ for all $x \in S$. We say that b_0 is a **least upper bound** of S if b_0 is an upper bound and $b_0 \leq b$ for any other upper bound b.

Though not all bounded sets have maximal elements, they all have least upper bounds, as the following theorem shows.

❏ **Theorem 2.5.1** Let S be a set of real numbers which is bounded above. Then S has a unique least upper bound.

Proof. We will prove the existence of a least upper bound; uniqueness is left as an exercise (problem 4). Let M be an upper bound for the set S and let b be an element of S. We shall construct a decreasing sequence, $\{a_n\}$, of upper bounds that converges to a least upper bound. Let $b_1 = b$, $c_1 = M$ and let I_1 be the interval $I_1 = [b_1, c_1]$. Choose $a_1 = M$. Note that a_1 is an upper bound and that there is an element of S, namely b, within $M - b$ of a_1. Let m_1 be the midpoint of the interval I_1. If there are no elements of S in the interval $[m_1, c_1]$, choose $a_2 = m_1$ and define $I_2 = [b_1, m_1]$. If there are elements of S in $[m_1, c_1]$, choose $a_2 = c_1$ and define $I_2 = [m_1, c_1]$. See Figure 2.5.1. Note that in both cases, a_2 is the right endpoint of I_2, a_2 is an upper bound for S, $a_2 \leq a_1$, and there is an element $x_2 \in S$ so that $|a_2 - x_2| \leq \frac{1}{2}(M - b)$.

Rename the endpoints of I_2 to be b_2 and c_2 and let m_2 be the midpoint of I_2. Choose a_3 to be m_2 if there is no point of S in $[m_2, c_2]$, and $a_3 = c_2$ otherwise. Continuing this constructive procedure, we obtain a decreasing sequence, $\{a_n\}$, of upper bounds for S. The nth term of the sequence, a_n, has the property that there is an element $x_n \in S$ so that $|a_n - x_n| \leq$

Figure 2.5.1

$2^{-n+1}(M - b)$. Since the sequence $\{a_n\}$ is decreasing and is bounded below by b, Theorem 2.4.3 guarantees that $\{a_n\}$ converges to a limit a.

To complete the proof we use two simple results that we have left as excercises. First, since a_n is an upper bound for S for each n, the limit a is also an upper bound for S (problem 2). Next, let $\varepsilon > 0$ be given, and choose n so that $2^{-n}(M - b) \leq \varepsilon/2$ and so that $|a_n - a| \leq \varepsilon/2$. Then,

$$|a - x_n| \leq |a - a_n| + |a_n - x_n| \leq \frac{\varepsilon}{2} + \frac{\varepsilon}{2},$$

so there is an element of S within ε of a. According to problem 3 this means that no upper bound for S could be less than a. Thus, a is a least upper bound for S. ❑

The least upper bound of the set S is also called the **supremum** of the set and denoted by $\sup S$. A similar proof shows that any set which is bounded below has a **greatest lower bound**. The greatest lower bound is called the **infimum** of the set and is denoted by $\inf S$. If a set S is not bounded above, we define $\sup S \equiv +\infty$. It is easy to see that if $\sup S = +\infty$ then there is a sequence of points $s_n \in S$ such that $s_n \to +\infty$ (problem 7). Similarly, if S is not bounded below we define $\inf S \equiv -\infty$.

Our discussion of completeness assumes the Archimedian property (for example, it was used implicitly in the proof of Theorem 2.4.3). The Archimedian property follows from Theorem 2.5.1.

❑ **Theorem 2.5.2** Let a and b be real numbers which satisfy $0 < a < b$. Then there is a positive integer n such that $b < na$.

Proof. We shall give a proof by contradiction. Assume that the Archimedian property fails for some a and b satisfying $0 < a < b$. That is, $na \leq b$ for all positive integers n. Thus b is an upper bound for the set $S \equiv \{na \mid n \in \mathbb{N}\}$. Let c be the least upper bound of S guaranteed by Theorem 2.5.1. Then, $c < c + a$ since $a > 0$, so $c - a < c$. Since c is the least upper of S, $c - a$ cannot be an upper bound. Thus, there is an element of S of the form ma, where m is a positive integer, such that $ma > c - a$. It follows that $c < (m+1)a$, but this is a contradiction since $(m+1)a$ is in S by definition and c is the least upper bound of S. Thus the Archimedian property holds for all a and b satisfying $0 < a < b$. ❑

The following result will be useful when we define the Riemann integral; see (10) in Section 3.3.

❏ **Theorem 2.5.3**   Let A and B be two sets of real numbers and suppose that for each $a \in A$ and $b \in B$ the inequality $a \leq b$ holds. Then $\sup A \leq \inf B$.

Proof.   Choose any $b \in B$. Since $a \leq b$ for all $a \in A$, it is clear that b is an upper bound for A. Since $\sup A$ is the least upper bound, we see that

$$\sup A \leq b.$$

Since this is true for all $b \in B$, the set B is bounded below by $\sup A$. Since $\inf B$ is by definition the greatest lower bound, we conclude that $\sup A \leq \inf B$.   ❏

We took as an axiom the statement that if a sequence of real numbers is a Cauchy sequence, then it converges to a limit. From that axiom we proved that bounded monotone sequences converge (Theorem 2.4.3) and that sets which are bounded above have least upper bounds (Theorem 2.5.1). We remark that Theorem 2.4.2, Theorem 2.4.3, and Theorem 2.5.1 are all equivalent (problem 10). That is, if one of them is true, then all three are true. Thus, we could have taken any one of them as an axiom and proved the other two.

Problems

1. Find the sup and inf of each of the following sets:

   (a) $\{\frac{1}{n} \mid n \in \mathbb{N}\}$.
   (b) $\{2 - \frac{1}{n} \mid n \in \mathbb{N}\}$.
   (c) $\{e^r \mid r \in \mathbb{Q}\}$.
   (d) $\{x^2 \mid 0 \leq x < 2\}$.

2. Let S be a set of real numbers and suppose that a_n is an upper bound for S for each n. Prove that if $a_n \to a$, then a is an upper bound for S.

3. Suppose that a set S of real numbers is bounded and let μ be an upper bound for S. Show that μ is the least upper bound of S if and only if for every $\varepsilon > 0$ there is an element of S in the interval $[\mu - \varepsilon, \mu]$.

4. Prove that least upper bounds are unique. That is, if μ_1 and μ_2 are both least upper bounds for the set S, then $\mu_1 = \mu_2$.

5. Suppose that a is an upper bound for the set S and that there is a sequence of elements $x_n \in S$ such that $x_n \to a$. Prove that a is the least upper bound.

6. (a) Let S_1 and S_2 be two bounded, nonempty sets of real numbers. Prove that

$$sup\{S_1 \cup S_2\} = max\{sup\{S_1\}, sup\{S_2\}\}.$$

   (b) Let S be the set consisting of all numbers of the form $a_1 + a_2$ where $a_1 \in S_1$ and $a_2 \in S_2$. Prove that $sup\{S\} = sup\{S_1\} + sup\{S_2\}$.

7. Show that if $\sup S = +\infty$, then there is a sequence of points $s_n \in S$ such that $s_n \to +\infty$.

8. Let S be a set of real numbers and suppose that $a > 0$. Prove that $\sup\{ax \mid x \in S\} = a \sup S$.

9. Let $\{a_n\}$ and $\{b_n\}$ be bounded sequences and define sets A, B, and C by $A = \{a_n\}$, $B = \{b_n\}$, and $C = \{a_n + b_n\}$. Prove that $\sup C \le \sup A + \sup B$. Give an example to show that strict inequality may hold.

10. (a) Assume that Theorem 2.5.1 is true and prove Theorem 2.4.2.

    (b) Assume that Theorem 2.4.3 is true and prove Theorem 2.4.2.

    (c) Conclude that Theorem 2.4.2, Theorem 2.4.3, and Theorem 2.5.1 are equivalent.

2.6 The Bolzano-Weierstrass Theorem

Suppose that we have a sequence $\{a_n\}$ of real numbers. We can make new sequences from the elements of $\{a_n\}$ by skipping terms:

$$a_2, a_4, a_6, a_8, \ldots$$

$$a_2, a_4, a_8, a_{16}, \ldots$$

$$a_1, a_{15}, a_{83}, a_{84}, a_{135}, \ldots$$

These new sequences are called, for obvious reasons, subsequences of the original sequence $\{a_n\}$. Notice that not only are each of the elements of the subsequence members of the original sequence but, in addition, they must occur in the same order as they did in the original sequence. These ideas are the basis of the following formal definition.

Definition. Let $\{a_n\}_{n=1}^{\infty}$ be a sequence of real numbers. Let $k \mapsto n(k)$ be a function from $\mathbb{N}$ to $\mathbb{N}$ having the property that $n(k + 1) > n(k)$ for all $k \in \mathbb{N}$. Then $\{a_{n(k)}\}_{k=1}^{\infty}$ is called a **subsequence** of $\{a_n\}_{n=1}^{\infty}$.

Normally, we use the simpler notation n_k instead of $n(k)$, and we write the subsequence as $\{a_{n_k}\}$, where it is assumed that k runs from 1 to ∞. Next, we introduce the notion of limit point.

Definition. Let $\{a_n\}$ be a sequence of real numbers. A number d is called a **limit point** of the sequence if, given $\epsilon > 0$ and an integer N, there exists an $n \geq N$ so that $|a_n - d| \leq \varepsilon$.

Intuitively, d is a limit point if the sequence keeps returning close to d. The notions of subsequence and limit point are related to each other, as the following proposition shows.

Proposition 2.6.1   Let $\{a_n\}$ be a sequence of real numbers. Then d is a limit point of the sequence if and only if there exists a subsequence $\{a_{n_k}\}$ of $\{a_n\}$ which converges to d.

Proof.   First suppose that there is a subsequence, $\{a_{n_k}\}$, which converges to a limit d. Let ε and a positive integer N be given. Since $a_{n_k} \to d$ as $k \to \infty$, we can choose a K so that

$$|a_{n_k} - d| \leq \varepsilon \quad \text{for all } k \geq K.$$

Since $n_k \to \infty$ as $k \to \infty$, we can choose a particular $k \geq K$ so that $n_k \geq N$. Thus, we have found an element of $\{a_n\}$, namely, a_{n_k}, so that $n_k \geq N$ and $|a_{n_k} - d| \leq \varepsilon$. Therefore d is a limit point.
    Conversely, suppose that d is a limit point. Then, there exists an n such that $|a_n - d| \leq 1$. Denote this n by n_1. Given n_1 and $\frac{1}{2}$, we can choose an $n_2 > n_1$ so that $|a_{n_2} - d| \leq \frac{1}{2}$. We continue in this way, defining successively larger integers $n_1 < n_2 < \ldots < n_k < \ldots$ so that

$$|a_{n_k} - d| \leq \frac{1}{k}.$$

It follows easily from this inequality that $a_{n_k} \to d$ as $k \to \infty$, so we have found a subsequence of $\{a_n\}$ that converges to d.   ❑

Example 1   Let $a_n = 2(-1)^n + \frac{1}{n}$. Then, since $\frac{1}{n} \to 0$ and $(-1)^n$ oscillates between $+1$ and -1, the sequence $\{a_n\}$ has two limit points, 2 and -2. According to Proposition 2.6.1, these limit points must be the limits of subsequences, and, indeed, they are:

$$a_{2k} = 2 + \frac{1}{2k} \to 2$$

$$a_{2k+1} = -2 + \frac{1}{2k+1} \rightarrow -2.$$

Example 2   Let $a_0 = .2$ and define the rest of the sequence $\{a_n\}$ by the recursive relation

$$a_{n+1} = (3.3)a_n(1 - a_n). \tag{13}$$

Using a hand calculator, we can compute approximately as many terms of the sequence as we like:

$a_0 =$	.2	$a_7 =$	.479631	$a_{14} =$	.823603
$a_1 =$	.528	$a_8 =$	.823631	$a_{15} =$	.479428
$a_2 =$	.822413	$a_9 =$	.479368	$a_{16} =$	.823603
$a_3 =$	.481965	$a_{10} =$	.823595	$a_{17} =$	.479427
$a_4 =$	.823927	$a_{11} =$	.479444	$a_{18} =$	.823603
$a_5 =$	.478736	$a_{12} =$	.823606	$a_{19} =$	.479427
$a_6 =$	.823508	$a_{13} =$	.479422	$a_{20} =$	.823603

The sequence appears to settle down to oscillating between the number .479427 and the number .823603. If this is the case, then the sequence would have two limit points, .479427 and .823603. The subsequence $\{a_{2n}\}$ would converge to .823603 and the subsequence $\{a_{2n+1}\}$ would converge to .479427. We use the word "would" because we have not *proved* that the sequence behaves in this way; we have found strong numerical evidence. One of the interesting things about the iteration (13) is that no matter what number between 0 and 1 we choose for a_0, the resulting sequence settles down to oscillating between the same two numbers, .479427 and .823603. This is discussed further in Section 2.7.

The proof of the following theorem, named after Bernhard Bolzano (1781 – 1848) and Karl Weierstrass (1815 – 1897), uses the idea of nested subintervals that was used in the proof of Theorem 2.4.3.

❏ **Theorem 2.6.2 (The Bolzano-Weierstrass Theorem)**   Every bounded sequence has a convergent subsequence.

Proof.   Let $\{a_n\}$ be the sequence. Since it is bounded, all of the terms are contained in an interval of the form $[-M, M]$ for some M. We will construct the subsequence and a nested family $\{I_k\}$ of subintervals whose lengths shrink to zero so that each I_k contains the tail of the subsequence.

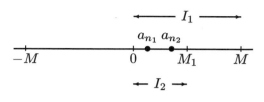

Figure 2.6.1

We first divide the interval into two parts at the midpoint. If $[0, M]$ contains infinitely many terms in the sequence $\{a_n\}$, we choose a_{n_1} to be one of those terms and choose $I_1 = [0, M]$. If $[0, M]$ contains only finitely many terms, then $[-M, 0]$ must contain infinitely many terms of $\{a_n\}$. In this case we let a_{n_1} be one of those terms and choose $I_1 = [-M, 0]$. In either case, I_1 has length M and contains a_{n_1} and infinitely many other terms of $\{a_n\}$. We now divide I_1 at its midpoint, making two subintervals. Either the right subinterval or the left subinterval (or both) must contain infinitely many points of $\{a_n\}$. Choose I_2 to be a subinterval with infinitely many points of $\{a_n\}$ and define a_{n_2} to be any member of $\{a_n\}$ in this subinterval whose index is greater than n_1. Note that the length of I_2 is $2^{-1}M$. See Figure 2.6.1.

Continuing in this manner we construct a nested family, $\{I_k\}$, of intervals, each containing all the succeeding ones. The length of I_k is $2^{-k+1}M$, and a term of the sequence a_{n_k} is in I_k. Since the intervals are nested, each I_k contains a_{n_j} for all $j \geq k$. Given $\varepsilon > 0$, we can choose K so that $2^{-K+1}M \leq \varepsilon$. If $j \geq K$ and $k \geq K$, both a_{n_j} and a_{n_k} are in I_K, so

$$|a_{n_j} - a_{n_k}| \leq 2^{-K+1}M \leq \varepsilon.$$

Thus the subsequence $\{a_{n_k}\}$ is a Cauchy sequence, and so by Theorem 2.4.2 it converges to a finite limit.   ❏

Corollary 2.6.3   If $\{a_n\}$ is a sequence of numbers in the closed interval $[b, c]$, then a subsequence of $\{a_n\}$ converges to a point in $[b, c]$.

Proof.   Since the sequence consists of numbers in the interval $[b, c]$ it is a bounded sequence and therefore, by the Bolzano-Weierstrass theorem, a subsequence $\{a_{n_k}\}$ converges to a finite limit, a. Suppose a is not in $[b, c]$. Then we can choose an $\epsilon > 0$ so that the intervals $[a - \epsilon, a + \epsilon]$ and $[b, c]$ are disjoint. Since $a_{n_k} \to a$, we know that a_{n_k} is in $[a - \epsilon, a + \epsilon]$ for k large enough. But this is a contradiction since all the a_n are in $[b, c]$ by hypothesis. Thus, a must be in $[b, c]$ too.   ❏

We shall see in the succeeding chapters that the Bolzano-Weierstrass theorem is extremely useful and arises in unexpected circumstances. Project 5 outlines an application to number theory.

Problems

1. Prove that if a sequence converges it has exactly one limit point. Is the converse true? Prove it or give a counterexample.

2. For each of the following sequences, find the limit points. For each limit point find a subsequence that converges to it:

   (a) $a_n = \frac{(-1)^n}{1+n}$.

   (b) $a_n = (-1)^n + \frac{1}{1+n}$.

   (c) $a_n = \sin \frac{n\pi}{4}$.

3. Suppose that the sequence $\{a_n\}$ converges to a and that d is a limit point of the sequence $\{b_n\}$. Prove that ad is a limit point of the sequence $\{a_n b_n\}$.

4. Let c be a limit point of $\{a_n\}$ and d be a limit point of $\{b_n\}$. Is $c + d$ necessarily a limit point of $\{a_n + b_n\}$? Prove it or give a counterexample.

5. Let $\{y_j\}_{j=1}^N$ be N given real numbers. Construct a sequence $\{a_n\}$ so that $\{y_j\}_{j=1}^N$ is the set of limit points of $\{a_n\}$ but $a_n \neq y_j$ for any n or j.

6. Consider the following sequence: $a_1 = \frac{1}{2}$; the next three terms are $\frac{1}{4}, \frac{1}{2}, \frac{3}{4}$; the next seven terms are $\frac{1}{8}, \frac{1}{4}, \frac{3}{8}, \frac{1}{2}, \frac{5}{8}, \frac{3}{4}, \frac{7}{8}; \dots$ and so forth. What are the limit points?

7. Let $\{d_n\}$ be a sequence of limit points of the sequence $\{a_n\}$. Suppose that $d_n \to d$. Prove that d is a limit point of $\{a_n\}$.

8. Let $\{I_k\}_{k=1}^\infty$ be a nested family of closed, finite intervals; that is, $I_1 \supseteq I_2 \supseteq \dots$. Prove that there is a point p contained in all the intervals, that is, $p \in \cap_{k=1}^\infty I_k$.

9. Suppose that $\{x_n\}$ is a monotone increasing sequence of points in $\mathbb{R}$ and suppose that a subsequence of $\{x_n\}$ converges to a finite limit. Prove that $\{x_n\}$ converges to a finite limit.

10. A sequence of points in the plane, $\{(x_n, y_n)\}$, is said to be **bounded** if there is an M so that $0 \leq |x_n| \leq M$ and $0 \leq |y_n| \leq M$ for all n. Explain why these definitions are geometrically reasonable. Prove that every bounded sequence in the plane has a convergent subsequence.

11. Let $\{(x_n, y_n)\}$ be a sequence of points in a rectangle $R \equiv [a, b] \times [c, d]$. Prove that $\{(x_n, y_n)\}$ has a subsequence which converges to a point of R.

12. Let $\{p_n\}$ be a sequence of points in $\mathbb{R}^2$. Use the notion of convergence introduced in problem 10 of Section 2.2 to

(a) Define what it means for a point $p \in \mathbb{R}^2$ to be a limit point of $\{p_n\}$.

(b) Prove that p is a limit point of $\{p_n\}$ if and only if $\{p_n\}$ has a subsequence which converges to p.

(c) Determine the limit points of the sequence $\{p_n\}$ if for each n, $p_n = ((-1)^n, \frac{2n}{n+1})$.

(d) Determine the limit points of $\{p_n\}$ if the polar coordinates of p_n are $r_n = 2 - \frac{1}{n}$ and $\theta_n = \frac{n\pi}{4}$.

13. Let $\{I_k\}_{k=1}^{\infty}$ be a family of closed finite intervals which has the property that the intersection of any finite subcollection of the intervals is nonempty. Prove that there is a point p contained in $\cap_{k=1}^{\infty} I_k$. Hint: consider the sets $J_k \equiv \cap_{n=1}^{k} I_n$ and use the idea of the proof of problem 8.

2.7 The Quadratic Map

In this section we consider sequences which are defined recursively by the relation

$$x_{n+1} = ax_n(1 - x_n), \tag{14}$$

where $a > 0$ and x_0 is a given number in the interval $[0, 1]$. If we define

$$F(x) \equiv ax(1 - x)$$

then the maximum of F on $[0, 1]$ occurs at $x = \frac{1}{2}$ and the maximum value is $\frac{a}{4}$. Thus, if $a \leq 4$, $F(x)$ is in $[0, 1]$ if x is in $[0, 1]$, so $\{x_n\}$ is a sequence of numbers between 0 and 1.

Sequences produced by iterative formulas such as (14) arise in a variety of applications. For example, in a simple model of limited population growth, x_n represents the fraction of some maximal population present in year n. Given an initial population x_0, one wants to understand the behavior of the whole sequence $\{x_n\}$, which is called the **orbit** of x_0 under the iteration. Will the population approach a limiting value? How does the behavior of the orbit depend on the initial condition x_0? How does the behavior of the orbit depend on the parameter a? These are the kinds of questions that one asks about most dynamical systems, both discrete and continuous. The importance of (14) is that, even though it looks so simple, it has many interesting properties usually associated with more complicated dynamical systems. We will see, for example, that as the

parameter a is changed, the behavior of the orbit $\{x_n\}$ changes dramati-
cally. When a is small, the orbits are quite regular, but as a becomes large
the orbits become very complicated.

First, we will consider the case $0 < a < 1$. If we choose a numerical
value for a and choose $x_0 \,\epsilon\, (0,1)$, then we can compute on a hand calculator as many iterates as we like. For each such numerical experiment,
we see that $x_n \to 0$ as $n \to \infty$. This gives us an idea about the behavior
of the orbits in the case $0 < a < 1$, but it is not a proof, of course, since we
are looking at only finitely many of the terms of each orbit and we cannot
perform the experiment for all possible choices of a and x_0. Furthermore,
because of round-off error the numerical calculations are only approximate. What if orbits that start close diverge from each other quickly?
In this case, the numerical experiments are likely to be inaccurate and
may be quite misleading. In addition, the numerical experiments do not
give us insight into the reason why the sequence $\{x_n\}$ converges to zero.
To gain understanding, we can construct the orbit approximately by a
geometric procedure.

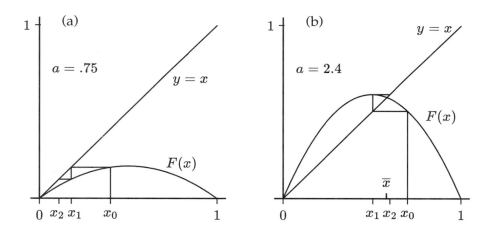

Figure 2.7.1

Given x_0, the horizontal line through the point $(x_0, F(x_0))$ intersects the
line $y = x$ in the point $(F(x_0), F(x_0))$. See Figure 2.7.1(a). Since $x_1 =
F(x_0)$, the first coordinate of this point is just x_1. The horizontal line
through $(x_1, F(x_1))$ intersects the line $y = x$ in the point $(F(x_1), F(x_1))$,
whose first coordinate is x_2. Continuing in this way, we can construct
approximately as many terms of the orbit $\{x_n\}$ as we like and see that
they get smaller and smaller because the graph of $F(x)$ lies below the

line $y = x$. We can also give an analytical proof that $x_n \to 0$.

❑ **Theorem 2.7.1**   Suppose that $0 < a \le 1$, $0 \le x_0 \le 1$, and the sequence $\{x_n\}$ satisfies (14). Then, $x_n \to 0$ as $n \to \infty$.

Proof.   If $x_0 = 0$ or $x_0 = 1$, all the rest of the terms in the orbit are zero so there is nothing more to prove. On the other hand, if $0 < x_n < 1$, then

$$x_{n+1} = a x_n (1 - x_n) < x_n$$

since $a \le 1$ and $1 - x_n < 1$. Thus, the sequence is strictly monotone decreasing and bounded below by zero. By Theorem 2.4.3, $\{x_n\}$ converges to some $\bar{x}$, and since each $x_n \ge 0$, we know that $\bar{x} \in [0, 1]$. Taking the limit of both sides of (14), using Theorems 2.2.4 and 2.2.5, we see that $\bar{x}$ satisfies

$$\bar{x} = a\bar{x}(1 - \bar{x}). \tag{15}$$

Since $a \le 1$ and $\bar{x} \in [0, 1]$, this can only be true if $\bar{x} = 0$. Thus, $x_n \to 0$ as $n \to \infty$.   ❑

Even in this very simple case we see the value of three different methods of investigation: numerical experimentation, geometric construction, and analytical proof. Let's try this same approach for $a > 1$. Set $a = 2.4$. If we try two different values for x_0, namely, $x_0 = .1$ and $x_0 = .85$, and iterate (14), we find

x_0	.1	.85
x_1	.216	.306
x_2	.406426	.509674
x_3	.578985	.599775
x_4	.585027	.576108
x_5	.582649	.586098
x_6	.583606	.582209
x_7	.583224	.58378
x_8	.583377	.583154
x_9	.583316	.583405
x_{10}	.58334	.583305

In both cases the sequences seem to be converging to the same number, something near .583. We know from the argument in the proof of

Theorem 2.7.1 that if $\{x_n\}$ converges, then the limit $\bar{x}$ must satisfy (15). Solving (15) algebraically, we find that there are two possibilities, $\bar{x} = 0$ and $\bar{x} = \frac{a-1}{a}$. Since $a = 2.4$, we calculate $\frac{a-1}{a} = .583333$, so now we see where the .583 came from. But why does the sequence converge to .583333 and not to zero? Let's look at the graph of F in Figure 2.7.1(b) and construct the successive points by the same geometrical method as above. The x coordinate of the point where the graph of F intersects the line $y = x$ is $\bar{x}$. If x_n is less than $\bar{x}$, then x_{n+1} will be larger than x_n because the graph of F is higher than the line $y = x$ for $x < \bar{x}$. This explains why the sequence cannot converge to zero. If x_n is greater than $\bar{x}$, then x_{n+1} will be less than x_n because the graph of F is lower than the line for $x > \bar{x}$. But that in and of itself doesn't prove that the sequence will converge to $\bar{x}$ since it could hop back and forth from one side of $\bar{x}$ to the other. In fact, in the case shown in Figure 2.7.1(b), it does hop back and forth, but, nevertheless, the sequence x_n converges to $\frac{a-1}{a}$. We shall give an analytical proof in the case $1 < a < 2\sqrt{2}$.

❏ **Theorem 2.7.2** Suppose that $1 < a < 2\sqrt{2}$ and that the sequence $\{x_n\}$ satisfies (14). Then, for all $0 < x_0 < 1$, the sequence $\{x_n\}$ converges to $\frac{a-1}{a}$ as $n \to \infty$.

Proof. Let $\bar{x} = \frac{a-1}{a}$. Recall that the function F achieves its maximum value at $x = \frac{1}{2}$ and that value is $\frac{a}{4}$. Thus, $x_n \leq \frac{a}{4}$ for $n \geq 1$. Now, choose a number δ small enough so that

$$a(1 - \delta) > 1 \quad \text{and} \quad \delta \leq F(\tfrac{a}{4}) = a\,\tfrac{a}{4}(1 - \tfrac{a}{4}).$$

We can choose such a δ satisfying the first inequality since $a > 1$. The reasons for this choice will become clear shortly. First of all, notice that the graph of F is concave downward since $F''(x) = -2a$. Thus on the closed interval $[\delta, \frac{a}{4}]$, the minimum value of F occurs at one of the endpoints. By our choice of δ above, $F(\delta) = a\delta(1 - \delta) > \delta$ and $F(\frac{a}{4}) \geq \delta$. Thus,

$$\delta < F(x) \leq \tfrac{a}{4} \qquad \text{if} \qquad \delta \leq x \leq \tfrac{a}{4}. \tag{16}$$

Therefore, if x_1 is in the interval $[\delta, \frac{a}{4}]$, then all the succeeding iterates will be in $[\delta, \frac{a}{4}]$. On the other hand, if any $x_j < \delta$, then

$$x_{j+1} = ax_j(1 - x_j) > a(1 - \delta)x_j.$$

Iterating this inequality shows that if $x_1, \ldots, x_k$ are all $< \delta$, then

$$x_k > (a(1 - \delta))^{k-1}x_1.$$

Since $a(1 - \delta) > 1$ and $x_1 > 0$, this gives the contradiction $x_k \geq \delta$ for k large. Therefore, for each x_0, some iterate, x_k must satisfy $x_k \geq \delta$, and, by what we proved above, all the succeeding iterates, x_n for $n \geq k$, will remain in $[\delta, \frac{a}{4}]$.

To prove the convergence, we subtract $\bar{x} = \frac{a-1}{a}$ from both sides of (14) and rearrange algebraically to obtain

$$(x_{n+1} - \bar{x}) \;=\; (1 - ax_n)(x_n - \bar{x}) \tag{17}$$

Since x_n is in $[\delta, \frac{a}{4}]$, $ax_n \geq a\delta$, and the hypothesis that $1 < a < 2\sqrt{2}$ implies that $ax_n \leq \frac{a^2}{4} < 2$. Thus,

$$-1 \;<\; 1 - \tfrac{a^2}{4} \;\leq\; 1 - ax_n \;\leq\; 1 - a\delta \;<\; 1.$$

Therefore, if we define

$$\beta \;\equiv\; \max\left\{|1 - \tfrac{a^2}{4}|,\, |1 - a\delta|\right\} \;<\; 1,$$

it follows from (17) that

$$|x_{n+1} - \bar{x}| \;\leq\; \beta|x_n - \bar{x}| \tag{18}$$

for all $n \geq k$. Iterating the inequality (18) gives

$$|x_n - \bar{x}| \;\leq\; \beta^{(n-k)}(x_k - \bar{x}),$$

which proves that $x_n \to \bar{x}$ as $n \to \infty$. ❏

Now let's investigate the behavior of the iteration for $a > 3$. If we set $a = 3.3$ and use a hand calculator to compute the first 20 iterations starting with three different initial values of x_0, we find the results in the following table:

x_0	.1	.5	.88
x_1	.297	.825	.34848
x_2	.68901	.476438	.749238
x_3	.707108	.823168	.620006
x_4	.683451	.480356	.777475
x_5	.713941	.823727	.570925
x_6	.673957	.479164	.8084
x_7	.725139	.823567	.511135
x_8	.657731	.479504	.824591

x_9	.742899	.823614	.477315
x_{10}	.6303	.479405	.823302
x_{11}	.768972	.8236	.480071
x_{12}	.586258	.479433	.823689
x_{13}	.800447	.823604	.479243
x_{14}	.527115	.479425	.823578
x_{15}	.822574	.823603	.479481
x_{16}	.481622	.479428	.823611
x_{17}	.823885	.823603	.479412
x_{18}	.478824	.479427	.823601
x_{19}	.82352	.823603	.479432
x_{20}	.479604	.479427	.823604

The sequences certainly don't look like they are converging! As n gets larger, they seem to alternate from close to .479 to close to .823. Amazingly, any other initial condition between 0 and 1 seems to give a sequence with the same properties. To understand what is happening, we look at the composition of F with itself, $F^{(2)}(x)$,

$$F^{(2)}(x) \equiv F(F(x)) = a^2 x(1-x)(1 - ax(1-x)),$$

which is a quartic polynomial. The graph of $F^{(2)}(x)$ is shown in Figure 2.7.2. In addition to $x = 0$, there are three other points, labeled p_1, $\bar{x}$, and p_2 in Figure 2.7.2, such that $F^{(2)}(x) = x$. These points are called **fixed points** of $F^{(2)}$. The point $\bar{x} = \frac{a-1}{a}$ is easy to understand. Since $\bar{x}$ is a fixed point of F, it is automatically a fixed point of $F^{(2)}$. Consider the point p_1. Since $F^{(2)}(F(p_1)) = F(F^{(2)}(p_1)) = F(p_1)$, the point $F(p_1)$ is a fixed point of $F^{(2)}$. Since the odd iterates of $F(p_1)$ equal p_1, $F(p_1)$ cannot be 0 or $\bar{x}$. If $F(p_1) = p_1$, then p_1 would be a fixed point of F. Since the only fixed points of F are 0 and $\bar{x}$, we conclude that $F(p_1) = p_2$, and likewise, $F(p_2) = p_1$. Thus, if $x_0 = p_1$ or $x_0 = p_2$, the iterates oscillate back and forth between p_1 and p_2. Such an orbit is said to have **period two**. The numerical evidence above suggests that any orbit starting with x_0 between 0 and 1 gets closer and closer as $n \to \infty$ to an orbit that oscillates between p_1 and p_2. This can be proven analytically if $x_0 \neq \bar{x}$

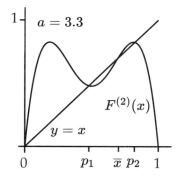

Figure 2.7.2

by showing that one of the two subsequences,

$$x_{2n+2} = F^{(2)}(x_{2n})$$

and

$$x_{2n+1} = F^{(2)}(x_{2n-1}),$$

converges to p_1 and the other to p_2 using techniques similar to those in Theorem 2.7.2. If $x_0 = \bar{x}$, then all the terms $x_n = \bar{x}$. Note that we would not normally be able to see this numerically because the infinite decimal expansion of $\bar{x}$ will be truncated by the calculator and the resulting orbit will get closer and closer to the period-two orbit discussed above.

And this is just the beginning! As a gets still larger, orbits of period $4, 8, \ldots$ and so on appear and the dynamics of the iterates becomes more interesting and difficult to analyze. When $a = 4$ the iterates are "chaotic" in the sense that most orbits wander everywhere, there is sensitive dependence of the orbit on the initial point, x_0, and the points with periodic orbits are dense. Further work in this direction would take us too far from our goal. Our intention here is just to show that investigating an iteration even as simple as (14) can lead to deep and interesting questions in analysis. For further development of the mathematics of discrete dynamical systems, see [9] and [8].

Problems

1. Consider the iteration

$$x_{n+1} = -\frac{1}{2}x_n.$$

   What happens to the iteration for different choices of x_0 in $\mathbb{R}$. Prove your conclusions.

2. Consider the iteration

$$x_{n+1} = x_n^2.$$

   What happens to the iteration for different choices of x_0 in $\mathbb{R}$. Prove your conclusions.

3. Consider the iteration

$$x_{n+1} = \sin x_n$$

   for $0 \leq x_0 \leq \pi$. Prove that $x_n \to 0$. Hint: derive the inequality $\sin x < x$ for $0 < x \leq \pi$ by using the Fundamental Theorem of Calculus.

4. Let a, b, and x_0 be positive numbers and consider the iteration

$$x_{n+1} = \frac{ax_n}{b + x_n}.$$

Find out all you can about the orbits by numerical experimentation and analytical proof.

5. Consider the iteration

$$x_{n+1} = x_n^3 - \lambda x_n,$$

where x_0 is any real number and $0 < \lambda < 1$. Find out all you can about the orbits by numerical experimentation and analytical proof.

The following discussion provides background for problems 6 – 9. For more information, see [25]. Consider a gene locus having two alleles, **a** and **b**. Individuals in a population of size N are said to have genetic types **aa**, **ab**, or **bb**, according to whether they have two **a** alleles, one of each, or two **b** alleles. Thus, there are $2N$ alleles in the population. We assume that the population reproduces synchronously and that each new individual receives two alleles chosen randomly from the gene pool of the previous generation. Let x_n denote the fraction of **a** alleles in the gene pool at the nth generation. Constants α, β, and γ, called fitness coefficients, represent the probabilities that each of the three genetic types will reach maturity and reproduce. Thus the total gene pool in the next generation will be proportional to

$$\alpha x_n^2 + 2\beta x_n(1 - x_n) + \gamma(1 - x_n)^2,$$

and the number of **a** alleles will be proportional to

$$\alpha x_n^2 + \beta x_n(1 - x_n).$$

Therefore,

$$x_{n+1} = \frac{\alpha x_n^2 + \beta x_n(1 - x_n)}{\alpha x_n^2 + 2\beta x_n(1 - x_n) + \gamma(1 - x_n)^2}.$$

6. (a) Show that if $0 < x_0 < 1$, then $0 < x_n < 1$ for all n.

   (b) Show that if $\alpha = \beta = \gamma$, then x_n remains constant in all generations.

7. Suppose that $\alpha > \beta > \gamma$. Prove that $x_n \to 1$ as $n \to \infty$.

8. Suppose that $\alpha < \beta < \gamma$. Prove that $x_n \to 0$ as $n \to \infty$.

9. Suppose that $\alpha < \beta$ and $\gamma < \beta$. Find numerical evidence that $\{x_n\}$ converges to a number $\bar{x}$ satisfying $0 < \bar{x} < 1$ and characterize $\bar{x}$ analytically.

Projects

1. The Fibonacci sequence, 1, 2, 3, 5, 8, 13, ..., is given by the recursive formula

$$F_{n+1} = F_n + F_{n-1},$$

where $F_1 = 1$ and $F_2 = 2$. Let $a_n = F_n/F_{n-1}$.

   (a) Suppose that $\{a_n\}$ converges to a limit. What must that limit be? Hint: divide the above equation by F_n to find an equation relating a_{n+1} and a_n.

   (b) Show that $3/2 \leq a_n \leq 2$ for all $n \geq 2$.

   (c) For each $n \geq 2$, prove that

   $$|a_{n+1} - a_n| \leq (2/3)^2 |a_n - a_{n-1}|.$$

   (d) Prove that for each $m \geq 2$,

   $$|a_{m+1} - a_m| \leq (2/3)^{2(m-1)} |a_2 - a_1|.$$

   (e) Use the inequality in (d) to show that $\{a_n\}$ is a Cauchy sequence and therefore converges to a limit. Hint: express

   $$a_n - a_m = (a_n - a_{n-1}) + (a_{n-1} - a_{n-2}) + \ldots + (a_{m+1} - a_m)$$

   and use the geometric series.

2. The purpose of this project is to derive the result of Section 2.3 more easily using the tools of linear algebra.

   (a) Show that 1 and $(1 - p - q)$ are the two eigenvalues of the matrix A.

   (b) Find a nonsingular matrix Q so that $A = QDQ^{-1}$, where D is the diagonal matrix

   $$D = \begin{pmatrix} 1 & 0 \\ 0 & 1 - p - q \end{pmatrix}.$$

   (c) Explain why $A^n = QD^nQ^{-1}$.

   (d) Prove that

   $$\lim_{n \to \infty} A^n \begin{pmatrix} p_0 \\ q_0 \end{pmatrix} = \begin{pmatrix} \frac{p}{p+q} \\ \frac{q}{p+q} \end{pmatrix}.$$

3. We are interested in the price of a commodity which is traded at regular intervals. We let Q_k denote the supply of the commodity, D_k the demand for the commodity, and p_k the price at the kth time. The demand depends on the current price,

$$D_k = a + bp_k,$$

and the supply depends on the previous price,

$$Q_k = c + dp_{k-1}.$$

(a) Explain why it is reasonable to take a, c, and d to be positive and b to be negative.

(b) Suppose we make the assumption that the supply is always equal to the demand. Find the difference equation satisfied by the sequence $\{p_k\}$.

(c) Suppose that the sequence of prices $\{p_k\}$ converges to a limiting price $\bar{p}$. What must $\bar{p}$ be?

(d) Find a condition on the coefficients so that you can prove that $p_k \to \bar{p}$. Why is it reasonable that the conditions depend on d and b?

(e) Use specific choices of the coefficients and find numerical evidence that the prices can oscillate wildly if the condition is not satisfied.

(f) Iterate the equation for p_k and use the partial sum of the geometric series to prove that the sequence does not converge if the coefficients do not satisfy the condition in (d).

4. The gas CO_2 is produced by the cellular metabolism of the body and is removed from the body when one exhales. There is a mechanism in the brain which senses the CO_2 level and sends signals to the breathing mechanism to take deeper breathes if the CO_2 level is too high. Let V_n be the volume of the nth breath and let C_n be the CO_2 concentration at the time of the nth breath. Let M be the (assumed) constant CO_2 concentration produced by the metabolism between each breath. Then, a simple model of this system is:

$$C_{n+1} = C_n - \alpha V_n + M \qquad (19)$$
$$V_{n+1} = \beta C_n \qquad (20)$$

The second equation says that the volume of the $n + 1$ breath is proportional to the CO_2 concentration at the time of the nth breath. The first equation says that CO_2 concentration C_n is lowered by an amount proportional to the volume of the nth breath and raised by M. Given the initial concentration C_0 and breath size V_0, we would like to determine the behavior of the sequences $\{C_n\}$ and $\{V_n\}$. Our discussion follows [10], where more information can be found.

(a) Show that the constant sequences $C_n = M/\alpha\beta$ and $V_n = M/\alpha$ solve (19) and (20).

(b) Define $c_n \equiv C_n - M/\alpha\beta$ and $v_n \equiv V_n - M/\alpha$. Show that the sequences $\{c_n\}$ and $\{v_n\}$ satisfy

$$c_{n+1} = c_n - \alpha v_n \qquad (21)$$
$$v_{n+1} = \beta c_n. \qquad (22)$$

(c) Show that $\{c_n\}$ satisfies the second order difference equation,

$$c_{n+1} - c_n + \alpha\beta c_{n-1} = 0. \qquad (23)$$

(d) Show that for appropriate choices of λ_1 and λ_2,

$$c_n = d_1 \lambda_1^n + d_2 \lambda_2^n$$

solves (23) for any choice of d_1 and d_2.

(e) For any choice of V_0 and C_0, show that d_1 and d_2 can be chosen so that

$$
\begin{aligned}
C_n &= d_1 \lambda_1^n + d_2 \lambda_2^n + M/\alpha\beta \\
V_n &= \beta(d_1 \lambda_1^n + d_2 \lambda_2^n) + M/\alpha
\end{aligned}
$$

solve the original problem.

(f) Prove that if $4\alpha\beta < 1$, then $C_n \to M/\alpha\beta$ and $V_n \to M/\alpha$ as $n \to \infty$, so that whatever the initial concentration and breath size, the breathing settles down to a steady rate.

(g) Show that if $1 < 4\alpha\beta < 4$, the concentrations exhibit oscillations which die out.

(h) Show that if α and β are large enough, the size of breaths and the CO_2 concentration can exhibit larger and larger oscillations. Can you explain this analytically?

5. We shall prove a theorem in number theory which uses the Bolzano-Weierstrass theorem. The proof is longer and more difficult than most of the proofs in Chapter 2. The project is to read and understand the proof, filling in all the details, and to prepare a lecture on it for presentation to the class.

For each positive integer n, there is a unique nonnegative integer j_n so that n lies in the half-open interval

$$2\pi j_n \leq n < 2\pi j_n + 2\pi.$$

Define

$$\alpha_n \equiv n - 2\pi j_n.$$

The number α_n, which is denoted $n(mod 2\pi)$ in number theory, is just the remainder when n is divided by 2π.

❑ **Theorem.**   The set of limit points of the sequence $\{\alpha_n\}$ is the entire interval $[0, 2\pi]$.

Proof.   Let $\varepsilon > 0$ and N be given. We will show that for any $x \in [0, 2\pi)$, there is an $n \geq N$ so that

$$|\alpha_n - x| \leq \varepsilon. \tag{24}$$

We first note that $n \neq m$ implies that $\alpha_n \neq \alpha_m$. If $\alpha_n = \alpha_m$, then

$$m - n = 2\pi(j_m - j_n),$$

but this would imply that π is a quotient of integers and therefore rational. Since π is irrational, we must have $\alpha_n \neq \alpha_m$. Now, we apply the Bolzano-Weierstrass theorem to the sequence $\{\alpha_n\}$, which is bounded because each α_n is in $[0, 2\pi)$. Since a subsequence converges, we can choose a large integer $m \geq N$ and a larger integer $n \geq m$ so that $|\alpha_n - \alpha_m| \leq \varepsilon$. Let $k \equiv n - m$. Then

$$k = \alpha_n - \alpha_m + 2\pi(j_n - j_m).$$

There are two cases, depending on whether $\alpha_n - \alpha_m$ is positive or negative. Suppose $\alpha_n - \alpha_m \equiv \delta > 0$. Then,

$$\alpha_k = \alpha_n - \alpha_m = \delta,$$

and since

$$2k = 2(\alpha_n - \alpha_m) + 4\pi(j_n - j_m),$$

we have $\alpha_{2k} = 2\delta$. Continuing in this manner, we see that the numbers $\alpha_k, \alpha_{2k}, \alpha_{3k}, \ldots$ equal $\delta, 2\delta, 3\delta, \ldots$, respectively. Each $x \in [0, 2\pi]$ is therefore within δ of one of the numbers $\alpha_k, \alpha_{2k}, \alpha_{3k}, \ldots$. Since $\delta \leq \varepsilon$, this proves (24). The other case is handled similarly. If $\delta < 0$, then $\alpha_k = 2\pi + \delta, \alpha_{2k} = 2\pi + 2\delta$, and so forth. As above, this proves that any x is within δ of one of the points $\alpha_k, \alpha_{2k}, \alpha_{3k}, \ldots$.   ❑

CHAPTER 3

The Riemann Integral

In this chapter we define what it means for a function to be continuous and we construct the Riemann integral. Continuity is defined in Section 3.1, and we prove several fundamental theorems about continuous functions on closed intervals in Section 3.2, for example, that they have maxima and minima. The Riemann integral for continuous functions on finite intervals is introduced in Section 3.3, where we show that the integral is well defined and has the properties that we expect. In Section 3.4, Taylor's theorem (from Section 4.3) is used to prove error estimates for numerical integration techniques. The integration theory is extended to functions with discontinuities in Section 3.5 and to unbounded continuous functions and functions on infinite intervals in Section 3.6.

3.1 Continuity

Throughout the first few chapters of this book we study functions which are defined on subsets of $\mathbb{R}$ and take values in $\mathbb{R}$. Recall that the set on which a function f is defined is called its domain and is denoted by $Dom(f)$. Normally the functions we deal with will be defined on intervals (open, closed, or half open) or unions of intervals. For example, the function $f(x) = (2 - x^2)^{-1}$ is naturally defined on the union of the intervals $(-\infty, -\sqrt{2}), (-\sqrt{2}, \sqrt{2})$, and $(\sqrt{2}, \infty)$, but not at the points $\sqrt{2}$ or $-\sqrt{2}$. Occasionally, functions are defined on more complicated sets.

Definition. A function f is said to be **continuous at a point** c in $Dom(f)$ if for every sequence $\{x_n\}$ of points of $Dom(f)$ which converges to c we have

$$\lim_{n\to\infty} f(x_n) \;=\; f(c). \tag{1}$$

f is said to be **continuous** if it is continuous at every c in $Dom(f)$.

The definition says that a function is continuous at a point if, as one gets closer to the point, the function values approach the function value at the point.

Example 1   Consider the function $f(x) = x^2$ whose domain is the whole real line $\mathbb{R}$. Let $c \in \mathbb{R}$ be given and suppose $x_n \to c$. Then,

$$\lim_{n \to \infty} f(x_n) = \lim_{n \to \infty} x_n^2$$

$$= (\lim_{n \to \infty} x_n)^2$$

$$= c^2$$

$$= f(c),$$

where we used Theorem 2.2.5 in the second step. Thus, f is continuous at c. Since c was arbitrary, f is continuous on $\mathbb{R}$.

The idea in Example 1 can easily be generalized to give a proof that all polynomials are continuous functions on $\mathbb{R}$ (problem 6 of Section 2.2). Let $p(x) = b_0 + b_1 x + b_2 x^2 + \ldots + b_N x^N$. We can take all of $\mathbb{R}$ as the domain of p since p is well defined for all real numbers. Choose any $c \in \mathbb{R}$ and suppose that $x_n \to c$. Then, using Theorems 2.2.3, 2.2.4, and 2.2.5,

$$\lim_{n \to \infty} p(x_n) = \lim_{n \to \infty} \{b_0 + b_1 x_n + b_2 x_n^2 + \ldots + b_N x_n^N\}$$

$$= \lim_{n \to \infty} b_0 + \lim_{n \to \infty} b_1 x_n + \lim_{n \to \infty} b_2 x_n^2 + \ldots + \lim_{n \to \infty} b_N x_n^N$$

$$= b_0 + b_1 \lim_{n \to \infty} x_n + b_2 \lim_{n \to \infty} x_n^2 + \ldots + b_N \lim_{n \to \infty} x_n^N$$

$$= b_0 + b_1 \lim_{n \to \infty} x_n + b_2 (\lim_{n \to \infty} x_n)^2 + \ldots + b_N (\lim_{n \to \infty} x_n)^N$$

$$= b_0 + b_1 c + b_2 c^2 + \ldots + b_N c^N$$

$$= p(c).$$

Thus p is continuous at $x = c$. Since c was arbitrarily chosen, we have shown that p is continuous on $\mathbb{R}$. Notice that p is therefore automatically continuous on any subset S contained in $\mathbb{R}$. Therefore, if a function is made up of pieces of polynomials (as in Example 2, below), the only points where the function might not be continuous are the points where the pieces match up, that is, where the definition formula of the function changes.

Example 2 Consider the following function defined on $[0, 2]$:

$$f(x) = \begin{cases} 2x & \text{if } 0 \le x < 1 \\ 2 - \frac{1}{2}x^2 & \text{if } 1 \le x \le 2. \end{cases}$$

Since f is made up of pieces of polynomials and polynomials are continuous, f is continuous on the interval $[0, 1)$ and on the interval $(1, 2]$. In Figure 3.1.1, it looks as if f is not continuous at $x = 1$ because the values of f as x approaches 1 from the left approach 2, while $f(1) = 1.5$. To see that the criterion in the definition is violated, let $x_n = 1 - \frac{1}{n}$. Then, $x_n \to 1$ as $n \to \infty$, but

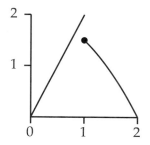

$$\begin{aligned} \lim_{n \to \infty} f(x_n) &= \lim_{n \to \infty} 2(1 - \tfrac{1}{n}) \\ &= 2 \\ &\ne f(1). \end{aligned}$$

Figure 3.1.1

Thus, f is continuous at all points of $[0, 2]$ except 1.

One of the nicest properties of continuous functions is that when they are added or multiplied the result is again a continuous function.

❑ **Theorem 3.1.1**   Let f and g be continuous functions and define $D = Dom(f) \cap Dom(g)$. Then,

(a) $f + g$ is continuous on D.

(b) For any constant κ, the function κf is continuous on $Dom(f)$.

(c) fg is continuous on D.

(d) $\frac{f}{g}$ is continuous at all $x \in D$ such that $g(x) \ne 0$.

Proof.   The proofs of all four parts follow easily from Theorems 2.2.3 – 2.2.6 in Section 2.2. For example, to prove (a), let c be a point of D. Suppose that $\{x_n\}$ is a sequence of points such that $x_n \in D$ and $x_n \to c$. Since

f is continuous at c, we know that $f(x_n) \to f(c)$. Since g is continuous at c we know that $g(x_n) \to g(c)$. Therefore, by Theorem 2.2.3,

$$\lim_{n \to \infty} (f(x_n) + g(x_n)) = \lim_{n \to \infty} f(x_n) + \lim_{n \to \infty} g(x_n)$$

$$= f(c) + g(c).$$

Thus, by definition, $f + g$ is continuous at c. Since c was an arbitrary point of D, we have proven (a). The other parts are proved similarly. ❑

We will see later that $\sin x$, $\cos x$, e^{ax}, and many other elementary functions are continuous on $\mathbb{R}$. Because of Theorem 3.1.1, any algebraic combination of these functions and polynomials is a continuous function on $\mathbb{R}$, except possibly where denominators vanish. In particular, parts (a) and (b) show that the set of continuous functions on a set D, denoted by $C(D)$, is a vector space; that is, linear combinations of continuous functions are again continuous functions. Vector spaces are defined formally in Section 5.8. The composition of continuous functions is also continuous.

❑ **Theorem 3.1.2**   Let f and g be continuous functions and define

$$D = \{x \mid x \in Dom(g) \text{ and } g(x) \in Dom(f)\}$$

Then $f \circ g$ is continuous on D.

Proof.   Let c be a point of D and suppose $x_n \in D$ and $x_n \to c$. By the definition of D, the numbers $g(x_n)$ and $g(c)$ are in $Dom(f)$. Further, since g is continuous, $g(x_n) \to g(c)$. Thus, since f is continuous, $f(g(x_n)) \to f(g(c))$, which proves that $f \circ g$ is continuous. ❑

The definition of D looks complicated but it merely expresses the fact that f can only be evaluated at points in $Dom(f)$. Therefore $g(x)$ must be in $Dom(f)$ and x must be in $Dom(g)$ in order to evaluate $f(g(x))$.

Example 3   Since $\sin x, \cos x, e^{ax}$, and polynomials are continuous on $\mathbb{R}$, all compositions such as

$$\sin(x^2 - 3x), \quad e^{3\cos x}, \quad \sin(\cos x)$$

are automatically continuous on $\mathbb{R}$.

ex: $g: [0,1] \to [2,3]$ $h \circ g$ cont, but not

$h: [2,3] \to [5,10]$ $g \circ h$

Example 4 Let's consider two examples where we have to be careful about domains. As you recall from calculus, the function $\ln x$ is defined for all $x > 0$. Later we will see that $\ln x$ is continuous. Thus $\ln(f(x))$ is defined only at those x where $f(x) > 0$. But on that set, $\ln(f(x))$ is automatically continuous if f is a continuous function, by Theorem 3.1.2. A slightly more complicated example is the function $\tan(\ln x)$. Since $\tan x = \frac{\sin x}{\cos x}$, the domain of the tangent function consists of all $x \neq \pi(2n+1)/2$ where n is an integer. Therefore, we must avoid those x such that $\ln x = \pi(2n+1)/2$, that is, the points of the form $e^{\pi(2n+1)/2}$. Of course, $\ln x$ is defined only if $x > 0$. Thus,

$$Dom(\tan(\ln x)) = \{x \mid x > 0 \text{ and } x \neq e^{\pi(2n+1)/2} \text{ for any } n\}.$$

By Theorem 3.1.2, $\tan(\ln x)$ is continuous on this domain.

Sometimes, it is convenient to have a criterion for continuity that does not involve sequences. We use this criterion in the next section when we define uniform continuity.

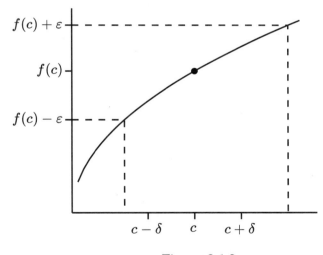

Figure 3.1.2

❑ **Theorem 3.1.3** A function $f(x)$ is continuous at a point c in $Dom(f)$ if and only if for each $\varepsilon > 0$ there is a $\delta > 0$ such that for all x in $Dom(f)$:

$$|x - c| \leq \delta \qquad \text{implies} \qquad |f(x) - f(c)| \leq \varepsilon. \qquad (2)$$

Proof.   First suppose that the $\varepsilon - \delta$ condition holds. We will show that f is continuous at c. Suppose that $x_n \in Dom(f)$ and $x_n \to c$. Let $\varepsilon > 0$ be given and choose δ so that (2) holds. Since $x_n \to c$, we can choose an N so that $|x_n - c| \leq \delta$ for $n \geq N$. Thus, by (2), we know that $|f(x_n) - f(c)| \leq \varepsilon$ for $n \geq N$. Thus, by the definition of convergence for sequences, the sequence $\{f(x_n)\}$ converges to $f(c)$. Therefore, f is continuous at c.

To show that continuity implies the $\varepsilon - \delta$ condition, we give a contra-positive proof. That is, assume that there exists an $\varepsilon > 0$ so that there is no δ so that (2) holds. We'll show that f is not continous at c. Let $\delta_n = \frac{1}{n}$. Since (2) does not hold, there is an $x_n \in Dom(f)$ satisfying

$$|x_n - c| \;\leq\; \delta_n \tag{3}$$

such that

$$|f(x_n) - f(c)| \;>\; \varepsilon. \tag{4}$$

From (3), it follows that $x_n \to c$ since $\delta_n \to 0$. But by (4), the sequence $\{f(x_n)\}$ cannot converge to $f(c)$ since each $f(x_n)$ is always a distance of more than ε away from $f(c)$. Therefore f is not continuous at c.   ❑

This theorem suggests a natural notation. For each c, we say that

$$\lim_{x \to c} f(x) \;=\; L \tag{5}$$

if and only if

$$\lim_{n \to \infty} f(x_n) \;=\; L$$

for all sequences $\{x_n\}$ in $Dom(f)$, with $x_n \neq c$, that converge to c. We assume in this definition that there is at least one such sequence (if not, (5) would be true for any L), which means that c can be approached from $Dom(f)$. According to the proof of Theorem 3.1.3, (5) is equivalent to saying that for each $\varepsilon > 0$ there is a $\delta > 0$ such that for all x in $Dom(f)$:

$$0 < |x - c| \leq \delta \quad \text{implies} \quad |f(x) - L| \leq \varepsilon.$$

Note that, in general, c need not be in $Dom(f)$ nor need L be a value of f. For example, the natural domain of $f(x) = x^{-1} \sin x$ is the set of all $x \neq 0$. We will show later that

$$\lim_{x \to 0} \frac{\sin x}{x} \;=\; 1.$$

Of course, if f is continuous and c is in $Dom(f)$, then $f(c) = L$.

Problems

1. Students are often told in calculus that continuous functions are those functions "whose graphs you can draw without picking your pencil up from the paper." Write a paragraph explaining the relation between this informal idea and the formal definition in this section.

2. Prove part (c) of Theorem 3.1.1.

3. Let $f(x)$ be a continuous function. Prove that $|f(x)|$ is a continuous function.

4. Where is the function $\ln(\sin x)$ defined and continuous?

5. Suppose that f is a continuous function on $\mathbb{R}$ such that $f(q) = 0$ for all $q \in \mathbb{Q}$. Prove that $f(x) = 0$ for all $x \in \mathbb{R}$.

6. Suppose that the function f is defined only on the integers. Explain why it is continuous.

7. Let $f(x) = 3x - 1$ and let $\varepsilon > 0$ be given. How small must δ be chosen so that $|x - 1| \leq \delta$ implies $|f(x) - 2| \leq \varepsilon$?

8. Let $f(x) = x^2$ and let $\varepsilon > 0$ be given.

   (a) Find a δ so that $|x - 1| \leq \delta$ implies $|f(x) - 1| \leq \varepsilon$.

   (b) Find a δ so that $|x - 2| \leq \delta$ implies $|f(x) - 4| \leq \varepsilon$.

   (c) If $n > 2$ and you had to find a δ so that $|x - n| \leq \delta$ implies $|f(x) - n^2| \leq \varepsilon$, would the δ be larger or smaller than the δ for parts (a) and (b)? Why?

9. Let $f(x) = 3x^3 - 2$ and let $\varepsilon > 0$ be given. Find a δ so that $|x - 1| \leq \delta$ implies $|f(x) - 1| \leq \varepsilon$.

10. Suppose that a function f is continuous at a point c and $f(c) > 0$. Prove that there is a $\delta > 0$ so that for all $x \in Dom(f)$,

$$|x - c| \leq \delta \quad \text{implies} \quad f(x) \geq \frac{f(c)}{2}.$$

11. Let $f(x) = \sqrt{x}$ with domain $\{x \mid x \geq 0\}$.

    (a) Let $\varepsilon > 0$ be given. For each $c > 0$, show how to choose δ so that $|x - c| \leq \delta$ implies $|\sqrt{x} - \sqrt{c}| \leq \varepsilon$. Hint: write

$$\sqrt{x} - \sqrt{c} = \frac{x - c}{\sqrt{x} + \sqrt{c}}.$$

    (b) Give a separate argument to show that f is continuous at zero.

12. (a) Sketch a rough graph of the function $f(x) = e^{-1/x^2}$ defined on the domain $\{x \mid x \neq 0\}$.

(b) Prove that $\lim_{x \to 0} e^{-1/x^2}$ exists.

(c) Show that f can be defined at zero in such a way that it is a continuous function on $\mathbb{R}$.

13. (a) Generate the graph of the function $f(x) = \sin \frac{1}{x}$ on the interval $(0, 1]$.

(b) Find sequences $\{\alpha_n\}$ and $\{\beta_n\}$ of numbers in $(0, 1]$, so that $\alpha_n \to 0$, $\beta_n \to 0$, and

$$\lim_{n \to \infty} \sin \frac{1}{\alpha_n} = 1, \qquad \lim_{n \to \infty} \sin \frac{1}{\beta_n} = -1.$$

(c) Conclude from part (b) that $\lim_{x \to 0} \sin \frac{1}{x}$ does not exist.

(d) Can f be defined at 0 so that f is continuous on $[0, 1]$?

(e) Can $g(x) = x \sin \frac{1}{x}$ be defined at 0 so that g is continuous on $[0, 1]$?

3.2 Continuous Functions on Closed Intervals

In the last section we proved several simple properties of continuous functions. We use the word "simple" because the proofs depend only on the idea of convergent sequence and the limit theorems in Section 2.2. In this section we prove several deeper properties of continuous functions which depend on the notion of Cauchy sequence and the completeness of the real numbers.

Definition.   A real-valued function f defined on a set S is said to be **bounded** if there exists a real number B so that

$$|f(x)| \leq B \qquad \text{for all } x \in S. \tag{6}$$

❏ **Theorem 3.2.1**   A continuous function on a closed finite interval is bounded.

Proof.   We give a proof by contradiction. Suppose that f is continuous on $[a, b]$ but no such B exists. Then for each n there is an $x_n \in [a, b]$ such that $|f(x_n)| > n$. Since the sequence $\{x_n\}$ is contained in $[a, b]$, the Bolzano-Weierstrass theorem and its corollary guarantee that $\{x_n\}$ has a subsequence $\{x_{n_k}\}$ that converges to a point c in $[a, b]$ as $k \to \infty$. Since f is continuous, $f(x_{n_k}) \to f(c)$ as $k \to \infty$. But this is impossible since the numbers $|f(x_{n_k})|$ diverge to $+\infty$. Therefore a B must exist so that (6) holds.   ❏

If f is a function on a set S, we define the **supremum** of f on the set S, denoted $\sup_S f$, to be

$$\sup_S f \equiv \sup\{f(x)\,|\,x \in S\}.$$

That is, $\sup_S f$ is just the supremum of the set of values of f on S. Similarly, we define the **infimum** of f on S by

$$\inf_S f \equiv \inf\{f(x)\,|\,x \in S\}.$$

Note that $\sup_S f$ could be $+\infty$ and $\inf_S f$ could be $-\infty$, depending on the function f and the set S. For example, if $f(x) = x^2$ and $S = \mathbb{R}$, then $\sup_S f = +\infty$ and $\inf_S f = 0$. Even in the case where $\sup_S f$ is finite, we do not necessarily know that there is a point c in S so that $f(c) = \sup_S f$, as the following example shows.

Example 1   Let $f(x) = x^2$ be defined on the set $S = (0, 2)$. See Figure 3.2.1. Then it is easy to see that $\sup_{(0,2)} f = 4$ and $\inf_{(0,2)} f = 0$, but there are no points in $(0, 2)$ where f takes on these values. Note that had we included the endpoints of the interval $(0, 2)$ in the domain of definition of f, then there would have been points (0 and 2) where the values of f were equal to $\sup_S f$ and $\inf_S f$.

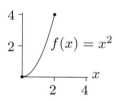

Figure 3.2.1

The example suggests that a continuous function on a *closed* interval assumes its supremum and infimum.

❏ **Theorem 3.2.2**   Let f be a continuous function on a closed interval $[a, b]$. Then, there exist points c and d in $[a, b]$ such that

$$f(c) = \sup_{[a,b]} f \quad \text{and} \quad f(d) = \inf_{[a,b]} f.$$

Proof.   According to Theorem 3.2.1, the set of values of the function f on $[a, b]$ is a bounded set, so $\sup_{[a,b]} f$ and $\inf_{[a,b]} f$ are finite numbers. Let

$M = \sup_{[a,b]} f$. For each n, there must be a point x_n in $[a, b]$ so that

$$M - \frac{1}{n} \leq f(x_n) \leq M, \tag{7}$$

because, otherwise, M wouldn't be the supremum of the values of f. Since the interval is finite, the Bolzano-Weierstrass theorem guarantees that the sequence $\{x_n\}$ has a subsequence $\{x_{n_k}\}$ that converges to a point c in $[a, b]$ as $k \to \infty$. Since f is continuous,

$$\lim_{k \to \infty} f(x_{n_k}) = f(c).$$

But, from (7), we see that

$$\lim_{k \to \infty} f(x_{n_k}) = M.$$

Thus, we must have $f(c) = M$ since a convergent sequence can have only one limit. The proof of the existence of d is similar.   ❏

❑ **Theorem 3.2.3 (The Intermediate Value Theorem)**   Let f be a continuous function on a closed interval $[a, b]$ such that $f(a) \neq f(b)$. Let y be a real number between $f(a)$ and $f(b)$; that is, either $f(a) < y < f(b)$ or $f(a) > y > f(b)$. Then there is a c in (a, b) such that $f(c) = y$.

Proof.   We shall consider the case $f(a) < y < f(b)$; the proof of the other case is similar. Let

$$S \equiv \{x \in [a, b] \mid f(x) < y\}$$

and define $c \equiv \sup S$. Note that S is nonempty since it contains a and S is bounded by b. Thus, c is a well defined element of $[a, b]$. Since f is continuous at a and $f(a) < y$, there is a $\delta > 0$ so that $f(x) < y$ if $a \leq x \leq a + \delta$. Thus $c > a$. We shall show that $f(c) = y$.

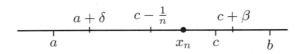

Figure 3.2.2

For each n we can choose an $x_n \in S$ that is in the interval $[c - \frac{1}{n}, c]$; see problem 3 of Section 2.5. Since $x_n \to c$ and f is continuous, $f(x_n) \to f(c)$.

See Figure 3.2.2. Thus, $f(c) \leq y$ since $f(x_n) < y$ for all n. If $f(c) < y$, then if $\beta > 0$ is small enough, $f(c+\beta) < y$ because f is continuous. Since this contradicts the definition of c, we conclude that $f(c) = y$. ❏

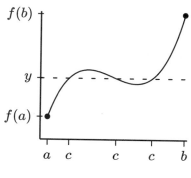

Figure 3.2.3

    The Intermediate Value Theorem says something that seems almost obvious. If the graph of a continuous function f starts below the height y and ends above it, then there must be a point where it crosses the horizontal line at height y. As in Figure 3.2.3, there can be more than one such point. If the result is "obvious," why is it so hard to prove? Consider the function $f(x) = x^2 - 2$ on the interval $[0, 2]$. Since $f(0) = -2$ and $f(2) = 2$, the Intermediate Value Theorem guarantees that there is a number $c \in [0, 2]$ so that $c^2 - 2 = 0$; that is, $c = \sqrt{2}$. Since $\sqrt{2}$ is irrational, if our number system contained only rational numbers, then the "graph" of $x^2 - 2$ would cross the x-axis without meeting it. Thus, the completeness of $\mathbb{R}$ must be involved in a deep way in the proof of the Intermediate Value Theorem. It is, since we used the fact that a bounded set has a least upper bound, which is equivalent to the axiom of completeness.

Corollary 3.2.4 Let f be a continuous function on a closed interval $[a, b]$ and define $m \equiv \inf_{[a,b]} f$ and $M \equiv \sup_{[a,b]} f$. Then, the range of f is the interval $[m, M]$.

Proof. According to Theorem 3.2.2, there exist c and d in $[a, b]$ so that $f(c) = M$ and $f(d) = m$. By Theorem 3.2.3, for every $y \in [m, M]$ there is an x between c and d so that $f(x) = y$. Thus, the range of f contains $[m, M]$. By the definitions of M and m, the range cannot contain any points outside $[m, M]$, so the range equals $[m, M]$. ❏

We now study more deeply the criterion for continuity introduced in Theorem 3.1.3. In order to prove continuity of f at c, one must show that, given $\varepsilon > 0$, we can choose a δ so that (2) holds. In general, the size of δ may depend on what point c we are looking at. Intuitively, we will have to choose δ smaller if the function f is rising or falling faster.

Definition.   A continuous function f defined on $Dom(f)$ is said to be **uniformly continuous** if for each $\varepsilon > 0$ there is a $\delta > 0$ so that for all x in $Dom(f)$ and all c in $Dom(f)$

$$|x - c| \leq \delta \qquad \text{implies} \qquad |f(x) - f(c)| \leq \varepsilon. \qquad (8)$$

Thus, to show that a function is uniformly continuous one must be able to find a δ that is independent of c.

Example 2   Consider the function $f(x) = x^2$, with domain equal to the whole real line, $\mathbb{R}$. Let $\varepsilon > 0$ be given and choose a point c. Then,

$$|f(x) - f(c)| \leq |x - c||x + c|.$$

Choose a δ and suppose that x satisfies $|x - c| \leq \delta$. Then,

$$|f(x) - f(c)| \leq \delta|x + c|.$$

We want to see how small δ must be so that the right-hand side is less than ε. Since x is within δ of c, we know that $|x| \leq |c| + \delta$. Therefore if $\delta \leq 1$, we have

$$\begin{aligned} |f(x) - f(c)| &\leq \delta(|x| + |c|) \\ &\leq \delta(|c| + \delta + |c|) \\ &\leq \delta(2|c| + 1). \end{aligned}$$

Now we can see how to choose δ. If $\delta < 1$ and

$$\delta \leq \frac{\varepsilon}{2|c| + 1},$$

then

$$|x - c| \leq \delta \qquad \text{implies} \qquad |f(x) - f(c)| \leq \varepsilon.$$

Thus we have proven that $f(x)$ is continuous at c, and we see explicitly how the choice of δ depends on c. As $|c|$ gets larger, we have to choose

δ smaller. This makes sense because the slope of f gets steeper as x gets larger. So, to make the difference $|f(x) - f(c)|$ small, we have to choose δ smaller as c gets larger. Notice, however, that $\delta = \frac{\varepsilon}{2|N|+1}$ is a suitable δ for all c in $[-N, N]$. Thus $f(x) = x^2$ is uniformly continous on each interval $[-N, N]$ but not on the whole real line $\mathbb{R}$.

What we found in Example 2 is a general phenomenon.

❑ **Theorem 3.2.5**   Let f be a continuous function on a closed finite interval $[a, b]$. Then, f is uniformly continuous on $[a, b]$.

Proof.   We will give a proof by contradiction. Suppose that f is *not* uniformly continuous. Then, there exists an ε_0 so that we cannot choose any $\delta > 0$ with the property that

$$|x - c| \leq \delta \quad \text{implies} \quad |f(x) - f(c)| \leq \varepsilon_0$$

for all x and c in the interval. Set $\delta_n = 2^{-n}$. Then, for each n there are points x_n and c_n in $[a, b]$ so that

$$|x_n - c_n| \leq \delta_n \quad \text{and} \quad |f(x_n) - f(c_n)| \geq \varepsilon_0. \tag{9}$$

Since $[a, b]$ is a finite interval, the Bolzano-Weierstrass theorem guarantees that there exists a subsequence $\{x_{n_k}\}$ that converges to a point d in $[a, b]$. Furthermore,

$$\begin{aligned} |c_{n_k} - d| &\leq |c_{n_k} - x_{n_k}| + |x_{n_k} - d| \\ &\leq 2^{-n_k} + |x_{n_k} - d| \end{aligned}$$

so $c_{n_k} \to d$ also as $k \to \infty$. Since f is continuous, $f(x_{n_k}) \to f(d)$ and $f(c_{n_k}) \to f(d)$ as $k \to \infty$. This is impossible since $|f(x_{n_k}) - f(c_{n_k})| \geq \varepsilon_0$ for each k. Thus f is uniformly continuous on $[a, b]$. ❑

The following criterion guarantees uniform continuity (problem 7).

Definition.   A function f defined on a set $S \subseteq \mathbb{R}$ is said to be **Lipschitz continuous** on S if there exists an M so that

$$\frac{|f(x) - f(c)|}{|x - c|} \leq M$$

for all x and c in S such that $x \neq c$.

Problems

1. Prove that $f(x) = x^3 - 4x + 2$ has a zero in the interval $[0, 1]$.

2. Prove that all cubic polynomials have at least one real root.

3. Let f be a continuous function on a finite interval $[a, b]$. Suppose that $f(x) > 0$ for all x in $[a, b]$. Prove that there is an $\alpha > 0$ such that $f(x) > \alpha$ for all x in $[a, b]$.

4. Let f and g be continuous functions on the finite interval $[a, b]$. Suppose that $f(x) > g(x)$ for all x in $[a, b]$. Prove that there is an $\alpha > 0$ such that $f(x) > \alpha + g(x)$ for all x in $[a, b]$.

5. Let f and g be continuous functions on the finite interval $[a, b]$. Suppose that $f(x) < g(x)$ for all x in $[a, b]$. Prove that there is an $\alpha < 1$ such that $f(x) \le \alpha g(x)$ for all x in $[a, b]$. Hint: consider the sets $\{x \mid f(x) \ge (1 - \frac{1}{n})g(x)\}$.

6. Let f be a continuous function on a finite interval $[a, b]$. Suppose that $f(a) < f(b)$ and choose c and d so that $f(a) < c < d < f(b)$. Define

$$S \equiv \{x \mid c \le f(x) \le d\}.$$

   (a) Show by example that S need not be a single interval.

   (b) Suppose that $x \le y$ implies that $f(x) \le f(y)$ (such a function is called monotone increasing). Prove that in this case S is an interval.

7. Prove that if f is Lipschitz continuous on a set $S \subseteq \mathbb{R}$ then f is uniformly continuous on S.

8. In problem 13 of Section 3.1 we showed that $g(x) = x \sin \frac{1}{x}$ with domain $(0, 1]$ can be defined at zero in such a way that the extended function is continuous on $[0, 1]$. Is the extended function uniformly continuous on $[0, 1]$?

9. Show that the function $f(x) = \frac{1}{x}$ is not uniformly continuous on the interval $(0, \infty)$ but is uniformly continuous on any interval of the form $[\mu, \infty)$ if $\mu > 0$.

10. Prove that $f(x) = \sqrt{x}$ is uniformly continuous on $[0, \infty)$.

11. Show by example that a function can be uniformly continuous without being Lipschitz continuous. Hint: consider $f(x) = \sqrt{x}$.

12. A function f defined on $\mathbb{R}$ is said to be periodic with period p if $f(x+p) = f(x)$ for all $x \in \mathbb{R}$. Prove that if f is periodic and continuous, then f is uniformly continuous.

13. Use the result in Project 5 of Chapter 2 and the fact that $\sin x$ is a continuous function to prove that the set of limit points of the sequence $\{\sin n\}$ is the entire interval $[-1, 1]$.

3.3 The Riemann Integral

In this section we use many of the technical tools which we have developed and the uniform continuity of continuous functions on closed finite intervals to show the existence and properties of the Riemann integral. Let f be a bounded function on the interval $[a, b]$. We define a **partition,** P, of the interval to be any finite collection of points

$$x_0 < x_1 < x_2 < .... < x_N$$

such that $x_0 = a$ and $x_N = b$. For each of the N subintervals, $[x_{i-1}, x_i]$, we define

$$M_i \;\equiv\; \sup_x \{f(x) \,|\, x_{i-1} \leq x \leq x_i\}$$
$$m_i \;\equiv\; \inf_x \{f(x) \,|\, x_{i-1} \leq x \leq x_i\}.$$

Imagine that f is a positive continuous function such as the one pictured in Figure 3.3.1. Although we each have an intuitive idea of "the area under the curve between a and b," it is not clear how to define "area" technically. However we define area, it should have the property that

$$\sum_{i=1}^{N} m_i(x_i - x_{i-1}) \;\leq\; \text{area} \;\leq\; \sum_{i=1}^{N} M_i(x_i - x_{i-1})$$

because the sum on the left is the sum of the areas of the rectangles lying under the curve in Figure 3.3.1(a), and the sum on the right is the sum of the areas of the rectangles whose tops are over the curve in Figure 3.3.1(b). The sum on the left is called a **lower sum** and is denoted by $L_P(f)$ because it depends on f and on the partition P.

(a) $L_P(f)$

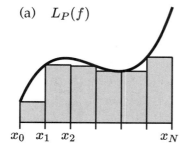

(b) $U_P(f)$

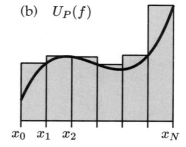

Figure 3.3.1

Similarly, the sum on the right is called an **upper sum** and is denoted by $U_P(f)$. For any bounded function f and any particular partition P we always know that $L_P(f) \le U_P(f)$ since $m_i \le M_i$ for each i.

Intuitively, upper sums should approximate the area from above and lower sums from below. And we should get better approximations to the "area" if we pick partitions with more points and shorter bases $(x_i - x_{i-1})$ for the rectangles. The following lemma shows that if we add points to a partition, then the upper sum cannot increase and the lower sum cannot decrease. The second lemma shows that, indeed, all upper sums are larger than all lower sums.

Lemma 1. Let Q be a partition which contains the points of P and some additional points. Then $L_P(f) \le L_Q(f)$ and $U_Q(f) \le U_P(f)$

Proof. We add the additional points to the partition P one at a time. Let Q_1 denote the partition with one new point y added in the jth interval:

$$x_0 < x_1 < \ldots < x_{j-1} < y < x_j < \ldots < x_N.$$

Then,

$$U_P(f) \;=\; \sum_{i=1}^{j-1} M_i(x_i - x_{i-1}) \;+\; M_j(x_j - x_{j-1}) \;+\; \sum_{i=j+1}^{n} M_i(x_i - x_{i-1})$$

$$U_{Q_1}(f) \;=\; \sum_{i=1}^{j-1} M_i(x_i - x_{i-1}) \;+\; \left(\sup_{x_{j-1} \le x \le y} f(x) \right)(y - x_{j-1})$$

$$+ \left(\sup_{y \le x \le x_j} f(x) \right)(x_j - y) \;+\; \sum_{i=j+1}^{n} M_i(x_i - x_{i-1}).$$

Since,

$$\left(\sup_{x_{j-1} \le x \le y} f(x) \right)(y - x_{j-1}) + \left(\sup_{y \le x \le x_j} f(x) \right)(x_j - y)$$

$$\le \;\; M_j(y - x_{j-1}) \;+\; M_j(x_j - y)$$

$$= \;\; M_j(x_j - x_{j-1}),$$

we conclude that $U_{Q_1}(f) \le U_P(f)$. Now we add a second point to Q_1 and make the same argument to show that $U_{Q_2}(f) \le U_{Q_1}(f)$ and so forth. Continuing in this way, we find that $U_Q(f) \leq U_P(f)$. The proof that $L_P(f) \leq L_Q(f)$ is similar. ❏

Lemma 2 Let P and Q be partitions. Then $L_P(f) \leq U_Q(f)$.

Proof. Consider the partition $P \cup Q$, made up of both the points of P and the points of Q. Then $L_{P \cup Q}(f) \leq U_{P \cup Q}(f)$. Since $P \cup Q$ contains P and $P \cup Q$ contains Q, Lemma 1 implies that

$$L_P(f) \; \leq \; L_{P \cup Q}(f) \; \leq \; U_{P \cup Q}(f) \; \leq \; U_Q(f).$$

Therefore, $L_P(f) \leq U_Q(f)$. ❏

Thus, each upper sum is larger than each lower sum, so by Theorem 2.5.3,

$$\sup_{P} \{L_P(f)\} \; \leq \; \inf_{P} \{U_P(f)\}. \tag{10}$$

Definition. A bounded function f on a finite interval $[a, b]$ is said to be **Riemann integrable** if

$$\sup_{P} \{L_P(f)\} \;\; = \;\; \inf_{P} \{U_P(f), \} \tag{11}$$

in which case we define the integral of f over $[a, b]$ by

$$\int_a^b f(x)\, dx \;\; \equiv \;\; \inf_{P} \{U_P(f)\}.$$

The Riemann integral is named after Georg Riemann (1826 – 1866).

Lemma 3 Let f be a bounded function on $[a, b]$. Suppose that for each $\varepsilon > 0$ there is a partition P so that

$$U_P(f) - L_P(f) \;\; \leq \;\; \varepsilon. \tag{12}$$

Then f is Riemann integrable.

Proof. Define $\alpha \equiv \inf_P \{U_P(f)\} - \sup_P \{L_P(f)\}$. Equation (10) shows that $\alpha \geq 0$. Suppose that $\alpha > 0$. For every partition P, $U_P(f) \geq \inf_P \{U_P(f)\}$ and $L_P(f) \leq \sup_P \{L_P(f)\}$, so $U_P(f) - L_P(f) \geq \alpha$. Thus, (12) can not hold for any partition if $\varepsilon < \alpha$. Since we are assuming (12), we conclude that $\alpha = 0$. This proves, by definition (11), that f is Riemann integrable.   ❏

❏ **Theorem 3.3.1**   A continuous function on a closed interval is Riemann integrable.

Proof.   Let f be continuous on $[a, b]$. We shall show that the criterion of Lemma 3 is satisfied. Since f is continuous on $[a, b]$, we know from Theorem 3.2.5 that it is uniformly continuous. Therefore, given $\varepsilon > 0$, we can choose $\delta > 0$ so that

$$|x - y| \leq \delta \quad \text{implies} \quad |f(x) - f(y)| \leq \frac{\varepsilon}{b - a}.$$

Let P be any partition such that the maximum subinterval length is less than or equal to δ. Since f is continuous, by Theorem 3.2.2 there are points c_i and d_i in each subinterval $[x_i, x_{i-1}]$ such that $f(c_i) = M_i$ and $f(d_i) = m_i$. Since $|c_i - d_i| \leq \delta$ for each i, we therefore know that $M_i - m_i \leq \varepsilon/(b - a)$. Thus,

$$\sum_{i=1}^{N} M_i(x_i - x_{i-1}) - \sum_{i=1}^{N} m_i(x_i - x_{i-1}) = \sum_{i=1}^{N} (M_i - m_i)(x_i - x_{i-1})$$

$$\leq \sum_{i=1}^{N} \frac{\varepsilon}{b - a}(x_i - x_{i-1})$$

$$= \varepsilon.$$

Since we have found a partition so that (12) holds, Lemma 3 guarantees that f is Riemann integrable.   ❏

Note that this theorem proves that the integral of a continuous function exists, but it doesn't tell us how to compute it. If the function f is the derivative of a function that we know, then we can evaluate the integral by using the Fundamental Theorem of Calculus (Section 4.2). If not, which is often the case, then we must use some method of approximation. If P is a partition and x_i^* is a point in the ith subinterval $[x_i, x_{i-1}]$, we call

$$\sum_{i=1}^{N} f(x_i^*)(x_i - x_{i-1})$$

a **Riemann sum** for f corresponding to the partition P. Note that there are many possible Riemann sums, depending on how we choose the points x_i^*. For example, if we choose each x_i^* to be the point in each $[x_i, x_{i-1}]$ where the maximum is taken on, then we get an upper sum. Similarly, if we choose x_i^* to be the point in each subinterval where the minimum is taken on, then we get a lower sum.

Corollary 3.3.2 Suppose that f is a continuous function on $[a, b]$, and let $\{P_k\}$ be a sequence of partitions such that the maximum length of the subintervals goes to zero as $k \to \infty$. Let S_k be any Riemann sum corresponding to P_k. Then

$$S_k \to \int_a^b f(x)\,dx \quad \text{as} \quad k \to \infty.$$

Proof. Let $\varepsilon > 0$ be given and choose δ as in the proof of Theorem 3.3.1. Choose K so that $k \geq K$ implies that the maximum subinterval length is less than δ. For all k,

$$L_{P_k}(f) \leq S_k \leq U_{P_k}(f)$$

and

$$L_{P_k}(f) \leq \int_a^b f(x)\,dx \leq U_{P_k}(f).$$

Therefore, since $U_{P_k}(f) - L_{P_k}(f) \leq \varepsilon$ for $k \geq K$, we conclude that

$$\left| S_k - \int_a^b f(x)\,dx \right| \leq \varepsilon$$

for $k \geq K$. This proves that $S_k \to \int_a^b f(x)\,dx$ as $k \to \infty$. $\qquad \Box$

In practice, the Riemann sums are evaluated by machine computation, and one is very interested in choosing the x_i^* so that the convergence of the S_k is rapid. This is the subject of Section 3.4. For the moment, we use Riemann sums to verify that the Riemann integral has the properties that we expect.

❏ **Theorem 3.3.3** Let f and g be continuous functions on the interval $[a, b]$ and let α and β be constants. Then

$$\int_a^b (\alpha f(x) + \beta g(x))\,dx = \alpha \int_a^b f(x)\,dx + \beta \int_a^b g(x)\,dx. \qquad (13)$$

Proof.   Let $\{P_k\}$ be a sequence of partitions such that the maximum length of the subintervals goes to zero as $k \to \infty$. For each k, we denote the points of the partition P_k by $x_0^k < x_1^k < \ldots < x_N^k$. Note that N depends on k. In each subinterval $[x_{i-1}^k, x_i^k]$ we choose a point, x_i^{k*} Then,

$$\sum_{i=1}^{N}(\alpha f(x_i^{k*}) + \beta g(x_i^{k*}))(x_i^k - x_{i-1}^k)$$

$$= \alpha \sum_{i=1}^{N} f(x_i^{k*})(x_i^k - x_{i-1}^k) + \beta \sum_{i=1}^{N} g(x_i^{k*})(x_i^k - x_{i-1}^k).$$

By Corollary 3.3.2, the sequence of Riemann sums on the left converges to the integral on the left of (14) as $k \to \infty$. Each of the sums on the right converges to the corresponding integral on the right of (13). Thus the result follows from Theorem 2.2.3 and Theorem 2.2.4.   ❑

❑ **Theorem 3.3.4**   Let f and g be continuous functions on the interval $[a, b]$ and suppose that $f(x) \leq g(x)$ for all $x \in [a, b]$. Then,

$$\int_a^b f(x)\,dx \quad \leq \quad \int_a^b g(x)\,dx. \tag{14}$$

Proof.   Let $\{P_k\}$ and x_i^{k*} be as above. Then,

$$\sum_{i=1}^{N} f(x_i^{k*})(x_i^k - x_{i-1}^k) \leq \sum_{i=1}^{N} g(x_i^{k*})(x_i^k - x_{i-1}^k).$$

Since the Riemann sums on the left converge to the integral on the left of (14) and the Riemann sums on the right converge to the integral on the right, we conclude from problem 3 of Section 2.4 that (14) holds.   ❑

❑ **Theorem 3.3.5**   Let f be a continuous function on the interval $[a, b]$. Then,

$$\left| \int_a^b f(x)\,dx \right| \quad \leq \quad \int_a^b |f(x)|\,dx. \tag{15}$$

Proof.   Let $\{P_k\}$ and x_i^{k*} be as above. By the triangle inequality,

$$\left| \sum_{i=1}^{N} f(x_i^{k*})(x_i^k - x_{i-1}^k) \right| \leq \sum_{i=1}^{N} |f(x_i^{k*})|(x_i^k - x_{i-1}^k).$$

The Riemann sums on the left converge to the integral on the left of (15) and the Riemann sums on the right converge to the integral on the right, so we conclude that (15) holds.   ❏

Theorems 3.3.4 and 3.3.5 can be combined to give the following fundamental integral estimate.

Corollary 3.3.6   Let f be a continuous function on the interval $[a, b]$. Then,

$$\left| \int_a^b f(x)\, dx \right| \leq (b - a) \sup_{[a,b]} |f(x)|. \tag{16}$$

❏ **Theorem 3.3.7**   Let f be a continuous function on the interval $[a, b]$ and suppose that $a \leq c \leq b$. Then,

$$\int_a^b f(x)\, dx = \int_a^c f(x)\, dx + \int_c^b f(x)\, dx. \tag{17}$$

The proof of Theorem 3.3.7 is left to the student (see problem 6). If $f(x) \geq 0$ for all $x \in [a, b]$, we define the **area** under f and above the x-axis between a and b to be $\int_a^b f(x)\, dx$. Note that by Theorem 3.3.4, the area is always nonnegative. If $f(x) \leq 0$, we define the area above f and below the x-axis to be $\int_a^b |f(x)|\, dx$. In this case also, the area is nonnegative.

In our definition of the Riemann integral $\int_a^b f(x)\, dx$, we assumed that $a < b$. If $a > b$ we define

$$\int_a^b f(x)\, dx \equiv - \int_b^a f(x)\, dx. \tag{18}$$

The Riemann integral on $\mathbb{R}^2$ is considered in project 4 of Chapter 4.

Problems

1. Let f be the function on $[0, 1]$ given by

$$f(x) = \begin{cases} 0 & \text{if } x \text{ is rational} \\ 1 & \text{if } x \text{ is irrational.} \end{cases}$$

   Explain why $U_P(f) = 1$ and $L_P(f) = 0$ for every partition P. Is f Riemann integrable?

2. Let f be the function on $[0, 1]$ given by

$$f(x) = \begin{cases} 1 & \text{if } x \neq \frac{1}{2} \\ 2 & \text{if } x = \frac{1}{2}. \end{cases}$$

   Prove that f is Riemann integrable and compute $\int_0^1 f(x)\, dx$. Hint: for each $\varepsilon > 0$, find a partition P so that $U_P(f) - L_P(f) \leq \varepsilon$ and use Lemma 3.

3. Suppose that f is Riemann integrable on $[a, b]$ and $f(x) \geq 0$ for all x.

   (a) Prove that $\int_a^b f(x)\, dx \geq 0$.

   (b) Prove that if $\int_a^b f(x)\, dx = 0$ and f is continuous, then $f(x) = 0$ for all x in $[a, b]$.

   (c) Find a counterexample which shows that the conclusion of part (b) may not hold if the hypothesis of continuity is removed.

4. Let $f(x) = 3x$ on the interval $[0, 1]$ and let $\varepsilon > 0$ be given.

   (a) Explain how to construct explicitly a partition P so that $U_P(f) - L_P(f) \leq \varepsilon$.

   (b) Compute $\int_0^1 f(x)\, dx$ without using the Fundamental Theorem of Calculus.

5. Let f be a continuous function on the interval $[a, b]$. Explain why $\int_a^b f(x)\, dx$ can be interpreted as "the sum of the areas above the x-axis minus the areas below." Explain why Theorem 3.3.4, Theorem 3.3.5, Corollary 3.3.6, and Theorem 3.3.7 make sense in terms of areas.

6. Prove Theorem 3.3.7.

7. Let $f(x) = x^2$ and let P be a partition of the interval $[1, 2]$ into subintervals of length δ. Compute $U_P(f)$ and $L_P(f)$ when $\delta = .5$, $\delta = .2$, and $\delta = .1$.

8. Let $f(x) = \frac{1}{x}$. Find a partition P of the interval $[1, 3]$ so that $U_P(f) - L_P(f) \leq 10^{-2}$. What can you conclude about $\int_1^3 \frac{1}{x}\, dx$?

9. Use Theorem 3.3.4 to prove that

$$e - 1 \leq \int_0^1 \sqrt{1 + xe^x}\, dx \leq \sqrt{2}(e - 1).$$

10. Let $f(x) = x + .1(\sin x)^3$. Estimate the integral $\int_0^2 f(x)\, dx$ from above and below.

11. Let f be the function in problem 10. Estimate $\int_1^2 e^{f(x)}\, dx$ from above and below.

12. Let f be a continuous function on $[a, b]$ and suppose that $\delta > 0$.

(a) Prove that $\int_a^b f(x)\,dx = \lim_{\delta \to 0} \int_{a+\delta}^b f(x)\,dx$.

(b) Prove that $\int_a^b f(x)\,dx = \lim_{\delta \to 0} \int_{a+\delta}^{b-\delta} f(x)\,dx$.

13. Let f be a continuous function on the interval $[a, b]$ and define

$$F(x) \equiv \int_a^x f(t)\,dt.$$

Prove that F is Lipschitz continuous on $[a, b]$.

14. Let f be a continuous function on the interval $[a, b]$. Suppose that for every $a_1 \in [a, b]$ and $b_1 \in [a, b]$ we know that

$$\int_{a_1}^{b_1} f(x)\,dx = 0.$$

Prove that $f(x) = 0$ for all x.

15. Let f be a continuous function on the interval $[a, b]$. Prove that there exists an $x \in [a, b]$ so that

$$f(x) = \frac{1}{b-a} \int_a^b f(x)\,dx.$$

Why does this make sense geometrically?

3.4   Numerical Methods

In the last section we saw that the Riemann integral of a continuous function can be approximated by a Riemann sum. Indeed, the integral was defined as the limit of such approximations. Sometimes one can use the Fundamental Theorem of Calculus to evaluate an integral exactly, but often some numerical method must be used. In this section we show how to use analytic techniques to derive error estimates for several different approximation schemes. Our purpose is not to find the "best" methods or estimates, but to show how analytical tools can be used.

Consider a particular rectangle in the approximation by the lower sum in Figure 3.3.1(a). See Figure 3.4.1(a) on page 98. The error is the area above the rectangle and below the curve. How big will this error be? In each interval we are approximating the function f by a constant function (whose height is the minimum of f in the interval). The error will, therefore, depend on how quickly f moves up from this constant value. That is, the error should depend on the derivative of f. If f has a small derivative, then the graph of f will be almost flat in a small interval and the approximation will be good. If f has a large derivative, then the graph of f will rise steeply and the approximation will be poor. Thus, what we want is an explicit estimate which says how good an approximation the Riemann sum is in terms of the size of the derivative of f. Throughout this section we use the properties of differentiation introduced in Chapter 4; in particular, we use the Mean Value Theorem from Section 4.2 and Taylor's theorem from Section 4.3.

From now on, we will always choose partitions which divide $[a, b]$ into subintervals of equal length. So, if there are N subintervals, then each has length

$$h \equiv x_i - x_{i-1} = \frac{b-a}{N}.$$

We choose points $x_i^* \in [x_{i-1}, x_i]$ and denote by

$$R_N \equiv \sum_{i=1}^{N} f(x_i^*)(x_i - x_{i-1}) = h \sum_{i=1}^{N} f(x_i^*)$$

a Riemann sum with N subintervals.

❏ **Theorem 3.4.1**   Let f be a continuously differentiable function on the interval $[a, b]$ and suppose that $|f'(x)| \leq M$ for all $x \in [a, b]$. Then,

$$\left| \int_a^b f(x)\, dx - R_N \right| \leq Mh(b - a).$$

Proof.   For every $x \in [x_i, x_{i-1}]$, the Mean Value Theorem guarantees that there is a ξ between x and x_i^* such that

$$\frac{f(x) - f(x_i^*)}{x - x_i^*} = f'(\xi)$$

so

$$\begin{aligned} |f(x) - f(x_i^*)| &= |f'(\xi)||x - x_i^*| \\ &\leq Mh. \end{aligned}$$

Therefore,

$$\left| \int_a^b f(x)\, dx - \sum_{i=1}^{N} f(x_i^*)(x_i - x_{i-1}) \right| = \left| \sum_{i=1}^{N} \int_{x_{i-1}}^{x_i} (f(x) - f(x_i^*))\, dx \right|$$

$$\leq \sum_{i=1}^{N} \left| \int_{x_{i-1}}^{x_i} (f(x) - f(x_i^*))\, dx \right|$$

$$\leq \sum_{i=1}^{N} \int_{x_{i-1}}^{x_i} |f(x) - f(x_i^*)|\, dx$$

$$\leq \sum_{i=1}^{N} \int_{x_{i-1}}^{x_i} Mh\, dx$$

$$= Mh(b - a). \qquad \qquad \Box$$

Notice that the theorem gives an upper bound for the error independent of how the points x_i^* in the Riemann sum are chosen. We can use the left endpoints, the right endpoints, the midpoints, or the points where f achieves maxima or minima. The actual error will depend on the x_i^*, and some choices may be better then others.

Example 1   Suppose that we wish to evaluate $\frac{1}{\sqrt{2\pi}} \int_{.5}^{1.5} e^{-x^2/2}\, dx$, which is just the probability that a standard normal random variable will take on a value between 0.5 and 1.5. Let $f(x) = e^{-x^2/2}$. Since $f'(x) = -xe^{-x^2/2}$ and $f''(x) = (x^2 - 1)e^{-x^2/2}$, it is easy to check that the maximal values of $|f'(x)|$ occur at $x = \pm 1$. Thus for all $x \, \epsilon \, [.5, 1.5]$, we have $|f'(x)| \leq e^{-1/2}$. Since $b - a = 1$, Theorem 3.4.1 guarantees that

$$\left| \int_{.5}^{1.5} e^{-x^2/2}\, dx - R_N \right| \leq he^{-1/2}$$

for any Riemann sum with N subintervals of length h. Therefore, if we want to be sure that the Riemann sum is within 10^{-6} of the true value of the integral, we need to choose $h \leq 10^{-6}\sqrt{e}$ or $N \geq 10^6/\sqrt{e}$.

The method of approximating integrals by general Riemann sums is called a **first-order** method because the error bound, $E(h) \equiv Mh(b - a)$, depends on the first power of h. You can see from the example why first-order methods aren't very good. To get an accuracy of h, we have to sum over approximately $1/h$ intervals (ignoring the $\sqrt{e}$). If $h = 10^{-6}$ this is definitely not something to do on a hand calculator! But there is another problem too. Any computer, whether it is a hand calculator, a PC, or a supercomputer, may make a small round-off error on each arithmetic operation. After a large number of arithmetic operations, the accumulated round-off errors may corrupt the answer one is trying to compute. For example, suppose that the round-off error is ε. To compute the Riemann sum R_N one has to evaluate the function N times, add the N values together, and multiply by h. Assuming that each evaluation and operation has a possible error of ε, the Riemann sum which we compute may differ by as much as $\varepsilon 2N$ from the actual Riemann sum. Thus our computation may differ from the real value of the integral by as much as

$$E(h) \;=\; Mh(b-a) \;+\; \frac{2\varepsilon(b-a)}{h}. \tag{19}$$

Of course, $E(h)$ is a bound for the error, not the error itself, which may be smaller. Notice that $E(h) \to \infty$ as $h \to 0$. In fact, $E(h)$ has a minimum at $h_0 = \sqrt{2\varepsilon/M}$ and that minimum is

$$E(h_0) \;=\; 2\sqrt{2M\varepsilon}(b-a).$$

Thus there is no point in choosing $h < h_0$ since the error may just get larger, and there is no way to choose h to guarantee that our approximation is closer than $2\sqrt{2M\varepsilon}(b-a)$ to the correct value of the integral.

To get better methods for computing $\int_a^b f(x)\,dx$ we should approximate f better on each subinterval. Since we need to be able to integrate the approximations explicitly, polynomials are natural candidates. In a Riemann sum, the function is approximated by a function which is constant on each subinterval, so it is natural to try the next simplest thing, linear approximations.

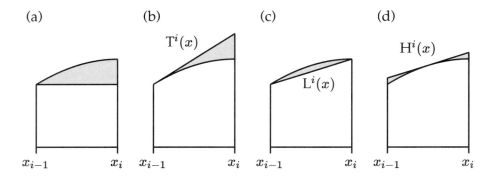

Figure 3.4.1. (a) A lower sum or left-hand endpoint approximation. (b) – (d): Different linear approximations. In all cases the error is shaded.

On each subinterval $[x_{i-1}, x_i]$, we shall approximate $f(x)$ by the first-order Taylor polynomial

$$T^i(x) \equiv f(x_{i-1}) + f'(x_{i-1})(x - x_{i-1}),$$

which has the same value as f at x_{i-1} and the same derivative as f at x_{i-1}. Let T_N denote the sum of the integrals of the functions $T^i(x)$ on the intervals $[x_{i-1}, x_i]$. That is,

$$
\begin{aligned}
T_N &\equiv \sum_{i=1}^{N} \int_{x_{i-1}}^{x_i} T^i(x)\, dx \\
&= h \sum_{i=1}^{N} f(x_{i-1}) + \sum_{i=1}^{N} f'(x_{i-1}) \int_{x_{i-1}}^{x_i} (x - x_{i-1})\, dx \\
&= h \sum_{i=1}^{N} f(x_{i-1}) + \frac{h^2}{2} \sum_{i=1}^{N} f'(x_{i-1}).
\end{aligned}
$$

❏ **Theorem 3.4.2** Let f be a twice continuously differentiable function on $[a, b]$ and suppose that $|f''(x)| \leq M$ for all $x \in [a, b]$. Then

$$\left| \int_a^b f(x)\, dx - T_N \right| \leq \frac{M h^2 (b - a)}{6}.$$

Proof. Suppose $x \in [x_{i-1}, x_i]$. By Taylor's theorem (Theorem 4.3.1), there is a $\xi \in [x_{i-1}, x]$ such that

$$f(x) - T^i(x) = \frac{f''(\xi)}{2!}(x - x_{i-1})^2.$$

Thus,

$$
\begin{aligned}
\left| \int_a^b f(x)\, dx - T_N \right| &= \left| \sum_{i=1}^{N} \int_{x_{i-1}}^{x_i} (f(x) - T^i(x))\, dx \right| \\
&\leq \sum_{i=1}^{N} \int_{x_{i-1}}^{x_i} |f(x) - T^i(x)|\, dx \\
&\leq \sum_{i=1}^{N} \int_{x_{i-1}}^{x_i} \frac{M}{2!}(x - x_{i-1})^2\, dx
\end{aligned}
$$

$$= \frac{M}{2!} \sum_{i=1}^{N} \frac{(x_i - x_{i-1})^3}{3}$$

$$= \frac{Mh^3 N}{6}$$

$$= \frac{Mh^2(b-a)}{6}. \qquad\qquad \square$$

Though natural, the approximation scheme of Theorem 3.4.2 is not used in practice because one must compute (symbolically or numerically) the derivative of f and evaluate it, as well as evaluate f itself. We can avoid this by choosing the linear approximation which connects $(x_{i-1}, f(x_{i-1}))$ and $(x_i, f(x_i))$ by a straight line

$$L^i(x) \; = \; f(x_{i-1}) \; + \; \frac{f(x_i) - f(x_{i-1})}{x_i - x_{i-1}}(x - x_{i-1}),$$

in each interval $[x_{i-1}, x_i]$. See Figure 3.4.1(c). This approximation scheme gives the trapezoidal rule (problem 3), which is also order 2 and has the error bound (20) where M is a bound for the second derivative of f. The proof of the error bound is outlined in project 4.

$$E(h) \;\; = \;\; \frac{Mh^2(b-a)}{12}. \qquad\qquad (20)$$

One can test the order of any scheme by trying it out on a function whose integral is known. If the scheme is order 1, then halving the size of h should halve the size of the error. If the scheme is order 2, then halving the size of h should cause the resulting error to be $\frac{1}{4}$ of the former error. If one tries this out on the special Riemann sum method

$$R_N \;\; = \;\; \sum_{i=1}^{N} f(\overline{x}_i)(x_i - x_{i-1}), \qquad \overline{x}_i \; = \; \frac{x_{i-1} + x_i}{2}, \qquad (21)$$

one finds that it is order 2. This is surprising since we proved in Theorem 3.4.1 that Riemann sum methods are order 1. Of course the error estimates are upper bounds for the error, so a particular Riemann sum method might have higher order. In fact, this method, called the Midpoint Rule for obvious reasons, is order 2. The following proof shows why.

❏ **Theorem 3.4.3** Let f be a twice continuously differentiable function on $[a, b]$ and suppose that $|f''(x)| \leq M$ for all $x \in [a, b]$. If R_N is given by (21), then

$$\left| \int_a^b f(x)\, dx - R_N \right| \leq \frac{Mh^2(b-a)}{24}. \tag{22}$$

Proof. On each interval we approximate f by the linear Taylor polynomial centered at the midpoint. Define

$$H^i(x) \equiv f(\bar{x}_i) + f'(\bar{x}_i)(x - \bar{x}_i), \qquad x \in [x_{i-1}, x_i].$$

See Figure 3.4.1(d). By Taylor's theorem, there is a $\xi \in [\bar{x}_i, x]$ such that

$$f(x) - H^i(x) = \frac{f''(\xi)}{2!}(x - \bar{x}_i)^2$$

and thus

$$\int_{x_{i-1}}^{x_i} |f(x) - H^i(x)|\, dx \leq \int_{x_{i-1}}^{x_i} \frac{M}{2!}(x - \bar{x}_i)^2\, dx \tag{23}$$

$$= \frac{M}{3}(h/2)^3. \tag{24}$$

Summing over the N intervals gives the estimate on the right side of (22). But why do the integrals of the $H^i(x)$ give the Midpoint Rule?

$$\int_{x_{i-1}}^{x_i} H^i(x)\, dx = f(\bar{x}_i) \int_{x_{i-1}}^{x_i} dx + f'(\bar{x}_i) \int_{x_{i-1}}^{x_i} (x - \bar{x}_i)\, dx.$$

The second term on the right is zero because $\bar{x}_i$ is the midpoint of $[x_{i-1}, x_i]$. Thus,

$$\sum_{i=1}^N \int_{x_{i-1}}^{x_i} H^i(x)\, dx = \sum_{i=1}^N f(\bar{x}_i)(x_i - x_{i-1}).$$

❏

Problems

1. If we approximate the integral in Example 1 by using the method of Theorem 3.4.2, how small should h be chosen so that the error is $\leq 10^{-4}$?

2. Show how to get an nth order scheme by generalizing the idea in Theorem 3.4.2.

3. Show that the linear approximation scheme with the $L^i(x)$ defined as in this section gives the trapezoid rule

$$\int_a^b f(x)\,dx \approx \frac{h}{2}(f(x_0) + 2f(x_1) + 2f(x_2) + \dots + 2f(x_{N-1}) + f(x_N)).$$

4. Use the integrals in problem 5 to provide experimental evidence that the Midpoint rule has order 2.

5. In each case below, compute how small h must be to guarantee that the error in the left-hand endpoint Riemann sum method and the error in the Midpoint rule are $\leq 10^{-4}$:

   (a) $\int_0^{\pi/2} \sin x \, dx$.

   (b) $\int_0^2 e^{2x} \, dx$.

   (c) $\int_2^{10} \ln x \, dx$.

6. Suppose that the error bound (including round-off) for a first-order numerical scheme method is given by (19).

   (a) Show that if h is small the error bound *increases* as we make h still smaller.

   (b) For what h is $E(h)$ smallest?

   (c) Suppose that the error bound for a second-order method is given by

   $$E_2(h) = Mh^2(b - a) + \frac{2\varepsilon(b - a)}{h}.$$

   Find the minimum of $E_2(h)$ and compare it to the minimum of $E(h)$ in the special case $M = (b - a) = 1$ and $\varepsilon = 10^{-9}$.

   (d) How small must h be for E_2 to be as small as the smallest value of E? Why is this important?

7. Let f be a continuous function on the interval $[a, b]$, which we divide into N segments of length $h = \frac{b-a}{N}$. Instead of approximating f on each subinterval $[x_{i-1}, x_i]$ by a linear function, we approximate f by the quadratic function that has the same values as f does at x_{i-1}, at x_i, and at the midpoint $\bar{x}_i$. Show that the sum of the integrals of these quadratic approximations is

$$S_N \equiv \sum_{i=1}^N \frac{h}{6}(f(x_{i-1}) + 4f(\bar{x}_i) + f(x_i)).$$

This is called **Simpson's rule**.

8. Provide numerical evidence that Simpson's rule is order 4 by using it to evaluate one of the integrals in problem 5.

9. If f is four times continuously differentiable and $|f^{(4)}(x)| \leq M$ for all $x \in [a, b]$, the error estimate for Simpson's rule is

$$\left| \int_a^b f(x)\, dx - S_N \right| \leq \frac{(b-a)(\frac{h}{2})^4 M}{180}.$$

For each of the integrals in problem 5, how small must h be so that the error in Simpson's rule is $\leq 10^{-4}$?

3.5 Discontinuities

Upper and lower sums were defined in Section 3.3 for any function on a finite interval $[a, b]$ that is bounded. We proved there that if f is continuous on $[a, b]$, then the inf of the upper sums equals the sup of the lower sums, and so, by definition, the Riemann integral exists. In this section we show that the Riemann integral also exists for special classes of functions which have points in their domains where they are not continuous. Such functions are called **discontinuous**, and the points are called **discontinuities**. We begin with an example (problem 1 of Section 3.3) which shows that if a function has too many discontinuities, the Riemann integral may not exist.

Example 1 Consider the function f on $[0, 1]$ defined by

$$f(x) = \begin{cases} 0 & \text{if } x \text{ is rational} \\ 1 & \text{if } x \text{ is irrational} \end{cases}$$

Let P be any partition. Then, because each interval in the partition contains both rational and irrational numbers (problem 11 of Section 1.1 and problem 11 of Section 1.4), the upper sum equals 1 and the lower sum equals 0. Since this is true for each partition P, the infimum of the upper sums is 1 and the supremum of the lower sums is 0. Thus the Riemann integral of f does not exist.

Some special classes of discontinuous bounded functions do have Riemann integrals, however. A function is said to be **monotone increasing** if $x < y$ implies that $f(x) \leq f(y)$. Monotone decreasing is defined analogously.

❏ **Theorem 3.5.1** Suppose that f is a bounded monotone function on $[a, b]$. Then f is Riemann integrable.

Proof. Suppose that f is monotone increasing; the proof for the decreasing case is similar. Let M be such that $|f(x)| \le M$ for all $x \in [a, b]$ and let $\varepsilon > 0$ be given. Let P be a partition so that the subintervals have equal length h with $h \le \varepsilon/2M$. Since f is monotone increasing, the value of f at the right-hand endpoint of each subinterval $[x_{i-1}, x_i]$ is the supremum of f over the interval and the value at the left-hand endpoint is the infimum of f. Thus,

$$U_P(f) \ = \ \sum_{i=1}^{N} f(x_i)(x_i - x_{i-1})$$

$$L_P(f) \ = \ \sum_{i=1}^{N} f(x_{i-1})(x_i - x_{i-1}),$$

and so,

$$U_P(f) - L_P(f) \ = \ \sum_{i=1}^{N}(f(x_i) - f(x_{i-1}))(x_i - x_{i-1})$$

$$= \ h\sum_{i=1}^{N}(f(x_i) - f(x_{i-1}))$$

$$= \ h(f(b) - f(a))$$

$$\le \ h(2M)$$

$$\le \ \varepsilon.$$

Thus, by Lemma 3 of Section 3.3, f is Riemann integrable.   ❏

If f is not monotone, the situation is much more difficult, and we will restict our attention to functions which are continuous except at finitely many points.

Example 2 Consider the function on the interval $[0, 1]$ defined by $f(0) = 0$ and $f(x) = sin\frac{1}{x}$ if $x > 0$. This function, whose graph is shown in Figure 3.5.1, is continuous on $(0, 1]$ but is not continuous at $x = 0$. This is not because we defined $f(0) = 0$. Since $\lim_{x \to 0} f(x)$ does not exist, no definition of f at $x = 0$ would make f a continuous function on $[0, 1]$. For each $\delta > 0$, f is continuous on the interval $[\delta, 1]$, so the integral $\int_\delta^1 f(x)\,dx$ certainly exists. If the integral of f on $[0, 1]$ does exist, it seems natural to guess that it should equal the limit of the numbers $\int_\delta^1 f(x)\,dx$ as $\delta \to 0$, if this limit exists.

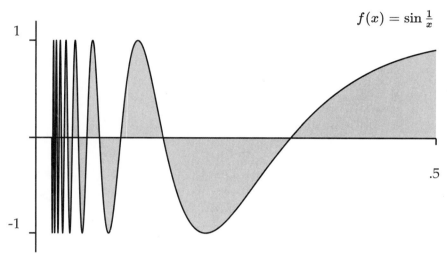

Figure 3.5.1

The following theorem answers the questions raised by Example 1.

❏ **Theorem 3.5.2**   Let f be a bounded function on $[a, b]$ that is continuous on (a, b). Then, the Riemann integral of f exists on $[a, b]$ and

$$\int_a^b f(x)\,dx \;=\; \lim_{\delta \to 0} \int_{a+\delta}^b f(x)\,dx. \tag{25}$$

Proof.   Choose M so that $|f(x)| \leq M$ for all x in $[a, b]$, and let $\varepsilon > 0$ be given. Choose δ small enough so that $\delta M \leq \frac{\varepsilon}{8}$. The reason for this choice will become clear later. Let P_δ be a partition of the interval $[a+\delta, b]$, which we label with $a + \delta = x_1 < x_2 < \ldots < x_N = b$. Any such partition P_δ can be extended to be a partition P of $[a, b]$ by adding the point $x_0 = a$. Let

$$M_i \equiv \sup_{x_{i-1} \leq x \leq x_i} f(x), \qquad m_i \equiv \inf_{x_{i-1} \leq x \leq x_i} f(x),$$

as in Section 3.3. Since f is continuous on $[a + \delta, b]$, we can choose the partition P_δ so that

$$\sum_{i=2}^{N} M_i(x_i - x_{i-1}) - \sum_{i=2}^{N} m_i(x_i - x_{i-1}) \ \leq \ \frac{\varepsilon}{4}. \tag{26}$$

Therefore,

$$\sum_{i=1}^{N} M_i(x_i - x_{i-1}) - \sum_{i=1}^{N} m_i(x_i - x_{i-1})$$

$$= \ (M_1 - m_1)\delta \ + \ \sum_{i=2}^{N} M_i(x_i - x_{i-1}) - \sum_{i=2}^{N} m_i(x_i - x_{i-1})$$

$$\leq \ \frac{\varepsilon}{4} + \frac{\varepsilon}{4}$$

by (26) and our choice of δ above. Thus, by Lemma 3 of Section 3.3, the Riemann integral of f exists on $[a, b]$. To see that $\int_a^b f(x)\,dx$ can be calculated by taking the limit (25), we estimate

$$\left| \int_a^b f(x)\,dx - \int_{a+\delta}^b f(x)\,dx \right| \tag{27}$$

$$\leq \ \left| \int_a^b f(x)\,dx \ - \ \sum_{i=1}^{N} M_i(x_i - x_{i-1}) \right|$$

$$+ \ M_1(x_1 - x_0) \ + \ \left| \sum_{i=2}^{N} M_i(x_i - x_{i-1}) - \int_{a+\delta}^b f(x)\,dx \right|$$

$$\leq \ \frac{\varepsilon}{2} + \frac{\varepsilon}{8} + \frac{\varepsilon}{4}.$$

Thus we have shown that, given $\varepsilon > 0$, we can choose a number $\delta > 0$ so that $|\int_a^b f(x)\,dx - \int_{a+\delta}^b f(x)\,dx| \leq \varepsilon$. This proves (25).   ❏

We remark that a special case of the hypotheses of Theorem 3.5.2 occurs when the function f is, in fact, continuous at a. In that case, we already know the limit formula (25); see problem 12 of Section 3.3. Notice that since the integral can be computed by the limit in (25), the value of $\int_a^b f(x)\,dx$ does not depend on the value of f at $x = a$.

Example 2 (revisited)   We now know that the Riemann integral $\int_a^b \sin \frac{1}{x} \, dx$ exists and the limit formula (26) suggests a way to compute it. Although the integral exists for the class of functions treated in Theorem 3.5.2, notice that we do not yet know that the properties of the integral (Theorems 3.3.3 – 3.3.7) are true. Assuming that the properties hold, we have

$$\int_0^1 \sin \frac{1}{x} \, dx \;=\; \int_0^\delta \sin \frac{1}{x} \, dx \;+\; \int_\delta^1 \sin \frac{1}{x} \, dx$$

by Theorem 3.3.7. Thus,

$$\left| \int_0^1 \sin \frac{1}{x} \, dx \;-\; \int_\delta^1 \sin \frac{1}{x} \, dx \right| \;=\; \left| \int_0^\delta \sin \frac{1}{x} \, dx \right|$$

$$\leq \; \delta$$

by Corollary 3.3.6 since $|\sin \frac{1}{x}| \leq 1$. This shows us how to compute $\int_a^b \sin \frac{1}{x} \, dx$ as closely as we like. Suppose that we want to know the value of $\int_0^1 \sin \frac{1}{x} \, dx$ to within 10^{-2}. If we choose $\delta = \frac{1}{2} 10^{-2}$ and replace $\int_0^1 \sin \frac{1}{x} \, dx$ by $\int_\delta^1 \sin \frac{1}{x} \, dx$, we make an error of less than $\frac{1}{2} 10^{-2}$. On the interval $[\delta, 1]$, the function $\sin \frac{1}{x}$ is infinitely often continuously differentiable, so we can use any of the numerical methods discussed in Section 3.4 to estimate $\int_\delta^1 \sin \frac{1}{x} \, dx$ to within $\frac{1}{2} 10^{-2}$.

Although the proof of Theorem 3.5.2 is a little complicated, the idea is very simple. Since the function is bounded, the terms in the upper and lower sums that come from the first interval $[a, a + \delta]$ can be made as small as we like by choosing δ small. On the rest of the interval, $[a + \delta, b]$, f is continuous, so the upper and lower sums there can be made as close as we like by picking the appropriate partition. We could have allowed f to be discontinuous at b too. In fact, if f is bounded and is continuous on $[a, b]$ except for finitely many points, then the Riemann integral of f exists. The proof, which uses exactly the same ideas as the proof of Theorem 3.5.2, is omitted.

❑ **Theorem 3.5.3**   Let f be a bounded function on $[a, b]$ that is continuous on $[a, b]$ except for finitely many points $a_1, a_2, \ldots, a_k$. Set $a_0 = a$ and $a_{k+1} = b$. Then, the Riemann integral of f exists on $[a, b]$ and

$$\int_a^b f(x) \, dx \;=\; \sum_{j=1}^{k+1} \int_{a_{j-1}}^{a_j} f(x) \, dx, \qquad (28)$$

where each integral on the right can be expressed as

$$\int_{a_{j-1}}^{a_j} f(x)\, dx \;=\; \lim_{\delta \searrow 0} \int_{a_{j-1}+\delta}^{a_j-\delta} f(x)\, dx. \tag{29}$$

Each of the integrals $\int_{a_{j-1}}^{a_j} f(x)\, dx$ is defined as the common value of the inf of the upper sums and the sup of the lower sums. Since this value can be computed by the limiting operation in (28), the value of $\int_a^b f(x)\, dx$ does not depend on the values of f at the points of discontinuity $a_1, a_2, \ldots, a_k$.

We want to define a particulary simple class of discontinuities called jump discontinuities. To do so, we first generalize the definition of $\lim_{x \to c} f(x)$ introduced in Section 3.1. Suppose that a function f is defined on an open interval (c, b). We say that f has a **limit from the right**, $f(c^+)$, at c, if for every sequence $x_n \to c$ with $x_n \in (c, b)$ we have

$$\lim_{n \to \infty} f(x_n) \;=\; f(c^+).$$

Note that each $x_n > c$. Similarly, suppose that f is defined on an open interval (a, c). We say that f has a **limit from the left**, $f(c^-)$, at c, if for every sequence $x_n \to c$ with $x_n \in (a, c)$ we have

$$\lim_{n \to \infty} f(x_n) \;=\; f(c^-).$$

We sometimes denote the limit from the left by $\lim_{x \nearrow c} f(x)$ and the limit from the right by $\lim_{x \searrow c} f(x)$. If f has both right-hand and left-hand limits at c and those limits are unequal, then f is said to have a **jump discontinuity** at c. Notice that nothing is said about the value of f at c. It may not even be defined there. However, if f is defined at c and

$$\lim_{x \nearrow c} f(x) \;=\; f(c) \;=\; \lim_{x \searrow c} f(x), \tag{30}$$

then f is continuous at c (problem 5).

Example 3   Consider the function

$$f(x) = \begin{cases} 2x & \text{if } 0 \le x < 1 \\ 1 & \text{if } x = 1 \\ 1/2 & \text{if } 1 < x < 2 \\ \frac{\pi}{2} - (x-2)^2 & \text{if } 2 \le x \le 3 \end{cases}$$

defined on $[0, 3]$. In the graph shown in Figure 3.5.2, we can see that f has jump discontinuities at $x = 1$ and $x = 2$. At $x = 1$, the limits from the left and right are 2 and $1/2$, respectively. At $x = 2$, the limits from the left and right are $1/2$ and $\pi/2$, respectively.

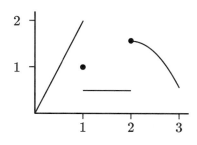

Figure 3.5.2

Definition. A function f defined on a finite interval $[a, b]$ is said to be **piecewise continuous** if it is continuous except at finitely many points at which it has jump discontinuities.

Suppose that f is piecewise continuous on $[a, b]$ with jump discontinuities at $a_1, a_2, \ldots, a_k$. Set $a_0 \equiv a$ and $a_{k+1} \equiv b$. Then f is continuous on each of the open intervals (a_{j-1}, a_j). Let f_j be the function on $[a_{j-1}, a_j]$ that is equal to f on (a_{j-1}, a_j) and that is defined at a_{j-1} and a_j by the following limits:

$$f_j(a_{j-1}) \equiv \lim_{x \searrow a_{j-1}} f(x) \qquad \text{and} \qquad f_j(a_j) \equiv \lim_{x \nearrow a_j} f(x).$$

For example, in Example 3, there are three subintervals and three functions, f_1, f_2, and f_3. It is easy to check (problem 6) that the f_j is continuous on $[a_{j-1}, a_j]$. Therefore, by problem 12 of Section 3.3,

$$\int_{a_{j-1}}^{a_j} f_j(x)\, dx = \lim_{\delta \searrow 0} \int_{a_{j-1}+\delta}^{a_j - \delta} f_j(x)\, dx.$$

This means that for piecewise continuous functions, Theorem 3.5.3 can be written in a simpler form:

Corollary 3.5.4 Suppose that f is piecewise continuous on $[a, b]$ with jump discontinuities at $a_1, a_2, \ldots, a_k$. Set $a_0 \equiv a$ and $a_{k+1} \equiv b$ and define the functions f_j as above. Then f is Riemann integrable and

$$\int_a^b f(x)\,dx \ = \ \sum_{j=1}^{k+1} \int_{a_{j-1}}^{a_j} f_j(x)\,dx. \tag{31}$$

The distinction that is being made here is as follows. In Theorem 3.5.3 the integrals $\int_{a_{j-1}}^{a_j} f(x)\,dx$ exist because the inf of the upper sums is equal to the sup of the lower sums. This doesn't help us to evaluate the integral since taking such infs and sups is difficult. Therefore, an important part of Theorem 3.5.3 is that $\int_{a_{j-1}}^{a_j} f(x)\,dx$ can be expressed as the limit (28). For each $\delta > 0$, $\int_{a_{j-1}+\delta}^{a_j-\delta} f(x)\,dx$ is the integral of a continuous function on a closed interval and so can be evaluated by ordinary means (analytical or numerical). If, however, f is piecewise continuous, then one does not need the limiting operation in (28). The integral $\int_{a_{j-1}}^{a_j} f(x)\,dx$ equals $\int_{a_{j-1}}^{a_j} f_j(x)\,dx$, which is an integral of a continuous function on a finite interval. Thus, in Example 3,

$$\int_0^3 f(x)\,dx \ = \ \int_0^1 2x\,dx \ + \ \int_1^2 \frac{1}{2}\,dx \ + \ \int_2^3 \frac{\pi}{2} - (x-2)^2\,dx$$

and each of the integrals on the right can be evaluated by elementary means.

We know that bounded monotone functions and functions which are bounded and continuous except for finitely many points are Riemann integrable on finite closed intervals. On the other hand, from Example 1 we know that not all bounded functions have Riemann integrals. Thus, we have not characterized exactly which bounded functions have Riemann integrals. The proofs of Theorems 3.5.2 and 3.5.3 depend on the fact that the set of discontinuities is finite and thus the terms in upper and lower sums corresponding to small intervals about these points do not contribute much. The function in Example 1 is, however, discontinuous at every point in the interval $[0,1]$. This suggests that a bounded function on a finite interval should be Riemann integrable if the set of discontinuities is "small" in some sense. This is true, but a complete characterization, though interesting and important for historical reasons, is beyond the scope of this introductory text.

In Section 3.3, we proved the properties of the Riemann integral (Theorems 3.3.3 – 3.3.7) for continuous functions. In fact, they are true for all functions that are Riemann integrable. This can be proved for particular classes of functions that are Riemann integrable (see, for example, problems 9, 11, and 12) or directly from the definition of Riemann integrability (see problem 13).

Problems

1. Let f be the function on $[0, 1]$ given by

$$f(x) = \begin{cases} x & \text{if } x \text{ is rational} \\ 0 & \text{if } x \text{ is irrational} \end{cases}$$

   Show that f is not Riemann integrable.

2. Let f be the function on $[0, 1]$ given by

$$f(x) = \begin{cases} 0, & 0 \leq x < 1 \\ 1, & 1 \leq x < 2 \\ 2, & 2 \leq x \leq 3 \end{cases}$$

   (a) Prove that f is Riemann integrable without appealing to any theorems in this section.

   (b) Which theorems in this section guarantee that f is Riemann integrable?

   (c) What is $\int_0^3 f(x)\, dx$?

3. Let f be the function on $[0, 1]$ given by

$$f(x) = \begin{cases} 1, & 0 \leq x < \frac{1}{2} \\ x - \frac{1}{2} & \frac{1}{2} \leq x < 1 \end{cases}$$

   (a) Prove that f is Riemann integrable without appealing to any theorems in this section.

   (b) Which theorems in this section guarantee that f is Riemann integrable?

   (c) What is $\int_0^1 f(x)\, dx$?

4. Each of the following functions is well defined for $x > 0$. For each, explain whether Theorem 3.5.2 can be used to prove that the function is Riemann integrable on $[0, 1]$. Explain why the answers don't depend on how the functions are defined at $x = 0$:

   (a) $\sin^2 \frac{1}{x}$.

   (b) $\frac{1}{x} \sin \frac{1}{x}$.

   (c) $\ln x$.

(d) $\frac{\sin x}{x}$. Hint: derive the inequality $\sin x < x$ for $0 < x \leq \pi$ by using the Fundamental Theorem of Calculus.

5. Suppose that f is defined in an open interval containing c and that (30) holds. Prove that f is continuous at c; that is, show that for *every* sequence $x_n \to c$ we have $\lim_{n \to \infty} f(x_n) = f(c)$.

6. Suppose that f is continuous on an open interval (a, b) and that f has a right-hand limit at a and a left-hand limit at b. Define an extension of f to $[a, b]$ by setting $f_e(x) \equiv f(x)$ for $a < x < b$ and $f_e(a) \equiv f(a^+)$, $f_e(b) \equiv f(b^-)$.

   (a) Show that f_e is continuous on $[a, b]$.

   (b) Show that if f_e has any other values at a or b, then f_e would not be continuous on $[a, b]$.

7. Suppose that f is monotone increasing on $[a, b]$. Prove that any discontinuities that f has are jump discontinuities.

8. Show by example that a monotone increasing function on a finite interval can have infinitely many jump discontinuities.

9. Use Corollary 3.5.4 to prove that the properties of the Riemann integral (Theorems 3.3.3 – 3.3.7) hold for the integrals of piecewise continuous functions on finite intervals.

10. Prove Theorem 3.5.3 by following the ideas of the proof of Theorem 3.5.2.

11. Let f be bounded on $[a, b]$ and continuous except for finitely many points. Let P_n be a sequence of partitions so that the maximal subinterval length goes to zero as $n \to \infty$, and let S_n be a Riemann sum corresponding to P_n. Prove that $S_n \to \int_a^b f(x)\,dx$ as $n \to \infty$.

12. Use the result of problem 11 to show that the properties of the Riemann integral (Theorems 3.3.3 – 3.3.7) hold for functions which are bounded on $[a, b]$ and are continuous except for finitely many points.

13. Use the definition of Riemann integrability to show directly that if f and g are Riemann integrable on $[a, b]$, then $f + g$ is Riemann integrable on $[a, b]$ and

$$\int_a^b f(x) + g(x)\,dx \ = \ \int_a^b f(x)\,dx \ + \ \int_a^b g(x)\,dx.$$

3.6 Improper Integrals

In the last section we investigated the Riemann integral for bounded functions with discontinuities. In this section we extend the Riemann integral to some unbounded functions and to some functions on infinite intervals. We begin with an example.

Example 1 Consider the function $f(x) = 1/\sqrt{x}$ whose graph is shown in Figure 3.6.1. Note that f is continuous on $(0,1]$ but diverges to ∞ as $x \searrow 0$. The Riemann integral as defined in Section 3.3 certainly does not exist since $U_P(f) = \infty$ for every partition P because the supremum of f is infinite over any interval of the form $[0, \delta)$ as long as $\delta > 0$. However, our experience in Section 3.5 suggests another approach. For every $\delta > 0$ the Riemann integral of f exists on $[\delta, 1]$ because f is continuous on $[\delta, 1]$. And since f is so simple, we can compute it explicitly:

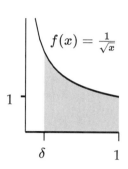

Figure 3.6.1

$$\int_\delta^1 \frac{1}{\sqrt{x}}\, dx \ = \ 2 - 2\sqrt{\delta}.$$

Since the right-hand side has a limit as $\delta \to 0$, namely 2, we could *define* the integral of $\frac{1}{\sqrt{x}}$ on $[0,1]$ to be that limit:

$$\int_0^1 \frac{1}{\sqrt{x}}\, dx \ \equiv \ \lim_{\delta \searrow 0} \int_\delta^1 \frac{1}{\sqrt{x}}\, dx \ = \ 2.$$

The example gives us the idea for the following definition.

Definition. Suppose that f is a continuous function on a half-open finite interval $(a, b]$. If the limit of $\int_{a+\delta}^b f(x)\, dx$ exists and is finite as $\delta \searrow 0$, we define

$$\int_a^b f(x)\, dx \ \equiv \ \lim_{\delta \searrow 0} \int_{a+\delta}^b f(x)\, dx.$$

and call it the **improper Riemann integral** of f on $[a, b]$. A similar definition holds if f is continuous but unbounded on $[a, b)$.

Example 2   Let us consider the family of functions $f_\alpha(x) = x^\alpha$ on the interval $[0, 1]$ for different values of α. For $\alpha \geq 0$, the function f_α is continuous on $[0, 1]$, so the Riemann integral exists. For $\alpha < 0$, f_α is continuous but unbounded on $(0, 1]$. There are three cases to consider. If $\alpha < 0$ and $\alpha \neq -1$,

$$\int_\delta^1 x^\alpha \, dx = \frac{1}{1 + \alpha} - \frac{\delta^{1+\alpha}}{1 + \alpha}.$$

In the case $-1 < \alpha < 0$, the right-hand side has limit $\frac{1}{1+\alpha}$ as $\delta \searrow 0$, so the improper Riemann integral exists and

$$\int_0^1 x^\alpha \, dx = \frac{1}{1 + \alpha}.$$

However, if $\alpha < -1$, then

$$\lim_{\delta \searrow 0} \left\{ \frac{1}{1 + \alpha} - \frac{\delta^{1+\alpha}}{1 + \alpha} \right\} = +\infty,$$

so the improper Riemann integral does not exist. The case $\alpha = -1$ is left as problem 3.

The functions in Example 2 were easy to analyze because the integrals could be computed explicitly. Often, this is impossible and one has to make estimates.

Example 3   Consider the integral $\int_0^1 \frac{\cos x}{\sqrt{1-x}} \, dx$. The problem here is near $x = 1$, where the function is unbounded. Define

$$I_\delta \equiv \int_0^{1-\delta} \frac{\cos x}{\sqrt{1 - x}} \, dx.$$

We want to show that $\lim_{\delta \searrow 0} I_\delta$ exists. This means that for any sequence $\delta_n \to 0$ with $\delta_n > 0$, the sequence $\{I_{\delta_n}\}$ has a limit and the limit is independent of the sequence $\{\delta_n\}$ chosen. We did not emphasize this in Example 2 since the limits there were simple and explicit. Here we will be more careful. Let $\{\delta_n\}$ be such a sequence and consider the two terms I_{δ_n} and I_{δ_m}. Suppose $\delta_n \leq \delta_m$. Then,

$$|I_{\delta_n} - I_{\delta_m}| = \left| \int_0^{1-\delta_n} \frac{\cos x}{\sqrt{1 - x}} \, dx - \int_0^{1-\delta_m} \frac{\cos x}{\sqrt{1 - x}} \, dx \right|$$

$$= \left| \int_{1-\delta_m}^{1-\delta_n} \frac{\cos x}{\sqrt{1-x}} \, dx \right|$$

$$\leq \int_{1-\delta_m}^{1-\delta_n} \frac{|\cos x|}{\sqrt{1-x}} \, dx$$

$$\leq \int_{1-\delta_m}^{1-\delta_n} \frac{1}{\sqrt{1-x}} \, dx$$

$$= 2\sqrt{\delta_m} - 2\sqrt{\delta_n}.$$

Since $\delta_n \searrow 0$ and $\sqrt{x}$ is continuous, $\sqrt{\delta_n} \to 0$ as $n \to \infty$. Thus, given $\varepsilon > 0$, we can choose N so that $n \geq N$ and $m \geq N$ implies $|\sqrt{\delta_m} - \sqrt{\delta_n}| \leq \varepsilon/2$. By the above estimate, it follows that $|I_{\delta_n} - I_{\delta_m}| \leq \varepsilon$ for $n \geq N$ and $m \geq N$. Thus, $\{I_{\delta_n}\}$ is a Cauchy sequence and therefore converges since the real numbers are complete.

We have shown that $\{I_{\delta_n}\}$ converges for any sequence $\delta_n \searrow 0$. If $\gamma_n \searrow 0$ is another such sequence, then the same estimate as above shows that

$$|I_{\delta_n} - I_{\gamma_n}| \leq 2|\sqrt{\delta_n} - \sqrt{\gamma_n}|$$

for all n. Since the right-hand side converges to zero as $n \to \infty$, the left-hand side must converge to zero too; that is,

$$\lim_{n \to \infty} I_{\delta_n} = \lim_{n \to \infty} I_{\gamma_n}.$$

Thus, the improper integral $\int_0^1 \frac{\cos x}{\sqrt{1-x}} \, dx$ exists. To compute it approximately, one can use the idea in Example 2 (revisited) of Section 3.5.

This same idea allows us to define the improper Riemann integral on semi-infinite intervals.

Definition. Let f be a continuous function on the semi-infinite interval $[a, \infty)$. If $\lim_{b \to \infty} \int_a^b f(x) \, dx$ exists and is finite, we define

$$\int_a^\infty f(x) \, dx \equiv \lim_{b \to \infty} \int_a^b f(x) \, dx.$$

and call it the **improper Riemann Integral** of f on $[a, \infty)$. A similar definition holds if f is continuous on a semi-infinite interval of the form $(-\infty, b]$.

Example 4   Let us consider the function $f(x) = 1/x^\alpha$ for $\alpha > 0$. If $\alpha = 2$, the graph is shown in Figure 3.6.2. Suppose that $\alpha > 1$.

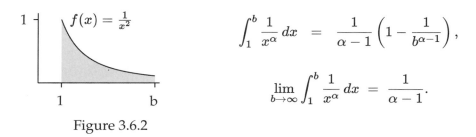

$$\int_1^b \frac{1}{x^\alpha}\, dx \;=\; \frac{1}{\alpha - 1}\left(1 - \frac{1}{b^{\alpha-1}}\right),$$

$$\lim_{b\to\infty} \int_1^b \frac{1}{x^\alpha}\, dx \;=\; \frac{1}{\alpha - 1}.$$

Figure 3.6.2

Thus, the improper Riemann integral exists. Similarly, it is not hard to see that the improper integral does not exist in the case $0 < \alpha \leq 1$.

Sometimes one can show that improper Riemann integrals exist even though one can not evaluate the limit explicitly.

Example 5   Consider the improper Riemann integral $\int_1^\infty \frac{\sin x}{x}\, dx$. Using integration by parts (problem 12 of Section 4.2), we see that

$$\int_1^b \frac{\sin x}{x}\, dx \;=\; \cos(1) \;-\; \frac{\cos b}{b} \;+\; \int_1^b \frac{\cos x}{x^2}\, dx.$$

The first two terms on the right clearly have nice limits as $b \to \infty$. To handle the third term notice that if c and d are both large and $c \leq d$, then

$$\left| \int_1^d \frac{\cos x}{x^2}\, dx - \int_1^c \frac{\cos x}{x^2}\, dx \right| \;=\; \left| \int_c^d \frac{\cos x}{x^2}\, dx \right|$$

$$\leq \int_c^d \frac{|\cos x|}{x^2}\, dx$$

$$\leq \int_c^d \frac{1}{x^2}\, dx$$

$$= \frac{1}{c} - \frac{1}{d}.$$

This estimate can be used similarly to the way in which the estimate was used in Example 3 to show that $\int_1^{c_n} \frac{\cos x}{x^2}\, dx$ converges to a finite limit as $c_n \to \infty$, and the limit is independent of the sequence $\{c_n\}$ chosen (problem 7).

There are subtleties in improper Riemann integrals. Choose μ small enough so that $\sin x \geq \frac{1}{2}$ if $x \in [\frac{\pi}{2}, \frac{\pi}{2} + \mu]$. Then,

$$\sum_{n=1}^{N} \int_{\frac{\pi}{2}+2n\pi}^{\frac{\pi}{2}+2n\pi+\mu} \frac{\sin x}{x} \, dx \;\geq\; \frac{1}{2} \sum_{n=1}^{N} \int_{\frac{\pi}{2}+2n\pi}^{\frac{\pi}{2}+2n\pi+\mu} \frac{dx}{x}$$

$$\geq \frac{1}{2} \sum_{n=1}^{N} \frac{1}{\frac{\pi}{2} + 2n\pi} \int_{\frac{\pi}{2}+2n\pi}^{\frac{\pi}{2}+2n\pi+\mu} dx$$

$$= \frac{\mu}{\pi} \sum_{n=1}^{N} \frac{1}{1 + 4n}$$

$$\geq \frac{\mu}{8\pi} \sum_{n=1}^{N} \frac{1}{n}.$$

Since the harmonic diverges (see Section 6.2), we see that the total amount of area in the positive bumps of the function $x^{-1} \sin x$ is infinite. Similarly there is an infinite amount of area in the negative bumps. Yet the improper integral $\int_1^\infty \frac{\sin x}{x} \, dx$ exists because when one takes the limit of

$$\int_1^b \frac{\sin x}{x} \, dx$$

as $b \to \infty$ there are cancelations between the successive positive bumps and negative bumps. This is analogous to an infinite series that is conditionally but not absolutely convergent.

Problems

1. Prove that the following functions have improper Riemann integrals on the interval $[0, 1]$:

   (a) $\frac{1+x^2}{\sqrt{x}}$.

   (b) $\frac{\cos 2x}{x^{3/4}}$.

   (c) $\frac{1}{\sqrt{x}(x+1)}$.

2. Determine which of the following functions have improper Riemann integrals on the interval $[0, 1]$.

   (a) $\ln x$. Hint: integrate by parts.

(b) $\frac{\sin x}{x^{3/2}}$. Hint: see problem 4(d) in Section 3.5.

3. Use the integral formula for the natural logarithm (see Example 2 of Section 4.3) and properties of the logarithm to prove that the improper Riemann integral $\int_0^1 \frac{1}{x}\, dx$ does not exist.

4. Prove that the improper Riemann integral $\int_0^\infty e^{-x}\, dx$ exists.

5. Determine whether the improper Riemann integral $\int_2^\infty \frac{1}{(\ln x)^2}\, dx$ exists.

6. Prove that the improper Riemann integral $\int_0^\infty e^{-x^2/2}\, dx$ exists. Hint: for large x, estimate $e^{-x^2/2}$ by e^{-x}.

7. Complete the proof of Example 5 by using the estimate to show that if $c_n \to \infty$, then the improper integral $\int_1^{c_n} \frac{\sin x}{x}\, dx$ converges as $n \to \infty$ and the limit is independent of the choice of $\{c_n\}$.

8. Suppose that f and g are continuous functions on the interval $(a, b]$ and assume that the improper Riemann integrals $\int_a^b f(x)\, dx$ and $\int_a^b g(x)\, dx$ both exist. Prove that the improper Riemann integral $\int_a^b f(x) + g(x)\, dx$ exists and equals $\int_a^b f(x)\, dx + \int_a^b g(x)\, dx$.

9. Suppose that f is an unbounded continuous function on the open interval (a, b). We say that the improper Riemann integral $\int_a^b f(x)\, dx$ exists if there is a $c \in (a, b)$ so that the improper integrals $\int_a^c f(x)\, dx$ and $\int_c^b f(x)\, dx$ exist (one of them might be proper). In this case we define

$$\int_a^b f(x)\, dx \;=\; \int_a^c f(x)\, dx \;+\; \int_c^b f(x)\, dx.$$

Show that if this is true for one $c \in (a, b)$ then it is true for all $c \in (a, b)$ and the value of $\int_a^b f(x)\, dx$ is independent of the choice of c.

10. Suppose that f is a continuous function on $\mathbb{R}$. We say that the improper Riemann integral $\int_{-\infty}^\infty f(x)\, dx$ exists if there is a $c \in \mathbb{R}$ so that the improper integrals $\int_{-\infty}^c f(x)\, dx$ and $\int_c^\infty f(x)\, dx$ exist. In this case we define

$$\int_{-\infty}^\infty f(x)\, dx \;=\; \int_{-\infty}^c f(x)\, dx \;+\; \int_c^\infty f(x)\, dx.$$

Show that if this is true for one $c \in \mathbb{R}$ then it is true for all $c \in \mathbb{R}$ and the value of $\int_{-\infty}^\infty f(x)\, dx$ is independent of the choice of c.

11. Which of the following improper Riemann integrals exist?

(a) $\int_0^1 \left(\frac{1}{\sqrt{x}} + \frac{1}{\sqrt{1-x}} \right) dx$.

(b) $\int_{-\infty}^\infty \frac{1}{1+x^2}\, dx$.

(c) $\int_{-\infty}^{\infty} e^{-x^2}\, dx$.

(d) $\int_{-\infty}^{\infty} x\, dx$.

12. Suppose that f is a continuous function on $\mathbb{R}$ such that the improper integral $\int_{-\infty}^{\infty} |f(x)|\, dx$ exists.

   (a) Show that the improper integral $\int_{-\infty}^{\infty} f(x)\, dx$ exists.

   (b) Show that if g is continuous and bounded on $\mathbb{R}$, then the improper integral $\int_{-\infty}^{\infty} g(x)f(x)\, dx$ exists.

Projects

1. Suppose that f is continuous on the open interval (a, b). The purpose of this project is to show that f can be extended to be a continuous function on $[a, b]$ if and only if f is uniformly continuous on (a, b). It is clear from Theorem 3.2.5 that if f can be extended it must be uniformly continuous. So, suppose f is uniformly continuous on (a, b).

   (a) Prove that f is bounded on (a, b).

   (b) Let $a_n \to a$. Prove that $\{f(a_n)\}$ has a subsequence which converges to a limit L.

   (c) Define $f(a) = L$ and prove that f is continuous at a.

   (d) Do the same at the other endpoint and conclude that f has a continuous extension. Is the extension unique?

   (e) Give examples of bounded continuous functions on open intervals which do not have continuous extensions.

2. The purpose of this project is to derive integral expressions for certain physical or geometric quantities.

   (a) Let f be a continuously differentiable function on the interval $[a, b]$. As in Section 3.4, divide up $[a, b]$ into N subintervals $[x_{i-1}, x_i]$ of equal length. Approximate the length of the graph of f over $[a, b]$ by writing down the sum of the lengths of the straight line segments between the successive points $(x_i, f(x_i))$. Rearrange your sum so that it is a Riemann sum on $[a, b]$. As $N \to \infty$, the Riemann sums converge to an integral. What is the integral? Why is this a reasonable way to define the arc length of the graph of f over $[a, b]$?

   (b) A metal bar is lying on the x-axis between $x = 0$ and $x = 2$. The density of the bar at x is $\rho(x)$ units of mass per unit length. Find and justify an expression for the total mass of the bar.

   (c) Consider the problem of finding a point $\bar{x}$ so that if one puts one's finger under the bar at $\bar{x}$ it will just balance. The point $\bar{x}$ is called the **center of mass.** A unit of mass m at a distance d from $\bar{x}$ exerts a downward moment proportional to m times d. The point $\bar{x}$ will be such that the moments on the left are equal to the moments on the right. Find an integral expression for $\bar{x}$.

(d) Let f be a positive continuously differentiable function on the interval $[a, b]$. Derive the formulas for the volume and surface area of the figure obtained by revolving the graph of f around the x-axis.

3. This project is a cautionary tale about numerical methods. Define the number E_n for each nonnegative integer $n \geq 0$ by

$$E_n \equiv \int_0^1 x^n e^x \, dx.$$

Suppose that we are interested in E_{10} but don't want to evaluate the integral numerically.

(a) Show that $E_n = e - nE_{n-1}$.

(b) Show that $0 \leq E_n \leq e$ for each n and that $E_0 = e - 1$.

(c) Use the recursion relation to compute $E_1, E_2, ..., E_{10}$. Ignore the fact that your calculator has an "e" key and use the (good) approximate value of 2.718 for e. Do you think you've found a good value for E_{10}? Do your numbers satisfy the inequality in (b)? Explain what happened by deriving a recursion relation for the difference between your approximate E_n (call it $\bar{E}_n$) and the real E_n.

(d) Rewrite the recursion relation in (a) to express E_{n-1} in terms of E_n. Pick any number between -100 and $+100$ for E_{20} and use the recursion relation to compute $E_{19}, E_{18}, \ldots, E_{10}$.

(e) Do you think this value for E_{10} is right? See if it is right by numerically estimating the integral for E_{10}. How can this be?

(f) Explain what is going on!

4. The purpose of this project is to prove that the trapezoid rule has error bound $E(h) = \frac{Mh^2(b-a)}{12}$. We use the linear approximation $L^i(x)$ on $[x_{i-1}, x_i]$ as defined in Section 3.4.

(a) Let $x \in (x_{i-1}, x_i)$ and define

$$g_i(t) \equiv f(t) - L^i(t) - (f(x) - L^i(x)) \frac{(t - x_{i-1})(t - x_i)}{(x - x_{i-1})(x - x_i)}$$

Verify that $g(x_i) = g(x) = g(x_{i-1}) = 0$.

(b) Use Rolle's theorem (Theorem 4.2.2) twice to conclude that there exists a $\xi \in (x_{i-1}, x_i)$ so that $g''(\xi) = 0$.

(c) Prove that $f(x) - L^i(x) = \frac{1}{2}f''(\xi)(x - x_{i-1})(x - x_i)$.

(d) Prove the error bound for the trapezoid rule.

CHAPTER 4

Differentiation

In this chapter we introduce the notion of derivative, develop its properties, and explain the relationship between differentiation and integration. Some of the results, though probably not the proofs, will be familiar from calculus. In Section 4.3 we prove Taylor's theorem, a fundamental result because it often allows one to approximate functions by polynomials. We have already used Taylor's theorem in discussing numerical integration techniques in Section 3.4. Further applications appear in Section 4.4 and throughout the book. In Section 4.5, we show how to compute the derivatives of inverse functions, and the result is used to justify the usual change-of-variables techniques in integration. Finally, in Section 4.6 we show how some of the ideas of Chapters 3 and 4 can be extended to functions of two variables.

4.1 Differentiable Functions

Definition. A function f is said to be **differentiable** at a point $x \in Dom(f)$ if the limit of the difference quotient

$$\frac{f(x+h) - f(x)}{h}$$

exists as $h \to 0$, in which case we call the limit the **derivative** of f at x and denote it by

$$f'(x) \equiv \lim_{h \to 0} \frac{f(x+h) - f(x)}{h}. \tag{1}$$

We will also use the standard notation $\frac{df}{dx} = f'(x)$.

It is assumed in the definition that an open interval about x is contained in the domain of f. Recall the definition of $\lim_{h \to}$ from Section 3.1. Saying that (1) is true means that

$$f'(x) \;=\; \lim_{n \to \infty} \frac{f(x + h_n) - f(x)}{h_n}$$

for every sequence $\{h_n\}$ such that $h_n \neq 0$ and $x + h_n \in Dom(f)$ for each n and $h_n \to 0$. Equivalently, given $\varepsilon > 0$, there is a $\delta > 0$ such that

$$\left| f'(x) - \frac{f(x + h) - f(x)}{h} \right| \;\leq\; \varepsilon \qquad \text{if } 0 < |h| < \delta.$$

Note that f' is itself a function since the limit will, in general, depend on x. Since x is required to be in the domain of f we always have $Dom(f') \subseteq Dom(f)$.

Example 1   To illustrate the definition, we'll check that $f(x) = x^2$ is differentiable everywhere. Let x be given, then

$$\frac{f(x + h) - f(x)}{h} \;=\; \frac{(x + h)^2 - x^2}{h}$$

$$=\; 2x + h.$$

Since the limit of $(2x + h)$ exists as $h \to 0$ and equals $2x$, we conclude from the definition that f' exists at x and $f'(x) = 2x$.

The difference quotient (1) is the slope of the straight line through the points $(x, f(x))$ and $(x+h, f(x+h))$ on the graph of f; see Figure 4.1.1(a). Thus, intuitively, $f'(x)$ is the limit of these slopes, that is, the slope of the line tangent to the graph of f at $(x, f(x))$.

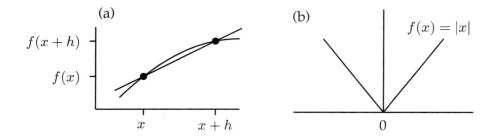

Figure 4.1.1

Example 2 Figure 4.1.1(b) shows an example of a function, $f(x) = |x|$, which is not differentiable at $x = 0$. In order to show that the difference quotient does not have a limit, we just need to exhibit a sequence $\{h_n\}$ satisfying $h_n \to 0$ such that the limit of

$$\frac{f(0 + h_n) - f(0)}{h_n}$$

as $n \to \infty$ does not exist. Let $h_n = \frac{(-1)^n}{n}$. Then $h_n \to 0$ as $n \to \infty$. However, when n is odd, h_n is negative and

$$\frac{f(0 + h_n) - f(0)}{h_n} = -1.$$

When n is even, h_n is positive and

$$\frac{f(0 + h_n) - f(0)}{h_n} = 1.$$

Thus the sequence of quotients does not converge as $n \to \infty$. The difference quotient does have a limit from the left and a limit from the right at $x = 0$, but the limits are not the same.

It is useful to think of $f'(x)$ in three ways. It is the limit of the difference quotient (1). It is the rate of change of the function f at x. It is the slope of the tangent line to the graph of f at the point $(x, f(x))$.

❑ **Theorem 4.1.1** If f is differentiable at x, then f is continuous at x.

Proof. We write

$$f(x + h) - f(x) = \frac{f(x + h) - f(x)}{h} h.$$

Since f is differentiable at x, the quotient on the right converges to a finite limit. Therefore, since $h \to 0$, the right-hand side converges to zero by Theorem 2.2.5. Thus, the left-hand side converges to zero too, which proves that f is continuous at x. ❑

❑ **Theorem 4.1.2** Suppose that f and g are differentiable at x. Then,

(a) For any constants α and β, $\alpha f + \beta g$ is differentiable at x and

$$(\alpha f + \beta g)' = \alpha f' + \beta g'.$$

(b) The product fg is differentiable at x and $(fg)' = f'g + fg'$.

(c) If $g(x) \neq 0$, then f/g is differentiable at x and

$$\left(\frac{f}{g}\right)' = \frac{gf' - fg'}{g^2}.$$

Proof.   By definition, $(\alpha f + \beta g)(x) = \alpha f(x) + \beta g(x)$. Thus,

$$\frac{(\alpha f + \beta g)(x + h) - (\alpha f + \beta g)(x)}{h}$$

$$= \alpha \frac{f(x + h) - f(x)}{h} + \beta \frac{g(x + h) - g(x)}{h}.$$

Since f and g are differentiable at x, the difference quotients for f and g on the right side have limits $f'(x)$ and $g'(x)$, respectively. By Theorem 2.2.3 and Theorem 2.2.4, the whole right-hand side converges to $\alpha f'(x) + \beta g'(x)$. Thus, the left-hand side has a limit, and the limit is $\alpha f'(x) + \beta g'(x)$. This proves (a).

To prove (b), we write

$$\frac{f(x + h)g(x + h) - f(x)g(x)}{h}$$

$$= \frac{f(x + h) - f(x)}{h}\, g(x + h) + \frac{g(x + h) - g(x)}{h}\, f(x).$$

As above, the difference quotients for f and g on the right side have limits $f'(x)$ and $g'(x)$ respectively. Furthermore, by Theorem 4.1.1, $g(x + h) \to g(x)$ as $h \to 0$. Thus, by Theorems 2.2.3 – 2.2.5, the right-hand side has a limit, and the limit is $f'(x)g(x) + f(x)g'(x)$. Therefore, the left-hand side has a limit, and the limit is $f'(x)g(x) + f(x)g'(x)$.

We omit the proof of (c).   ❏

Example 3   To show how useful this theorem is, we will prove that polynomials are differentiable at all x and derive the usual formula for the derivative. First, it is easy to see that if $f(x) = \alpha$ for all x, then the difference quotient is zero for all h, so $f'(x) = 0$. It is also easy to use the definition to see that if $f(x) = x$, then $f'(x) = 1$. To differentiate $f(x) = x^2$, notice that $x^2 = x \cdot x$. So, applying part (b) of the theorem, we see that $f(x) = x^2$ is differentiable and

$$(x \cdot x)' = x \cdot 1 + 1 \cdot x = 2x.$$

Of course, we knew this already by explicitly evaluating the limit of the difference quotient, but this gives us the idea for a general proof. Suppose we know that $(x^n)' = nx^{n-1}$ for some $n \geq 1$. Then, by part (b) of Theorem 4.1.2,

$$
\begin{aligned}
(x^{n+1})' &= (x \cdot x^n)' \\
&= (x)'(x^n) + x(x^n)' \\
&= x^n + x \cdot nx^{n-1} \\
&= (n+1)x^n.
\end{aligned}
$$

Therefore, by induction, we have proven that $(x^n)' = nx^{n-1}$ for all $n \geq 1$. Using part (a) of the theorem repeatedly, we obtain

$$
(\alpha_0 + \alpha_1 x + \alpha_2 x^2 + \ldots + \alpha_m x^m)' = \alpha_1 + 2\alpha_2 + \ldots + m\alpha_m x^{m-1},
$$

which is the usual formula for the derivative of a polynomial. To see what we gained by using Theorem 4.1.2, try to prove directly that if $f(x) = \alpha_0 + \alpha_1 x + \alpha_2 x^2 + \ldots + \alpha_m x^m$, then the limit of the difference quotient exists.

❏ **Theorem 4.1.3 (The Chain Rule)** Suppose that f is differentiable at x and g is differentiable at $f(x)$. Then $g \circ f$ is differentiable at x and $(g(f(x))' = g'(f(x))f'(x)$.

Proof. We define a new function H on

$$
Dom(H) \equiv \{y \,|\, y \in Dom(g) \text{ and } y \neq f(x)\}
$$

by the formula

$$
H(y) \equiv \frac{g(y) - g(f(x))}{y - f(x)}. \tag{2}
$$

Let $y = f(x) + h$. Then

$$
H(f(x) + h) = \frac{g(f(x) + h) - g(f(x))}{h} \rightarrow g'(f(x))
$$

as $h \rightarrow 0$ since g is differentiable at $f(x)$. Therefore, if we extend the domain of H to include $f(x)$ and define $H(f(x)) \equiv g'(f(x))$, then H is continuous at $f(x)$. Multiply both sides of (2) by $(y - f(x))$, substitute $y = f(x + h)$, and divide both sides by h. The result is

$$\frac{g(f(x+h)) - g(f(x))}{h} \; = \; H(f(x+h)) \, \frac{f(x+h) - f(x)}{h}.$$

Since f is differentiable at x, $\frac{1}{h}(f(x+h) - f(x)) \to f'(x)$ as $h \to 0$. Since f is continuous at x, $f(x+h) \to f(x)$; so, by the continuity of H, we see that $H(f(x+h)) \to H(f(x))$. Thus, the limit of the right-hand side exists and is equal to $g'(f(x))f'(x)$. It follows that the left-hand side has a limit and the limit is $g'(f(x))f'(x)$. Therefore, by definition, $g \circ f$ is differentiable at x and $(g(f(x)))' = g'(f(x))f'(x)$.   ❏

We will often use $\sin x$, $\cos x$, and the exponential function in examples even though we will not formally define them until Chapter 6. We assume that the reader is familiar with them, knows that they are differentiable, and knows their derivatives. The chain rule allows us to assert that complicated functions like $\sin(e^{2x})$ are differentiable (and it gives a formula for the derivative!) without our having to take the limit of the difference quotient.

Definition.   If f is differentiable at every point of an open interval (a, b), we say that f is differentiable *on* (a, b). If, in addition, f' is continuous on (a, b), we say that f is **continuously differentiable** on (a, b).

Recall from Section 3.2 that continuous functions on closed intervals have very strong properties, so it is natural to try to extend this definition to closed intervals. We say that f is differentiable on $[a, b]$ if it is differentiable on (a, b) and, in addition, the right-hand limit of the difference quotient exists at a and the left-hand limit of the difference quotient exists at b. Right-hand and left-hand limits are defined in Section 3.5.

Definition.   If f is differentiable on $[a, b]$ and f' is continuous on $[a, b]$, we say that f is **continuously differentiable** on $[a, b]$.

It is convenient to give names to sets of functions which satisfy different hypotheses. We denote the set of continuous functions on $[a, b]$ by $C[a, b]$ and the continuously differentiable functions on $[a, b]$ by $C^{(1)}[a, b]$. Similarly, $C(\mathbb{R})$ and $C^{(1)}(\mathbb{R})$ denote the continuous and continuously differentiable functions on $\mathbb{R}$. Note that these sets of functions are vector spaces in the sense that linear combinations of functions in these sets are again in the sets (by Theorems 3.1.1 and 4.1.2). They are also algebras because products of functions in these spaces are again in the spaces.

If a function f is differentiable in an open interval about x and its derivative f' is differentiable at x, we say that f is twice differentiable at x and denote the second derivative by

$$f''(x) \equiv \frac{d}{dx}\left(\frac{df}{dx}\right)$$

or $\frac{d^2 f}{dx^2}$. Higher derivatives are defined analogously; we will often denote the nth derivative by $f^{(n)}(x)$. As above, we denote by $C^{(n)}[a,b]$ the set of functions which are n times continuously differentiable on $[a,b]$, and by $C^{(\infty)}[a,b]$ the functions which are infinitely often continuously differentiable. The spaces $C^{(n)}(\mathbb{R})$ and $C^{(\infty)}(\mathbb{R})$ are defined analogously. We showed in Example 3 that polynomials are differentiable everywhere and that their derivatives are again polynomials. Thus all polynomials are in $C^{(\infty)}(\mathbb{R})$. Since $\sin x$, $\cos x$, and e^x are continuous and $(\sin x)' = \cos x$, $(\cos x)' = -\sin x$, and $(e^x)' = e^x$, these functions are continuously differentiable. By differentiating repeatedly, one can easily show by induction that they are in $C^{(\infty)}(\mathbb{R})$.

Problems

1. Prove that $f(x) = x^3$ is differentiable everywhere by showing that the limit of the difference quotient exists.

2. Let $n > 0$ be a positive integer. For all $x \neq 0$, prove that $f(x) = \frac{1}{x^n}$ is differentiable everywhere and $f'(x) = \frac{-n}{x^{n+1}}$, by showing that the limit of the difference quotient exists.

3. Let p and q be integers, $q \neq 0$. Suppose that $f(x) = x^{\frac{p}{q}}$ is differentiable for $x > 0$. Prove that

$$\frac{d}{dx}x^{\frac{p}{q}} = \frac{p}{q}x^{\frac{p}{q}-1}.$$

   Hint: differentiate $f(x)^q = x^p$.

4. Where are the following functions differentiable?

   (a) $|\sin x|$.

   (b) $\sin|x|$.

5. Let $p(x)$ be a polynomial and suppose that x_o is a real root; that is, $p(x_o) = 0$. When will $|p(x)|$ be differentiable at x_o?

6. Is $f(x) = \sqrt{x}$ continuously differentiable on the interval $(0, 1)$? On the interval $[0, 1]$?

7. Suppose that we assume that e^{ax}, $\sin x$, and $\cos x$ are continuously differentiable with derivatives ae^{ax}, $\cos x$, and $-\sin x$, respectively. Prove that $e^{\sin x} \epsilon C^{(\infty)}(\mathbb{R})$.

8. Suppose that $f \epsilon C^{(1)}[a, b]$. Prove that f is Lipschitz continuous on $[a, b]$. Hint: use the Mean Value Theorem.

9. Suppose that f is differentiable at x. Prove that

$$\lim_{h \to 0} \frac{f(x + h) - f(x - h)}{2h} = f'(x).$$

10. Suppose that $0 \leq f(x) \leq x^2$ for all $x \epsilon \mathbb{R}$.

   (a) Prove that f is differentiable at $x = 0$ and $f'(0) = 0$.

   (b) Give an example of a function which satisfies the hypothesis but which is not continuous for $x \neq 0$. Is the function continuous at $x = 0$?

11. Suppose that $f(x) \geq 0$ for all $x \epsilon \mathbb{R}$. Assume that $f(x)^2$ is differentiable. Is $f(x)$ necessarily differentiable?

12. Let $f(x) = x^2 \sin(1/x)$ on the set $E = (-\infty, 0) \cup (0, \infty)$.

   (a) Explain why f is continuously differentiable on E.

   (b) Compute the derivative of f on E.

   (c) Define $f(0) = 0$ and use the difference quotient to show that f is differentiable at zero.

   (d) Show that if $f(0) = 0$, then f is differentiable but not continuously differentiable on $\mathbb{R}$.

13. Let g be the function $g(x) = |x|$ on the interval $[-1, 1)$. Extend g to the whole real line by requiring that $g(x + 2) = g(x)$ for all x.

   (a) Draw the graph of g.

   (b) Where is g continuous? Where is g differentiable?

   (c) Define $g_n(x) \equiv g(4^n x)$. Where is g_n continuous? Where is g_n differentiable?

   Remark: These functions are used in Example 2 of Section 6.3 where we prove the existence of a continuous function that is nowhere differentiable.

14. A function is called **piecewise smooth** on $[a, b]$ if it is infinitely often continuously differentiable at all but finitely many points, $x_1, x_2, \ldots, x_N$, of $[a, b]$ at which f and all its derivatives are continuous or have jump discontinuities. Which of the following functions are piecewise smooth?

(a) $f(x) = |x|$ on $[-1, 1]$.

(b) The functions in parts (a) and (b) of problem 4 defined on $[-2\pi, 2\pi]$.

(c) $f(x) = \sqrt{x}$ on $[1, 2]$.

(d) The function f of problem 12 on $[-1, 1]$.

(e) The function g of problem 13 on $[-1, 1]$.

4.2   The Fundamental Theorem of Calculus

Before proving the Fundamental Theorem of Calculus, we prove three theorems which relate the values of a function f to values of the derivative. Each theorem contains a simple geometric idea. The first theorem says that at an interior maximum point, the tangent line must be horizontal. See Figure 4.2.1(a).

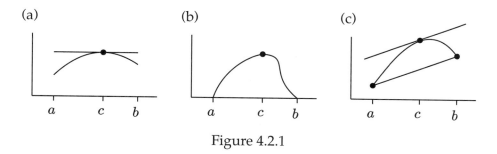

Figure 4.2.1

❑ **Theorem 4.2.1**   Suppose that f is continuous on the finite interval $[a, b]$. Let c be a point where f attains its maximum. If $a < c < b$ and f is differentiable at c, then $f'(c) = 0$.

Proof.   Suppose $f'(c) > 0$. Since $\frac{1}{h}(f(c + h) - f(c)) \to f'(c)$, we can choose a δ so that

$$\frac{f(c + h) - f(c)}{h} \geq f'(c)/2 \quad \text{if } |h| \leq \delta.$$

Thus for $0 < h \leq \delta$,

$$f(c + h) \geq f(c) + hf'(c)/2 > f(c),$$

which contradicts the hypothesis that f achieves its maximum at c. A similar proof shows that $f'(c) < 0$ is also impossible.   ❑

A similar result holds for the point where f achieves its minimum (problem 1).

The next theorem says that if f is zero at two different points then it must achieve a maximum or a minimum somewhere in between.

❏ **Theorem 4.2.2 (Rolle's Theorem)** Suppose that f is continuous on the finite interval $[a, b]$, differentiable on (a, b), and $f(a) = 0 = f(b)$. Then there is a point c satisfying $a < c < b$ such that $f'(c) = 0$.

Proof.   If $f(x) = 0$ for all $x \epsilon (a, b)$, then $f'(x) = 0$ for all x, so we can choose c to be any point in the interval. Otherwise, there must be a point x_o such that $|f(x_o)| \neq 0$. If $f(x_o) > 0$, then, by Theorem 3.2.2, f achieves a positive maximum at some point c in the interval $[a, b]$. The point c cannot equal a or b since f is zero there, so $a < c < b$. Thus, by Theorem 4.2.1, $f'(c) = 0$. See Figure 4.2.1(b). If $f(x_o) < 0$, a similar proof, using the analogue of Theorem 4.2.1 for minima, gives the result.   ❏

The next theorem states that there must be a point on the graph of a function between $(a, f(a))$ and $(b, f(b))$ where the slope of the tangent line equals the slope of the straight line between $(a, f(a))$ and $(b, f(b))$. See Figure 4.2.1(c).

❏ **Theorem 4.2.3 (The Mean Value Theorem)** Suppose that f is continuous on the finite interval $[a, b]$ and differentiable on (a, b). Then there is a point c satisfying $a < c < b$ such that

$$f'(c) = \frac{f(b) - f(a)}{b - a}.$$

Proof.   The idea of the proof is simple; we subtract from f the function whose graph is the straight line and then use Rolle's theorem. Define

$$g(x) \equiv f(x) - \left\{ f(a) + \frac{f(b) - f(a)}{b - a} (x - a) \right\}. \qquad (3)$$

Then $g(a) = 0 = g(b)$, so Rolle's theorem implies that there is a point c satisfying $a < c < b$ such that $g'(c) = 0$. Differentiating both sides of (3) and evaluating at c, we obtain the result.   ❏

Notice that, in general, the point c depends on a and b. Though Rolle's theorem and the Mean Value Theorem are very simple, they are extremely useful and important.

Example 1 Consider $\sin x$ on the interval $[0, x]$. Since $(\sin x)' = \cos x$, the Mean Value Theorem tells us that for each x there is a point $c(x)$ satisfying $0 \leq c(x) \leq x$ such that

$$\frac{\sin x}{x} = \frac{\sin x - \sin 0}{x - 0} = \cos c(x).$$

As $x \to 0$, $c(x) \to 0$ since $0 < c(x) < x$. Thus, because $\cos x$ is continuous,

$$\lim_{x \searrow 0} \frac{\sin x}{x} = \lim_{x \searrow 0} \cos c(x) = 1.$$

Since $\sin x$ is an odd function, the limit from the left is the same. Thus,

$$\lim_{x \to 0} \frac{\sin x}{x} = 1.$$

❏ **Theorem 4.2.4 (The Fundamental Theorem, Part I)** Let f be a continuously differentiable function on a finite interval $[a, b]$. Then,

$$\int_a^b f'(x)\, dx = f(b) - f(a). \tag{4}$$

Proof. Let P be a partition of $[a, b]$ into N subintervals. By the Mean Value Theorem, there is ξ_i in each subinterval $[x_{i-1}, x_i]$ such that

$$f(x_i) - f(x_{i-1}) = f'(\xi_i)(x_i - x_{i-1}).$$

Thus,

$$f(b) - f(a) = \sum_{i=1}^{N} (f(x_i) - f(x_{i-1})) \tag{5}$$

$$= \sum_{i=1}^{N} f'(\xi_i)(x_i - x_{i-1}). \tag{6}$$

The sum (4) is a Riemann sum for $\int_a^b f'(x)\, dx$. By Corollary 3.3.2, the Riemann sum converges to $\int_a^b f'(x)\, dx$ as the maximal length of the subintervals gets smaller since f' is continuous. But each of the Riemann sums equals $f(b) - f(a)$, so the limit of the Riemann sums equals $f(b) - f(a)$.
❏

Part I of the Fundamental Theorem is the basis for analytical integration. Given an integral $\int_a^b g(x)dx$, one tries to guess a G so that $G'(x) = g(x)$.

If one is able to, then $\int_a^b g(x)dx = G(b) - G(a)$, by the Fundamental Theorem.

❑ **Theorem 4.2.5 (The Fundamental Theorem, Part II)** Let f be a continuous function on a finite interval $[a, b]$. Define

$$F(x) \equiv \int_a^x f(t)\,dt.$$

Then F is continuously differentiable on $[a, b]$, and $F'(x) = f(x)$.

Proof.   Let $x \in [a, b)$. We shall show that the difference quotient for F has the right-hand limit $f(x)$ at x. Suppose $h > 0$. Then

$$\frac{F(x+h) - F(x)}{h} = \frac{1}{h}\int_a^{x+h} f(t)\,dt - \frac{1}{h}\int_a^x f(t)\,dt$$

$$= \frac{1}{h}\int_x^{x+h} f(t)\,dt.$$

For the moment, fix h and let m and M denote the minimum and maximum of f on the interval $[x, x + h]$. Then, by Theorem 3.3.4,

$$m \leq \frac{1}{h}\int_x^{x+h} f(t)\,dt \leq M.$$

Since f takes on the values m and M, the Intermediate Value Theorem guarantees that there is a c in the interval $[x, x + h]$ so that

$$f(c) = \frac{1}{h}\int_x^{x+h} f(t)\,dt.$$

Now, c depends on h, but since $c \in [x, x+h]$ we must have $c \searrow x$ as $h \to 0$. Since f is continuous, we know that $f(c) \to f(x)$ as $h \searrow 0$. Thus the difference quotient for $F(x)$ converges to $f(x)$ as $h \searrow 0$. A similar proof shows that if $x \in (a, b]$, then the left-hand limit of the difference quotient of F converges to $f(x)$ as $h \nearrow 0$. Hence F is differentiable on $[a, b]$ and its derivative is f. Since f is continuous, F is continuously differentiable on $[a, b]$. ❑

Note that Part II answers a very natural question. Does every continuous function have an antiderivative? That is, given f, is there an F so that $F' = f$?

All five theorems proven in this section have versions with weaker hypotheses.

Problems

1. State and prove the analogue of Theorem 4.2.1 for the point where f achieves its minimum.

2. Let $f(x) = e^{-x^2}$. Find a formula for a function F so that $F'(x) = f(x)$.

3. Let g be twice continuously differentiable on $[a, b]$. Suppose that there are three distinct points, x_1, x_2, x_3, in $[a, b]$ so that $g(x_1) = g(x_2) = g(x_3) = 0$. Prove that there is a c satisfying $a < c < b$ such that $g''(c) = 0$. Hint: use Rolle's theorem twice.

4. Prove that
$$\lim_{x \to 0} \frac{\cos x - 1}{x} = 0.$$
Hint: use the Mean Value Theorem.

5. Suppose that f is continuously differentiable on $[a, b]$ and $f'(x) = 0$ for all x. Prove that f is a constant function.

6. Suppose that f is continuously differentiable on $[a, b]$ and $f'(x) > 0$ for all x. Prove that f is strictly monotone increasing on $[a, b]$; that is, if $x < y$, then $f(x) < f(y)$. $\quad 0 < y - x \qquad 0 < f(y) - f(x) = \int_x^y f'(t) dt$

7. Let f and g be in $C^{(1)}(\mathbb{R})$ and suppose that $f(0) = g(0)$ and $f'(x) \le g'(x)$ for all $x \ge 0$. Prove that $f(x) \le g(x)$ for all $x \ge 0$.

8. Suppose that f is continuously differentiable on $[a, b]$ and that $f(a) = 2$ and $|f'(x)| \le .3$ for all $x \in [a, b]$. What can you say about $f(b)$?

9. Let f be a piecewise continuous function on a finite interval $[a, b]$ with jump discontinuities at $a_1, a_2, ..., a_K$. Define $F(x) \equiv \int_a^x f(t) \, dt$. Prove that F is differentiable except at the points $a_1, a_2, ..., a_K$ and $F'(x) = f(x)$.

10. Suppose that f is continuous and g is continuously differentiable on $\mathbb{R}$. Prove that $\int_0^{g(x)} f(t) \, dt$ is a continuously differentiable function of x and compute its derivative.

11. Let f be twice continuously differentiable on $[a, b]$ and suppose that $f'(c) = 0$ for some $c \in (a, b)$. Suppose $f''(c) > 0$. Prove that

    (a) there is a $\delta > 0$ such that $f''(x) > 0$ for all $x \in (c - \delta, c + \delta)$.

    (b) $f'(x) < 0$ for $x \in (c - \delta, c)$ and $f'(x) > 0$ for $x \in (c, c + \delta)$.

    (c) $f(x) > f(c)$ for $x \in (c - \delta, c + \delta)$ such that $x \ne c$; that is, c is a local minimum.

(d) the hypothesis $f''(c) \geq 0$ is not sufficient to guarantee the conclusion of part (c).

12. Suppose that f and g are continuously differentiable on $[a, b]$. Prove that

$$\int_a^b f'(x)g(x)\,dx = f(b)g(b) - f(a)g(a) - \int_a^b f(x)g'(x)\,dx.$$

Hint: use the Fundamental Theorem of Calculus.

13. Suppose that f is a continuous function on $\mathbb{R}$ that satisfies

$$f(x) \;=\; 5 + 2\int_0^x f(t)\,dt.$$

Prove that $f \in C^{(1)}(\mathbb{R})$ and express f' in terms of f. Then find f.

14. Suppose that f is a continuous function on an interval about 0 that satisfies

$$f(x) \;=\; 5 + 2\int_0^x f(t)^2\,dt.$$

Prove that $f \in C^{(1)}(\mathbb{R})$ and express f' in terms of f. Then find f. Hint: see project 1.

4.3   Taylor's Theorem

It is natural to try to approximate functions by polynomials because polynomials are easy to integrate and easy to differentiate. Suppose that we are given a function f defined on some finite interval $[a, b]$ and we want to approximate it near some given point x_o in the interval.

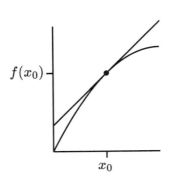

If f is a differentiable function, it is reasonable to make the linear approximation $f(x) \approx f(x_o) + f'(x_o)\,(x - x_o)$. We use the symbol $\approx$ to mean "approximately equal to." The function on the right is the straight line that has the same value at x_o as f and the same derivative at x_o as f. For x close to x_o the linear approximation should be pretty good. But how close and how good? If f itself were a straight line near x_o, that is, if f' were constant, then the approximation would be perfect. But

$f(x_0)$

x_0

Figure 4.3.1

if the slope $f'(x)$ is changing, then the graph of f will curve away from the straight line approximation. See Figure 4.3.1. Since the rate of change of $f'(x)$ is given by $f''(x)$, our intuition suggests that the size of the second derivative of f will determine how good the linear approximation is. We shall see below that this intuition is correct.

If we want to make better approximations, we could use a quadratic approximation

$$f(x) \approx f(x_o) + f'(x_o)(x - x_o) + \frac{f''(x_o)}{2!}(x - x_o)^2,$$

which has the same value, the same derivative and the same second derivative as f at x_o. Continuing in this way, we define the **n^{th} Taylor polynomial** of f at x_o by

$$T^{(n)}(x, x_o) \equiv f(x_o) + f'(x_o)(x - x_o) + \frac{f''(x_o)}{2!}(x - x_o)^2 +$$

$$\dots + \frac{f^{(n)}(x_o)}{n!}(x - x_o)^n.$$

$T^{(n)}(x, x_o)$ is a polynomial in x, but the coefficients depend on x_o. Note that f must be differentiable n times at x_o in order to define $T^{(n)}(x, x_o)$.

Example 1   Let $f(x) = \cos x$ and $x_o = 0$. Then, $f(0) = 1, f'(0) = 0, f''(0) = -1, f^{(3)}(0) = 0$, and $f^{(4)}(0) = 1$, so the first five Taylor polynomials are:

$$T^{(0)}(x, 0) = 1$$
$$T^{(1)}(x, 0) = 1$$
$$T^{(2)}(x, 0) = 1 - \frac{x^2}{2!}$$
$$T^{(3)}(x, 0) = 1 - \frac{x^2}{2!}$$
$$T^{(4)}(x, 0) = 1 - \frac{x^2}{2!} + \frac{x^4}{4!}.$$

The graphs of $\cos x$, $T^{(0)}(x, 0)$, $T^{(2)}(x, 0)$, $T^{(4)}(x, 0)$, and $T^{(6)}(x, 0)$ are shown in Figure 4.3.2.

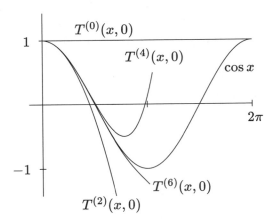

Figure 4.3.2

❑ **Theorem 4.3.1 (Taylor's Theorem)** Suppose that f is n times continuously differentiable on the interval $[a, b]$ and suppose that $f^{(n+1)}$ exists on (a, b). Let x_o be a point of $[a, b]$. Then for all x in $[a, b]$, $x \neq x_o$, there is a ξ between x and x_o such that

$$f(x) \;=\; T^{(n)}(x, x_o) \;+\; \frac{f^{(n+1)}(\xi)}{(n+1)!}\,(x - x_o)^{n+1}. \tag{7}$$

Proof.   Fix an x in $[a, b]$ and let α be the number so that

$$f(x) \;=\; T^{(n)}(x, x_o) \;+\; \alpha\,(x - x_o)^{n+1}. \tag{8}$$

Define a new function g on $[a, b]$ by

$$g(t) \;=\; f(t) \;-\; T^{(n)}(t, x_o) \;-\; \alpha\,(t - x_o)^{n+1}.$$

Differentiating $n + 1$ times yields

$$g^{(n+1)}(t) \;=\; f^{(n+1)}(t) \;-\; \alpha\,(n+1)!.$$

Hence, we just need to show that there is a ξ between x and x_o such that $g^{(n+1)}(\xi) = 0$, for then $\alpha = f^{(n+1)}(\xi)/(n+1)!$. Because the first n derivatives of $T^{(n)}$ equal those of f at x_o, we have

$$g(x_o) \;=\; g'(x_o) \;=\; g^{(2)}(x_o) \;=\; ... \;=\; g^{(n)}(x_o) \;=\; 0.$$

Now, since $g(x) = 0$ (by the way we chose α in (8)), Rolle's theorem implies that there is an x_1 between x and x_o so that $g'(x_1) = 0$. Since $g'(x_o) = 0$ too, Rolle's theorem implies that there is an x_2 between x_1 and x_o so that $g^{(2)}(x_2) = 0$. Continuing in this manner, we find an x_{n+1} so that $g^{(n+1)}(x_{n+1}) = 0$. Setting $\xi = x_{n+1}$, we conclude that (7) holds.

❏

Notice that the Mean Value Theorem is just Taylor's theorem for the case $n = 0$.

Example 1 (revisited) Let's see how good an approximation $T^{(4)}(x, 0)$ is to $\cos x$. Since the fifth derivative of $\cos x$ is $-\sin x$,

$$\left| \cos x - \left\{ 1 - \frac{x^2}{2!} + \frac{x^4}{4!} \right\} \right| \le \left| \frac{\sin \xi}{5!} x^5 \right|$$

$$\le \frac{|x|^5}{5!}.$$

Thus, the approximation is very good near the origin but gets much worse as x grows, as we saw in Figure 4.3.2.

Example 2 (the natural logarithm) The function $\ln x$, is defined for $x > 0$ by

$$\ln x = \int_1^x \frac{1}{t}\, dt.$$

By the Fundamental Theorem of Calculus, $\ln x$ is continuously differentiable and $(\ln x)' = 1/x$. Thus $\ln x$ is infinitely often continuously differentiable on $(0, \infty)$. The first three derivatives evaluated at $x = 1$ are $f'(1) = 1$, $f''(1) = -1$, and $f^{(3)}(1) = 2$. We shall use the third-order Taylor polynomial about the point $x_o = 1$,

$$T^{(3)}(x, 1) = (x - 1) - \frac{1}{2}(x - 1)^2 + \frac{1}{3}(x - 1)^3,$$

to approximate $\ln 1.2$. Note that $\ln 1.2$ can also be evaluated by computing the defining integral approximately using the methods of Section 3.4 (problem 6). Since the fourth derivative of $\ln x$ is $-\frac{3!}{x^4}$, Taylor's theorem says that

$$\ln x - T^{(3)}(x, 1) = -\frac{1}{4\xi^4}(x - 1)^4$$

for some ξ between 1 and x. We are interested in $x = 1.2$. Since $\xi \in (1, 1.2)$, we know that $\frac{1}{\xi} \leq 1$, so

$$| \ln 1.2 - T^{(3)}(1.2, 1)| \leq \frac{1}{4}(.2)^4 = .0004.$$

Thus, the value $T^{(3)}(1.2, 1) = .1827$ is within .0004 of $\ln 1.2$.

Taylor's Theorem can also be used to evaluate certain difficult limits.

❏ **Theorem 4.3.2 (l'Hospital's Rule)**   Suppose that f and g are continuously differentiable on an interval containing the point x_o and that $f(x_o) = 0 = g(x_o)$. Suppose also that $g'(x_o) \neq 0$. Then,

$$\lim_{x \to x_o} \frac{f(x)}{g(x)} = \frac{f'(x_o)}{g'(x_o)}.$$

Proof.   Since $g'(x_o) \neq 0$, both g and g' are nonzero in a small interval about x_o except for the root of g at x_o (problem 12). Using the Mean Value Theorem, we can therefore compute

$$\lim_{x \to x_o} \frac{f(x)}{g(x)} = \lim_{x \to x_o} \frac{f(x) - f(x_o)}{g(x) - g(x_o)}$$

$$= \lim_{x \to x_o} \frac{f'(\xi_1)}{g'(\xi_2)}$$

$$= \frac{f'(x_o)}{g'(x_o)}$$

since f' and g' are continuous. We used the fact that ξ_1 and ξ_2 are between x_o and x, so $\xi_1 \to x_o$ and $\xi_2 \to x_o$ as $x \to x_o$.   ❏

Suppose that f is infinitely often continuously differentiable in an open interval about x_o. In that case we can define the nth Taylor polynomial for all n. This raises the question of whether the infinite series

$$\sum_{n=0}^{\infty} \frac{f^{(n)}(x_o)}{n!}(x - x_o)^n$$

converges and, if it does, whether it converges to $f(x)$. We study the convergence of power series in Section 6.4, where we shall see that this question is deeper than it looks.

Problems

1. Compare the graph of $\ln x$ to the graphs of the first five Taylor polynomials of $\ln x$ about $x = 1$.

2. Compare the graph of $\frac{1}{\sqrt{2\pi}} e^{-x^2/2}$ to the graphs of the first few Taylor polynomials about $x = 0$.

3. Use l'Hospital's rule to evaluate the following limits:

    (a) $\lim_{x \to 0} \frac{\sin 5x}{x}$.

    (b) $\lim_{x \to 1} \frac{x^5 - 1}{x^3 - 1}$.

    (c) $\lim_{x \to 0} \frac{e^{3x} - 1}{x}$.

4. Compute

$$\lim_{x \to 0} \frac{1}{\sin x} \int_0^{\sin 2x} \cos 5t \, dt.$$

5. Verify all the calculations in Example 2.

6. Use one of the methods in Section 3.4 to estimate the integral for $\ln 1.2$ and compare the result with that obtained in Example 2.

7. Use Taylor polynomials at $x = x_o$ to approximate $\sqrt{9.2}$ and compare the result with what you get using your hand calculator. How do you think the hand calculator makes the computation?

8. Suppose that f is twice continuously differentiable on $[0, 1]$ and that $f(0) = 1$, $f'(0) = 2$, and $|f''(x)| \leq .3$ for all $x \in [0, 1]$. Compute $\int_0^1 f(x) \, dx$ as best you can.

9. Let $f(x) = \frac{\sin x - x}{x^3}$.

    (a) Find the third-order Taylor polynomial for $\sin x$ about $x = 0$.

    (b) Use the error estimate in Taylor's theorem to compute $\lim_{x \to 0} f(x)$.

10. (a) Prove that $\ln a^2 = 2 \ln a$ for all $a > 0$. Hint: write

$$\int_1^{a^2} \frac{1}{x} \, dx = \int_1^a \frac{1}{x} \, dx + \int_a^{a^2} \frac{1}{x} \, dx$$

and use a change of variables in the second integral. Change of variables is justified in Section 4.5.

(b) Prove that $\ln a^n = n \ln a$ for all $a > 0$ and all integers $n \geq 0$.

11. (a) Prove that $\ln x$ is strictly monotone increasing and that $\ln x \to \infty$ as $x \to \infty$. Hint: the harmonic series diverges.

(b) Prove that there is a unique number e such that $\ln e = 1$.

(c) Find the first few Taylor polynomials of $f(x) = \ln(1 + x)$ about $x = 0$.

(d) Use Taylor's theorem to prove that $\lim_{n \to \infty} n \ln\left(1 + \frac{1}{n}\right) = 1$.

(e) Prove that $\lim_{n \to \infty} \left(1 + \frac{1}{n}\right)^n = e$.

12. Suppose that g is continuously differentiable in an interval about x_o and $g(x_o) = 0$. If $g'(x_o) \neq 0$, show that there is a small interval about x_o in which neither g nor g' vanish except for the root of g at the point x_o.

13. Suppose that f and g are twice continuously differentiable on an interval containing the point x_o and that $f(x_o) = f'(x_o) = 0 = g'(x_o) = g(x_o)$. Suppose that $g''(x_o) \neq 0$. Prove that

$$\lim_{x \to x_o} \frac{f(x)}{g(x)} = \frac{f''(x_o)}{g''(x_o)}.$$

14. Evaluate $\lim_{x \to 0} \frac{\cos x - 1}{x^2}$.

4.4 Newton's Method

Finding the roots of functions, that is, the values of x so that $f(x) = 0$, is important in both pure and applied mathematics. The most familiar use is finding the roots of the derivative of a function in order to determine possible local maxima and minima. Even if $f(x)$ is a relatively simple function like a polynomial, this is not an easy question. The quadratic formula allows one to find the roots of quadratic polynomials, and more complicated formulas allow one to write down analytic expressions for the roots of cubic and quartic polynomials. But it can be proven that there are no general formulas for quintic and higher order polynomials. Newton's method, discovered by Isaac Newton, is based on a simple geometric idea. Consider the graph of a function $f(x)$ near one of its roots.

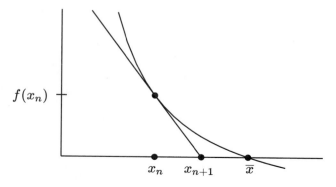

Figure 4.4.1

Let x_n denote the nth guess for the root. We will explain how to construct the $n+1$ guess. If $f(x_n) = 0$, we have found the root and we can stop. Otherwise, go to the point $(x_n, f(x_n))$ on the graph of f. Construct the tangent line to the graph at this point; it has slope $f'(x_n)$. Then x_{n+1} is defined to be the point where the tangent line crosses the x-axis. See Figure 4.4.1. To find the formula for x_{n+1}, consider the triangle whose vertices are at $(x_n, 0), (x_{n+1}, 0)$, and $(x_n, f(x_n))$. The height is $f(x_n)$ and the base is $x_{n+1} - x_n$, so the slope of the hypotenuse (which is a piece of the tangent line) is the ratio. That is,

$$f'(x_n) = \frac{-f(x_n)}{x_{n+1} - x_n}.$$

Solving for x_{n+1}, we find

$$x_{n+1} = x_n - \frac{f(x_n)}{f'(x_n)}. \qquad (9)$$

Example 1 Let's try out Newton's method on the polynomial $f(x) = 2 - x^2$. Of course we know that the roots are $\pm\sqrt{2}$, but let's suppose that we don't know that. From a very rough graph, we can see that there is a positive root somewhere between 1 and 2. If we choose as our initial guess $x_0 = 1$ and apply Newton's method, we find

$$\begin{aligned} x_0 &= 1.0 \\ x_1 &= 1.5 \\ x_2 &= 1.41667 \\ x_3 &= 1.41422 \\ x_4 &= 1.41421 \end{aligned}$$

If we choose as our initial guess $x_0 = 1.9$ and apply Newton's method, we find

$$
\begin{aligned}
x_0 &= 1.9 \\
x_1 &= 1.47632 \\
x_2 &= 1.41552 \\
x_3 &= 1.41421 \\
x_4 &= 1.41421
\end{aligned}
$$

In both cases the sequence converges extremely rapidly to the positive root $\sqrt{2} \approx 1.41421$.

It is clear, even in a simple case like this, that there are pitfalls in Newton's method. First, notice that that if we start with $x_0 < 0$, then the sequence of iterates will converge to the root $-\sqrt{2}$ rather than to the root $\sqrt{2}$. Second, if we had been unlucky enough to start with the initial guess $x_0 = 0$, then the tangent line has zero slope and so doesn't intersect the x-axis anywhere; hence the method breaks down immediately. Intuition suggests that we can avoid these pitfalls if we start close enough to the root that we want to find. The following theorem shows that under reasonable hypotheses the intuition is correct.

❏ **Theorem 4.4.1**   Suppose that $f(\overline{x}) = 0$ and that f is twice continuously differentiable in an interval $(\overline{x} - \delta, \overline{x} + \delta)$ containing $\overline{x}$. Suppose that $f'(\overline{x}) \neq 0$. Then, if x_0 is chosen sufficiently close to $\overline{x}$, the iterates in Newton's method will converge to $\overline{x}$.

Proof.   Note that we are not sure yet that the iterates given by formula (9) even exist: for how do we know the iterates stay in the interval or that $f'(x_n) \neq 0$ for each of the x_n? For the moment we assume that everything is all right and make an estimate. That estimate will tell us how close we need to choose x_0 to $\overline{x}$. From (9),

$$
x_{n+1} - \overline{x} \;=\; x_n - \overline{x} - \frac{f(x_n)}{f'(x_n)} \tag{10}
$$

$$
= \; (x_n - \overline{x}) - \frac{(f(x_n) - f(\overline{x}))}{f'(x_n)} \tag{11}
$$

$$
= \; (x_n - \overline{x}) - \frac{f'(\overline{x})(x_n - \overline{x}) + f''(\tau_n)(x_n - \overline{x})^2/2!}{f'(x_n)} \tag{12}
$$

$$= (x_n - \overline{x}) \left\{ \frac{f'(x_n) - f'(\overline{x})}{f'(x_n)} + \frac{f''(\tau_n)(x_n - \overline{x})}{2! \, f'(x_n)} \right\} \tag{13}$$

$$= (x_n - \overline{x}) \left\{ \left(\frac{1}{f'(x_n)} \right) \left(f''(\overline{\tau}_n) + \frac{f''(\tau_n)}{2!} \right) (x_n - \overline{x}) \right\} \tag{14}$$

We used Taylor's theorem in (12), and the Mean Value Theorem applied to f' in (14). Hence τ_n and $\overline{\tau}_n$ are between $\overline{x}$ and x_n. Notice that if the expression in curly brackets has absolute value less than 1, then each iterate will be strictly closer to $\overline{x}$ than the previous one. We are now ready to say how close is close. Since f' is continuous and $f'(\overline{x}) \neq 0$ by hypothesis, we can choose a $\delta_1 \leq \delta$ such that

$$|f'(x)| \geq \frac{|f'(\overline{x})|}{2} \qquad \text{for all } x \in [\overline{x} - \delta_1, \overline{x} + \delta_1].$$

Equivalently,

$$\frac{1}{|f'(x)|} \leq \frac{2}{|f'(\overline{x})|} \qquad \text{for all } x \in [\overline{x} - \delta_1, \overline{x} + \delta_1].$$

Next, since f'' is continuous, there is an M so that $|f''(x)| \leq M$ for all $x \in [\overline{x} - \delta_1, \overline{x} + \delta_1]$. Thus,

$$\left| \left(\frac{1}{f'(x_n)} \right) \left(f''(\overline{\tau}_n) + \frac{f''(\tau_n)}{2!} \right) (x_n - \overline{x}) \right| \leq \frac{3M}{|f'(\overline{x})|} |x_n - \overline{x}| \leq \frac{1}{2}$$

if

$$|x_n - \overline{x}| \leq (6M)^{-1} |f'(\overline{x})|. \tag{15}$$

Therefore, when (15) holds,

$$|x_{n+1} - \overline{x}| \leq \frac{1}{2} |x_n - \overline{x}|. \tag{16}$$

Hence, if x_0 is within δ_1 of $\overline{x}$ and satisfies (15) for $n = 0$, all the succeeding x_n will be successively closer to $\overline{x}$ and (16) will hold for all n. Iterating (16), we obtain

$$|x_{n+1} - \overline{x}| \leq \left(\frac{1}{2} \right)^n |x_0 - \overline{x}|, \tag{17}$$

which proves that the iterates converge to $\overline{x}$. $\qquad \square$

We have proven the above theorem to illustrate how to use Taylor's formula to get convergence. One can show convergence under much weaker hypotheses. For example, it is not necessary that $f'(\overline{x}) \neq 0$. Notice that if we define $C \equiv 3M/|f'(\overline{x})|$, then it follows from (14) and the estimate above (15) that

$$|x_{n+1} - \overline{x}| \leq C|x_n - \overline{x}|^2.$$

This estimate shows that the convergence in Newton's method is extremely rapid once one is close enough to $\overline{x}$. For example, suppose that $C = 10$ and that $|x_n - \overline{x}| \leq 10^{-3}$. Then, $|x_{n+1} - \overline{x}| \leq 10 \times (10^{-3})^2 = 10^{-5}$. This explains why the convergence in Example 1 is so rapid. In contrast, another technique, the bisection method, discussed in problems 6 and 7, is reliable but slow.

For functions with many roots and many points where the derivative is zero, one must start close to a particular root in order to converge to it. The following example shows that, even for a beautiful function with only one root, the iterates of Newton's method may diverge if we don't start close enough to the root.

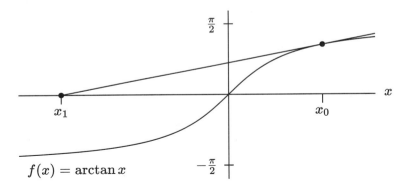

Figure 4.4.2

Example 2 Let $f(x) = \arctan x$. The graph of $f(x)$ is shown in Figure 4.4.2. It is not hard to check that f and the root $\overline{x} = 0$ satisfy the hypotheses of Theorem 4.4.1. So if x_0 is close enough to zero, the sequence $\{x_n\}$ produced by Newton's method will converge to zero. On the other hand, if one starts far enough away from zero, the sequence of iterates in Newton's method diverges. You can see the reason by drawing the appropriate tangent lines on the graph. If one starts with a large positive x_0, then x_1 will be even further away from zero, and so forth. We can prove this divergence analytically too. Choose a number b so that $b = \tan 1$; that is, $1 = \arctan b$. Since $\arctan x$ is monotone increasing and odd, we know that

$$|\arctan x| > 1 \qquad \text{if } |x| > b.$$

Using the inequality $1 + x^2 \geq 2x$, we see that

$$\left|\frac{f(x_n)}{f'(x_n)}\right| = \left|(1 + x_n^2)\arctan x_n\right| > 2|x_n|$$

if $|x_n| > b$. It follows from (9) that

$$|x_{n+1}| > |x_n|$$

if $|x_n| > b$. Thus if we begin with $|x_0| > b$, the successive iterates will get further and further away from the origin.

Example 3   One important use of Newton's method is to solve equations which cannot be solved analytically. For example, suppose that $0 < a < 1$ and that we want to find the x such that

$$a = e^x(1 - x). \tag{18}$$

Let $f(x) = e^x(1-x)$. Since $f(0) = 1$ and $f(1) = 0$, the Intermediate Value Theorem guarantees that there is an x between 0 and 1 so that (18) holds. In fact, since f is strictly monotone decreasing, there is a *unique* solution x for each $a \in (0, 1)$. Unfortunately, no amount of algebraic manipulation (or taking logs) solves (18) explicitly for x in terms of a. However, given any particular a, we can find x by using Newton's method to find the root of $e^x(1 - x) - a$.

When one uses the method of separation of variables (Project 1) to solve ordinary differential equations, the solution, $y(t)$, is typically determined by an implicit relation $g(y(t), t) = 0$, which often cannot be solved to find y explicitly in terms of t. However, one can use Newton's method and the idea in Example 3 to find y approximately for different values of t. See Project 2.

Finally, we note that finding the roots of functions of one variable is relatively easy. First we generate the graph of the function. This enables us to make a good guess for each of the roots, and Newton's method does the rest. Finding the common roots of several functions of several variables is much more difficult.

Problems

1. Find the roots of $f(x) = x^4 - 6x^2 + 2x + 1$.

2. Let $a = 0.5$. Find the x which satisfies (18).

3. Find the values of x so that $\cos x = x$.

4. What is the property of the function $\arctan x$ which makes Newton's method diverge if x_0 is large? Find other functions for which Newton's method can diverge if one starts too far from the root.

5.   (a) Which hypothesis of the Theorem 4.4.1 is not satisfied by the function $f(x) = x^2$?

    (b) For different choices of x_0, provide numerical evidence that Newton's method converges rapidly to zero anyway.

    (c) Prove that for $f(x) = x^2$ and any x_0, the iterates in Newton's method converge to zero.

6. Suppose that f is a continuous function on an interval $[a, b]$. If $f(a) < 0$ and $f(b) > 0$, then by the Intermediate Value Theorem, f has a root between a and b. If m is the midpoint of $[a, b]$ and $f(m) \neq 0$, then there is a root of f either in $[a, m]$ or in $[m, b]$, depending on the sign of $f(m)$. We carry out the same procedure on this new interval, and so forth. For obvious reasons, this is called the **bisection method**.

    (a) If $\{y_n\}$ is the sequence of midpoints of the intervals in the bisection method, prove that $\{y_n\}$ converges to a number c.

    (b) Prove that $f(c) = 0$.

    (c) Compute in terms of a and b how large n must be to guarantee that $|y_n - c| \leq 10^{-6}$.

7. Let $f(x) = x^2 - 2$. Let $\{x_n\}$ be the sequence of iterates generated by Newton's method starting with $x_0 = 1$. Let $\{y_n\}$ be the sequence of iterates generated by the bisection method starting with the interval $[0, 2]$.

    (a) How many iterates are required in each method to find $\sqrt{2}$ to within 10^{-6}?

    (b) Explain carefully why Newton's method is so much faster.

8. Explain why the factor $(1/2)^n$ in (17) can be made to be $(1/10)^n$ if we start even closer to the root.

9. Theorem 4.4.1 guarantees that Newton's method will converge if we start in the interval $[\bar{x} - \delta_1, \bar{x} + \delta_1]$. How do we know what δ_1 is? Suppose that $f(x)$ is twice continuously differentiable on $(\bar{x} - \delta, \bar{x} + \delta)$ and $|f''(x)| \leq M$. Suppose that $f(\bar{x}) = 0$ and $f'(\bar{x}) = c \neq 0$. Prove that δ_1 in the proof of Theorem 4.4.1 can be chosen to be $\frac{c}{2M}$. Hint: use the Fundamental Theorem of Calculus to relate $f'(x)$ and $f''(x)$.

10. For λ larger than 2.6 (approximately), the function $f(x) = x^3 - \lambda x^2 + x + 1$ has three real roots, which we denote $r_1(\lambda)$, $r_2(\lambda)$, and $r_3(\lambda)$. Use Newton's method to generate good graphs of r_1, r_2, and r_3 on the interval $[2.6, 4]$.

4.5   Inverse Functions

A function f is said to be **strictly monotone increasing** if $x < y$ implies $f(x) < f(y)$. Strictly monotone decreasing is defined similarly. It is easy to see from the Fundamental Theorem of Calculus that if f is continuously differentiable and $f'(x) > 0$, then f is strictly monotone increasing, although this condition is not necessary (problem 1). The importance of strict monotonicity is that such a function is automatically one-to-one; that is, if $x \neq y$ then $f(x) \neq f(y)$. Thus, a strictly monotone function f has an inverse function, denoted f^{-1}, satisfying $f^{-1}(f(x)) = x$ for all $x \in Dom(f)$, and $f(f^{-1}(y)) = y$ for all $y \in Dom(f^{-1}) = Ran(f)$.

❑ **Theorem 4.5.1**   Suppose that f is a strictly monotone function on an interval $[a, b]$.

(a) If $Ran(f)$ is an interval, then f is continuous.

(b) If f is continuous, then f^{-1} is continuous.

Proof. We will assume that f is strictly monotone increasing. The proof for the other case is similar. Suppose that $c \in (a, b)$. Then, $f(x) < f(c) < f(z)$ if $x < c < z$, so

$$L \equiv \sup_{x<c} f(x) \leq f(c) \leq \inf_{z>c} f(x) \equiv R$$

Suppose $L < R$. By monotonicity $f(x) \leq L$ for $x < c$ and $f(z) \geq R$ for $z > c$. Thus $Ran(f)$ would contain at most one point, $f(c)$, in the interval (L, R). This is impossible since then $Ran(f)$ would not be an interval which would contradict the hypothesis. Therefore $L = f(c) = R$. Now suppose $c_n \to c$ and let ε be given. Since, $f(c) = \sup_{x<c} f(x)$, we can choose $x_1 < c$ so that $f(c) - f(x_1) \leq \varepsilon$. Similarly we can choose $z_1 > c$ so that $f(z_1) - f(c) \leq \varepsilon$. That is,

$$f(c) - \varepsilon \leq f(x_1) \leq f(c) \leq f(z_1) \leq f(c) + \varepsilon.$$

Since $c_n \to c$, we can choose N so that $x_1 \leq c_n \leq z_1$ if $n \geq N$. Therefore, by monotonicity,

$$f(c) - \varepsilon \leq f(x_1) \leq f(c_n) \leq f(z_1) \leq f(c) + \varepsilon$$

for $n \geq N$. This proves that $|f(c_n) - f(c)| \leq \varepsilon$ for $n \geq N$, so f is continuous. The proofs at the endpoints $c = a$ and $c = b$ are similar. This proves (a). Note that we needed only monotonicity, not strict monotonicity.

To prove (b) we recall that if f is continuous, then $Ran(f)$ is an interval (Corollary 3.2.4). Since f is strictly monotone, f^{-1} exists, and it is easy to check that f^{-1} is strictly monotone on the interval $Ran(f)$. Since the range of f^{-1} is $[a, b]$, part (a) implies that f^{-1} is continuous. ❏

❏ **Theorem 4.5.2**   Suppose that f is a strictly monotone continuous function on an interval $[a, b]$. If f is differentiable at a point x and $f'(x) \neq 0$, then f^{-1} is differentiable at $y = f(x)$ and

$$(f^{-1})'(y) \;=\; \frac{1}{f'(x)}. \tag{19}$$

Proof.   Suppose that x is in $[a, b]$ and satisfies the hypotheses. Let $y = f(x)$. Let $\lambda_n \to 0$ with $\lambda_n \neq 0$ and define $y_n = y + \lambda_n$. Then,

$$\frac{f^{-1}(y + \lambda_n) - f^{-1}(y)}{\lambda_n} \;=\; \frac{f^{-1}(y_n) - f^{-1}(y)}{y_n - y}$$

$$=\; \frac{f^{-1}(y_n) - x}{f(f^{-1}(y_n)) - f(x)}$$

$$=\; \frac{h_n}{f(x + h_n) - f(x)}$$

where we define $h_n \equiv f^{-1}(y_n) - x$. Note that by strict monotonicity, $h_n \neq 0$, and therefore $f(x + h_n) \neq f(x)$, again by strict monotonicity. Since f is continuous by hypothesis, f^{-1} is continuous by Theorem 4.5.1, which implies that $h_n \to 0$ as $n \to \infty$. Since f is differentiable at x, the limit of the right-hand side exists and equals $1/f'(x)$ by Theorem 2.2.6. Thus the limit of the left-hand side exists, which proves that f^{-1} is differentiable at y and that (19) holds. ❏

Corollary 4.5.3   If f is continuously differentiable on $[a, b]$ and $f'(x) \neq 0$ for all $x \in [a, b]$, then f^{-1} is continuously differentiable and (19) holds for all $x \in [a, b]$.

Proof. Since f is continuously differentiable and $f'(x) \neq 0$ for all x in $[a, b]$, f^{-1} is differentiable for all y in $Dom(f^{-1})$ and (19) holds. By hypothesis, f and f' are continuous and f^{-1} is continuous by Theorem 4.5.1. Therefore,

$$(f^{-1})'(y) = \frac{1}{f'(f^{-1}(y))}$$

is continuous by Theorems 3.1.1(d) and 3.1.2. ❏

Next, we prove two theorems which justify the usual techniques for changing variables in integrals. Both of these theorems have slick proofs that use the chain rule and the Fundamental Theorem of Calculus (problems 6 and 7). We give the proofs below because changing variables in Riemann sums is central to the definition of the integral over more complicated objects like curves and surfaces.

❏ **Theorem 4.5.4** Let f be a continuous function on a finite interval $[a, b]$. Suppose that ϕ is a strictly monotone continuously differentiable function on an interval $[c, d]$ such that $\phi(c) = a$ and $\phi(d) = b$. Then,

$$\int_a^b f(x) \, dx \;=\; \int_c^d f(\phi(t))\phi'(t) \, dt. \tag{20}$$

Proof. Let P be a partition of $[a, b]$ into n subintervals $[x_i, x_{i-1}]$. Let t_i be the point in $[c, d]$ so that $\phi(t_i) = x_i$. Since ϕ is strictly monotone, the points $\{t_i\}$ form a partition of $[c, d]$. On each interval $[t_i, t_{i-1}]$, the Mean Value Theorem guarantees the existence of a point t_i^* such that $\phi(t_i) - \phi(t_{i-1}) = \phi'(t_i^*) \, (t_i - t_{i-1})$. Let $x_i^* \equiv \phi(t_i^*)$. Then,

$$\sum_{i=1}^n f(x_i^*)(x_i - x_{i-1}) \;=\; \sum_{i=1}^n f(\phi(t_i^*))(\phi(t_i) - \phi(t_{i-1})) \tag{21}$$

$$=\; \sum_{i=1}^n f(\phi(t_i^*))\phi'(t_i^*) \, (t_i - t_{i-1}). \tag{22}$$

Now let the interval length in partition P get small. By Corollary 3.3.2, the Riemann sum on the left side of (21) converges to the left side of (20). On the other hand, since ϕ^{-1} is continuous, and therefore uniformly continuous, on $[a, b]$, the interval length of the $\{t_i\}$ partition also gets small. Thus, again by Corollary 3.3.2, the right-hand side of (22) converges to the right-hand side of (20). Thus (20) holds. ❏

❏ **Theorem 4.5.5**   Let f be a continuous function on a finite interval $[a, b]$. Suppose that ϕ is a continuously differentiable function such that $\phi'(x) \neq 0$ on $[a, b]$. Then,

$$\int_a^b f(\phi(x))\, dx \;=\; \int_{\phi(a)}^{\phi(b)} f(t)(\phi^{-1}(t))'\, dt. \tag{23}$$

Proof.   The proof is very similar to that above, so we just give a sketch here. Choose partition P as above and define $t_i = \phi(x_i)$. By the Mean Value Theorem, there are points t_i^* so that $\phi^{-1}(t_i) - \phi^{-1}(t_{i-1}) = (\phi^{-1})'(t_i^*)\,(t_i - t_{i-1})$. Define x_i^* so that $\phi(x_i^*) = t_i^*$. Then,

$$\sum_{i=1}^n f(\phi(x_i^*))(x_i - x_{i-1}) \;=\; \sum_{i=1}^n f(t_i^*)(\phi^{-1}(t_i) - \phi^{-1}(t_{i-1}))$$

$$\;=\; \sum_{i=1}^n f(t_i^*)(\phi^{-1})'(t_i^*)\,(t_i - t_{i-1}).$$

Letting the interval length in the partitions get small, we obtain (23). ❏

Problems

1. Suppose that f is continuously differentiable on $[a, b]$ and that $f'(x) > 0$. Prove that f is strictly monotone increasing. Give an example of a strictly monotone f whose derivative has a zero.

2. Suppose that f is continuous on $[a, b]$ and that f is one-to-one. Prove that f is either strictly monotone increasing or strictly monotone decreasing.

3. Prove that each the following functions is strictly monotone on $(0, \infty)$. Compute the inverse function and its derivative explicitly and verify that (19) holds.

    (a) $f(x) = x^3$.

    (b) $f(x) = \frac{1}{x}$.

    (c) $f(x) = \frac{1}{\sqrt{\ln x}}$.

4. Let $f(x) = x^4 + x^2 + 1$. Compute $(f^{-1})'(3)$ and $(f^{-1})'(21)$.

5. Prove that the function $\tan x$ is strictly monotone increasing on the interval $(-\frac{\pi}{2}, \frac{\pi}{2})$. Use Theorem 4.5.2 to compute a formula for the derivative of $\arctan x$.

6. Give a different proof of Theorem 4.5.4 as follows. Let $F(u) = \int_a^u f(x)\,dx$ and define $G(t) \equiv F(\phi(t))$. Now, apply the Fundamental Theorem of Calculus to G'. Note that this proof does not require ϕ to be monotone.

7. Give a different proof of Theorem 4.5.5 as follows. Let G be such that $G' = f \circ \phi$. Now, compute $(G \circ (\phi)^{-1})'$ and use the Fundamental Theorem of Calculus.

8. The natural logarithm was defined in Example 2 of Section 4.3.

   (a) Explain why $\ln x$ defined on $(0, \infty)$ has an inverse function. Call it ϕ. What are the domain and range of ϕ? Is ϕ continuously differentiable?

   (b) Show by a change of variables that

   $$\int_b^{ab} \frac{dx}{x} = \int_1^a \frac{dx}{x}.$$

   (c) Prove that $\ln ab = \ln a + \ln b$ for all positive numbers a and b.

   (d) Prove that ϕ satisfies $\phi(x+y) = \phi(x)\,\phi(y)$ for all real numbers x and y. We shall see in Example 2 of Section 6.4 that $\phi(x) = e^x$.

4.6   Functions of Two Variables

In mathematics and its applications, functions which depend on several variables occur frequently. In this book we use such functions in the study of integral equations (Section 5.4), the calculus of variations (Section 5.5), ordinary differential equations (Chapter 7), complex analysis (Chapter 8), and partial differential equations (Section 9.1). Therefore, in this section, we show how two of the fundamental ideas which we have introduced, continuity and differentiation, can be extended to functions of several variables. For simplicity, we shall restrict our attention to functions of two variables that take values in the real numbers.

The Euclidean plane $\mathbb{R}^2 \equiv \mathbb{R} \times \mathbb{R}$ is the set of ordered pairs (x, y) where $x \in \mathbb{R}$ and $y \in \mathbb{R}$. We can add pairs, $(x_1, y_1) + (x_2, y_2) = (x_1 + x_2, y_1 + y_2)$, and multiply pairs by real numbers, $\alpha(x, y) = (\alpha x, \alpha y)$. These operations correspond to vector addition and scalar multiplication. We define the **Euclidean distance** from a point to the origin by $\|(x, y)\| \equiv \sqrt{x^2 + y^2}$. This notion of distance satisfies the triangle inequality (problem 10 of Section 2.2):

$$\|(x_1, y_1) + (x_2, y_2)\| \leq \|(x_1, y_1)\| + \|(x_2, y_2)\|.$$

If (x_1, y_1) and (x_2, y_2) are points in $\mathbb{R}^2$, we define the distance between them to be

$$\|(x_1, y_1) - (x_2, y_2)\| \equiv \sqrt{(x_1 - x_2)^2 + (y_1 - y_2)^2}.$$

Definition. Let $p_n = (x_n, y_n)$ be a sequence of points in the plane and let $p = (x, y)$. We say that $\{p_n\}$ **converges** to p, written $p_n \to p$, if $\|p_n - p\| \to 0$.

It is not hard to check that $p_n \to p$ if and only if $x_n \to x$ and $y_n \to y$ (problem 10 of Section 2.2). Now that we have a notion of convergence we can define continuity.

Definition.   A function f from $\mathbb{R}^2$ to $\mathbb{R}$ is said to be **continuous** at $p \in Dom(f)$ if, whenever $p_n \in Dom(f)$, $p \in Dom(f)$, and $p_n \to p$, it follows that

$$\lim_{n \to \infty} f(p_n) = f(p).$$

If f is continuous at every point of a set $E \subseteq \mathbb{R}^2$, then f is said to be **continuous on E**.

Example 1   Let $f(x, y) = 3x^2 y + y^3 + 1$ and suppose that $p_n = (x_n, y_n)$, $p = (x, y)$, and $p_n \to p$. Then

$$
\begin{aligned}
\lim_{n \to \infty} f(p_n) &= \lim_{n \to \infty} (3x_n^2 y_n + y_n^3 + 1) \\
&= 3(\lim_{n \to \infty} x_n)^2 (\lim_{n \to \infty} y_n) + (\lim_{n \to \infty} y_n)^3 + 1 \\
&= 3x^2 y + y^3 + 1 \\
&= f(p),
\end{aligned}
$$

where we used the limit theorems from Section 2.2 in the second step. Thus f is continuous on $\mathbb{R}^2$. The same idea can be used to prove that all polynomials in two variables are continuous functions on $\mathbb{R}^2$.

The three theorems in Section 3.1 have analogues for functions of two variables. The sum, product, and quotient of continuous functions are continuous except possibly where the denominator vanishes. If f is a continuous function from $\mathbb{R}^2$ to $\mathbb{R}$ and g is a continuous function on $\mathbb{R}$ with $Ran(f) \subseteq Dom(g)$, then $g \circ f$ is a continuous function on $Dom(f)$. Thus, a composition such as $\sin(x^2 + y^3)$ is continuous on the plane. Finally, f is continuous at p if and only if, given $\varepsilon > 0$, there is a $\delta > 0$ such that

$$\|q - p\| \leq \delta \quad \text{implies} \quad |f(q) - f(p)| \leq \varepsilon.$$

In all these cases, the proofs are virtually identical to those in Section 3.1 except that the absolute value, $|\cdot|$, in $\mathbb{R}$ is replaced by the Euclidean distance, $\|\cdot\|$. The reader is asked to give the proof of the third result in problem 1.

Definition.   A continuous function on a set $E \subseteq \mathbb{R}^2$ is said to be **uniformly continuous** if, given $\varepsilon > 0$, there exists a $\delta > 0$ such that

$$\|q - p\| \leq \delta \quad \text{implies} \quad |f(q) - f(p)| \leq \varepsilon \quad \text{for all } q \text{ and } p \text{ in } E.$$

Sets of the form $R = \{x \mid a_1 \leq x \leq b_1, \ a_2 \leq y \leq b_2\}$ are called closed rectangles. The following theorem generalizes Theorems 3.2.1, 3.2.2, and 3.2.5.

❏ **Theorem 4.6.1**   Let f be a continuous real-valued function on a closed rectangle R in the plane. Then,

(a) f is bounded on R.

(b) There exist points c and d in R so that

$$f(c) = \sup_{p \in R} f(p), \qquad f(d) = \inf_{p \in R} f(p).$$

(c) f is uniformly continuous on R.

Proof.   The proofs of all three parts follow closely the proofs in Section 3.2 except that they use the generalization of the Bolzano-Weierstrass theorem proved in problems 10 and 11 of Section 2.6. We will prove part (a), leaving parts (b) and (c) to the problems. To prove that f is bounded on R, we need to show that there is an M so that $|f(p)| \leq M$ for all $p \in R$. Suppose that this is not true. Then for each large integer n, there is a $p_n \in R$ such that $|f(p_n)| > n$. By the (generalization of the) Bolzano-Weierstrass theorem, the sequence $\{p_n\}$ has a subsequence $\{p_{n_k}\}$ that converges to a point $c \epsilon R$ as $k \to \infty$. Since f is continuous, $f(p_{n_k}) \to f(p_o)$ as $k \to \infty$. However, this is impossible since $f(p_{n_k}) \to \infty$. This contradiction shows that f must be bounded.   ❏

In the proof of Theorem 4.6.1, the only property of R that was used was that sequences in R have subsequences that converge to a point of R. Sets which have this property are called **compact** sets. Thus, continuous functions on compact sets have the three properties (a), (b), and (c). Compact sets are investigated in problems 5 and 6.

We are now ready to discuss differentiation. Let (x_o, y_o) be a point in the domain of a function of two variables, f, and suppose that all the points $(x_o + h, y_o)$ are in the domain of f for h small enough. Then, we define the **partial derivative of f with respect to x** at (x_o, y_o), written $\frac{\partial f}{\partial x}(x_o, y_o)$, by

$$\frac{\partial f}{\partial x}(x_o, y_o) \equiv \lim_{h \to 0} \frac{f(x_o + h, y_o) - f(x_o, y_o)}{h}$$

if the limit exists. Similarly, suppose that all the points $(x_o, y_o + h)$ are in the domain of f for h small enough. We define the **partial derivative of f with respect to y** at (x_o, y_o), written $\frac{\partial f}{\partial y}(x_o, y_o)$, by

$$\frac{\partial f}{\partial y}(x_o, y_o) \equiv \lim_{h \to 0} \frac{f(x_o, y_o + h) - f(x_o, y_o)}{h}$$

if the limit exists. We shall often use the simpler notation $f_x(x_o, y_o)$ and $f_y(x_o, y_o)$ for $\frac{\partial f}{\partial x}(x_o, y_o)$ and $\frac{\partial f}{\partial y}(x_o, y_o)$, respectively. In the language of calculus, we compute f_x by holding y fixed and differentiating with respect to x, and we compute f_y by holding x fixed and differentiating with respect to y.

Example 2   Suppose that $f(x, y) = 3x^2 y + y^3 + 1$. If we hold y fixed, then $f(x, y)$ is a polynomial in x and therefore differentiable. Similarly, if we hold x fixed, $f(x, y)$ is a polynomial in y and therefore differentiable. Thus f_x and f_y exist at all points (x, y) in the plane and

$$f_x(x, y) = 6xy, \qquad f_y(x, y) = 3x^2 + 3y^2.$$

The same argument shows that the partial derivatives of all polynomials in two variables exist.

Notice that we have not yet defined what it means for f to be "differentiable" at a point (x_o, y_o). It would be natural to think that the right condition is to require that both of the partial derivatives, f_x and f_y, exist. However, the following example shows clearly that this condition is not strong enough.

Example 3 Let $f(x, y)$ be the function such that $f(x, y) = 1$ if $x > 0$ and $y > 0$ and $f(x, y) = 0$ otherwise. Notice that the function f is identically zero on the x and y axes. It follows that the limits of the difference quotients for the partial derivatives of f exist and that $f_x(0, 0) = 0$ and $f_y(0, 0) = 0$. However, on any line through $(0, 0)$ besides the axes, the values of f jump from 0 to 1. See Figure 4.6.1. Thus, this function should not be considered "differentiable" at $(0, 0)$. In fact, f is not even continuous at the point $(0, 0)$.

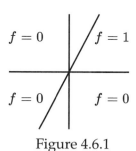

Figure 4.6.1

Definition. We say that f is **differentiable** at a point (x_o, y_o) of $\mathbb{R}^2$ if for some $\delta > 0$ the partial derivatives of f exist and are continuous in the disk $D_\delta(x_o, y_o) \equiv \{(x, y) \in \mathbb{R}^2 \mid (x - x_o)^2 + (y - y_o)^2 < \delta^2\}$. If this criterion holds at every point of a set E, then f is said to be **continuously differentiable** on E.

This criterion for differentiability is too strong since it implies differentiability in the entire disk $D_\delta(x_o, y_o)$. More delicate criteria can be given, but this one will be sufficient for our purposes.

❏ **Theorem 4.6.2** Suppose that f is differentiable at (x_o, y_o) and let a and b be any real numbers. Then,

$$\lim_{h \to 0} \frac{f(x_o + ah, y_o + bh) - f(x_o, y_o)}{h} = a\frac{\partial f}{\partial x}(x_o, y_o) + b\frac{\partial f}{\partial y}(x_o, y_o). \quad (24)$$

Proof. Let $D_\delta(x_o, y_o)$ be a disk in which the partials exist and are continuous. Choose h small enough so that the point $(x_o + ah, y_o + bh)$ is in the disk. See Figure 4.6.2. We now rewrite the difference quotient as

$$\frac{f(x_o + ah, y_o + bh) - f(x_o, y_o)}{h}$$

$$= \frac{f(x_o + ah, y_o + bh) - f(x_o + ah, y_o)}{h} + \frac{f(x_o + ah, y_o) - f(x_o, y_o)}{h}$$

$$= \frac{\partial f}{\partial y}(x_o + ah, \xi)\frac{bh}{h} + a\frac{f(x_o + ah, y_o) - f(x_o, y_o)}{ah}.$$

In the first term on the right, we used the Mean Value Theorem in y for the function $f(x_o + ah, y)$ on the interval $0 \le y \le bh$. See Figure 4.6.2. We can use the Mean Value Theorem because of the hypothesis that f is continuously differentiable in y for each fixed x. Since $0 \le \xi \le bh$ and the partial $\frac{\partial f}{\partial y}$ is continuous in the disk, the limit of the first term on the right as $h \to 0$ is $b\frac{\partial f}{\partial y}(x_o, y_o)$. Since the limit of the second term is $a\frac{\partial f}{\partial x}(x_o, y_o)$, we have proved the theorem. ❏

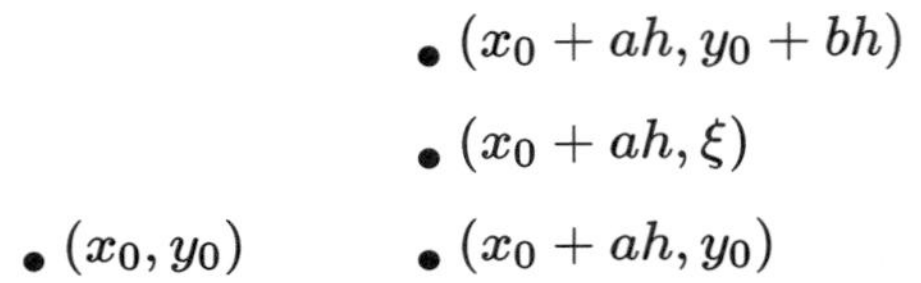

- $(x_0 + ah, y_0 + bh)$
- $(x_0 + ah, \xi)$
- (x_0, y_0)
- $(x_0 + ah, y_0)$

Figure 4.6.2

If $a^2 + b^2 = 1$, the limit in (24) is called the **directional derivative** in the direction (a, b). Finally, we prove a particular chain rule which we use in Section 5.5. Other chain rules are considered in the problems.

❏ **Theorem 4.6.3**   Let $x(t)$ and $y(t)$ be continuously differentiable functions of one variable and suppose that f is a continuously differentiable function of two variables. Define $g(t) \equiv f(x(t), y(t))$. Then, g is continuously differentiable and

$$g'(t) = \frac{\partial f}{\partial x}(x(t), y(t))x'(t) + \frac{\partial f}{\partial y}(x(t), y(t))y'(t). \qquad (25)$$

Proof.   Fix t and write the difference quotient for g as

$$\frac{f(x(t + h), y(t + h)) - f(x(t), y(t))}{h} =$$

$$\frac{f(x(t + h), y(t + h)) - f(x(t + h), y(t))}{h} + \frac{f(x(t + h), y(t)) - f(x(t), y(t))}{h}$$

$$= \frac{\partial f}{\partial y}(x(t+h), \xi_1) \frac{y(t+h) - y(t)}{h} + \frac{\partial f}{\partial x}(\xi_2, y(t)) \frac{x(t+h) - x(t)}{h}.$$

In each term we used the Mean Value Theorem, so ξ_1 is between $y(t)$ and $y(t+h)$ and ξ_2 is between $x(t)$ and $x(t+h)$. Thus, as $h \to 0$, we know that $\xi_1 \to y(t)$ and $\xi_2 \to x(t)$. Since $x(t)$ and $y(t)$ are differentiable and the partials of f are continuous, the limit of the right hand side exists and equals the right hand side of (25). Thus g is differentiable and (25) holds. Since the composition and product of continuous functions are continuous, g' is continuous.   ❏

A partial differential equation is an equation involving the partial derivatives of an unknown function. The investigation of partial differential equations was a primary stimulus for the rigorization of analysis in the 19th century, and such equations are used today in many models in the physical, biological, and social sciences. A simple example, the wave equation, is discussed in problems 12 and 13. The heat equation is discussed in Section 9.1.

Most of the concepts and theorems in this section have natural generalizations to real-valued functions defined on $\mathbb{R}^n \equiv \mathbb{R} \times \mathbb{R} \times \dots \times \mathbb{R}$. This is just the beginning of a difficult and important subject, the analysis of functions from $\mathbb{R}^n$ to $\mathbb{R}^m$. For more information, see [24] or [12]. We discuss some simple aspects of the integration of functions of two variables in project 4.

Problems

1. State and prove the analogue of Theorem 3.1.3 for real-valued functions of two variables.

2. Prove part (b) of Theorem 4.6.1.

3. Prove part (c) of Theorem 4.6.1.

4. Suppose that g and h are continuous functions on $\mathbb{R}$ and f is a continuous function on $\mathbb{R}^2$. For each t, define $y(t) = f(g(t), h(t))$. Prove that y is a continuous function on $\mathbb{R}$.

5. We wish to find criteria for a subset $E \subseteq \mathbb{R}^2$ to be compact. We know from problem 8 of Section 2.4 that every Cauchy sequence in $\mathbb{R}^2$ converges to a point of $\mathbb{R}^2$. We say that E is **closed** if every Cauchy sequence of points of E converges to a point of E.

(a) Use problem 10 in Section 2.6 to show that if E is closed and bounded, then E is compact.

(b) Show that if E is not closed and bounded, then E is not compact.

(c) Is $\{(x, y) \in \mathbb{R}^2 \mid x^2 + y^2 \leq 1\}$ compact?

(d) Is $\{(x, y) \in \mathbb{R}^2 \mid x^2 + y^2 < 1\}$ compact?

(e) Is $\{(x, y) \in \mathbb{R}^2 \mid x^2 + y^2 \geq 1\}$ compact?

6. Give an example of a noncompact set $E \subseteq \mathbb{R}^2$ and a continuous function f on E so that none of the three properties (a), (b), or (c) in Theorem 4.6.1 is true.

7. Show that the partial derivatives of

$$f(x, y) = \frac{xy}{\sqrt{x^2 + y^2}}$$

exist at $(0, 0)$ but that f is not differentiable there.

8. Let f be differentiable at (x, y). Show that the directional derivative of f at (x, y) is largest in the direction $(\bar{a}, \bar{b})$ where

$$\bar{a} = \frac{f_x}{\sqrt{f_x^2 + f_y^2}}, \qquad \bar{b} = \frac{f_y}{\sqrt{f_x^2 + f_y^2}}.$$

That is, it is largest in the direction of the gradient.

9. Let $H(x, p)$ be a differentiable function of two variables. For a given point (x_o, p_o) in the plane, let $x(t)$ and $p(t)$ be the solutions of the following system of differential equations:

$$x'(t) = \frac{\partial H}{\partial p}(x(t), p(t)), \qquad x(0) = x_o$$

$$p'(t) = -\frac{\partial H}{\partial x}(x(t), p(t)), \qquad p(0) = p_o.$$

This is called a **Hamiltonian system**, H is called the **Hamiltonian**, and the curve $(x(t), p(t))$ in the x-p plane is called an **orbit**.

(a) Prove that H is constant on orbits.

(b) Suppose that $H(x, p) = x^2 + p^2$. What do the orbits look like?

(c) Suppose that $H(x, p) = ax + bp$. What do the orbits look like?

10. Let g be a continuously differentiable function on $\mathbb{R}$ and let f be continuously differentiable on $\mathbb{R}^2$. Prove that $h(x, y) \equiv g(f(x, y))$ is continuously differentiable on $\mathbb{R}^2$ and compute its partial derivatives.

11. Let f be continuously differentiable on the plane and define

$$h(x,y) = \int_0^{f(x,y)} G(t)\,dt,$$

where G is a continuous function on $\mathbb{R}$. Prove that h is continuously differentiable and compute its partial derivatives.

12. Imagine a infinitely long elastic string lying along the x-axis. Suppose the string is set in motion at time $t = 0$, and let $u(x,t)$ denote the vertical displacement of the string at x at time t. According to a simple model for small displacements, u should satisfy the **wave equation**

$$u_{tt} - c^2 u_{xx} = 0,$$

where c is a constant determined by the elastic properties of the string. We assume that u also satisfies the initial conditions

$$u(x,0) = f(x), \qquad u_t(x,0) = g(x)$$

where f gives the initial displacement and g gives the initial velocity of the string at x. Assume that f and g are twice continuously differentiable. Verify that

$$u(x,t) = \frac{f(x+ct) + f(x-ct)}{2} + \frac{1}{2c}\int_{x-ct}^{x+ct} g(s)\,ds$$

solves the wave equation and the initial conditions. This solution was first written down by J. d'Alembert (1717–1783).

13. Let f be a twice continuously differentiable function of one variable. Define

$$u(t,x,y,z) \equiv \frac{f(t-r)}{r},$$

where $r \equiv \sqrt{x^2 + y^2 + z^2}$. Verify that u satisfies the wave equation in three dimensions:

$$u_{tt} - u_{xx} - u_{yy} - u_{zz} = 0.$$

Projects

1. A first-order ordinary differential equation with initial condition $y(0) = y_0$ is called **variables separable** if it can be written in the form

$$f(y)\frac{dy}{dt} = g(t).$$

In calculus, you probably learned that to solve the differential equation, you multiply both sides by "dt" to obtain

$$f(y)\,dy = g(t)\,dt.$$

You then integrate the left side with respect to y, obtaining $F(y)$, and integrate the right side with respect to t, obtaining $G(t)$. Setting $F(y) = G(t) + C$, you determine the constant C by the initial condition and then solve for y in terms of t. How can one integrate one side of an equality with respect to one variable and the other side with respect to another variable and expect the results to be equal?

(a) Using the chain rule and the Fundamental Theorem of Calculus, explain carefully what is really going on here.

(b) Use the method to solve $y' = \frac{e^{2t}}{y^2}$, $y(0) = 2$.

(c) Use the method to solve $y' = y^2$, $y(0) = 2$.

2. The purpose of this project is to show how Newton's method arises in solving differential equations.

(a) Show that the solution of the initial value problem

$$\frac{dy}{dt} = \frac{y}{y+1}, \quad y(0) = 1$$

satisfies

$$y(t) + \ln y(t) \quad = \quad t + 1. \tag{26}$$

(b) Try every trick in the book to find an expression for y in terms of t.

(c) Use the Intermediate Value Theorem to show that for each t there is a number $y(t)$ that satisfies (26).

(d) Prove that for each t, $y(t)$ is unique.

(e) For a number of different t between 0 and 4, use Newton's method to determine $y(t)$ and draw a sketch of the graph of $y(t)$.

3. Consider the function $f(x) = e^{-1/x^2}$.

(a) Explain why f is infinitely often differentiable on the set $(-\infty, 0) \cup (0, \infty)$.

(b) Show that f can be extended to be continuous on $\mathbb{R}$ by defining $f(0) = 0$.

(c) Show that f is continuously differentiable on $\mathbb{R}$ and $f'(0) = 0$.

(d) Prove that for $x \neq 0$ the nth derivative of f is of the form $p_n(1/x)e^{-1/x^2}$, where p_n is a polynomial.

(e) Prove that f is infinitely often continuously differentiable on $\mathbb{R}$ and $f^{(n)}(0) = 0$ for all n.

(f) What is the nth Taylor polynomial approximation to f around the point $x_o = 0$?

4. The purpose of this project is to develop some simple aspects of integration theory on $\mathbb{R}^2$. Let $R \equiv [a, b] \times [c, d]$ be a rectangle in the plane. We define a partition P of R to be a pair of partitions of $[a, b]$ and $[c, d]$, respectively:

$$a = x_0 < x_1 < x_2 < \ldots < x_N = b$$

$$c = y_0 < y_1 < y_2 < \ldots < y_M = d.$$

For each partition we define rectangles $R_{ij} \equiv [x_i, x_{i-1}] \times [y_{j-1}, y_j]$. Note that $R = \cup_{ij} R_{ij}$. Setting

$$M_{ij} \equiv \sup_{(x,y) \,\epsilon\, R_{ij}} \{f(x, y)\}, \qquad m_{ij} \equiv \inf_{(x,y) \,\epsilon\, R_{ij}} \{f(x, y)\},$$

we define upper and lower sums corresponding to partition P by:

$$U_P(f) \equiv \sum_{ij} M_{ij}(x_i - x_{i-1})(y_j - y_{j-1})$$

$$L_P(f) \equiv \sum_{ij} m_{ij}(x_i - x_{i-1})(y_j - y_{j-1}).$$

(a) Formulate and prove the analogues of Lemma 1 and Lemma 2 in Section 3.3.

If $\sup_P \{L_P(f)\} = \inf_P \{U_P(f)\}$, we say that f is **Riemann integrable** over R and define

$$\int_c^d \int_a^b f(x, y)\, dx\, dy \equiv \inf_P U_P(f).$$

(b) Prove that if f is continuous on R then f is Riemann integrable.

(c) Prove the analogue of Corollary 3.3.2.

(d) Prove that $G(y) \equiv \int_a^b f(x, y)\, dx$ is a continuous function of y on $[c, d]$.

(e) Prove that $\int_c^d \int_a^b f(x, y)\, dx\, dy$ can be computed by iterating the integrals; that is,

$$\int_c^d \int_a^b f(x, y)\, dx\, dy = \int_c^d \left(\int_a^b f(x, y)\, dx \right) dy.$$

Hint: write

$$\left| \int_c^d \int_a^b f(x,y)\, dx\, dy \; - \; \int_c^d \left(\int_a^b f(x,y)\, dx \right) dy \right|$$

$$\leq \left| \int_c^d \left(\int_a^b f(x,y)\, dx \right) dy \; - \; \sum_{ij} M_{ij}(x_i - x_{i-1})(y_j - y_{j-1}) \right|$$

$$+ \left| \sum_{ij} M_{ij}(x_i - x_{i-1})(y_j - y_{j-1}) \; - \; \sum_j (y_j - y_{j-1}) \int_a^b f(x, y_j)\, dx \right|$$

$$+ \left| \sum_j \left(\int_a^b f(x, y_j)\, dx \right)(y_j - y_{j-1}) \; - \; \int_c^d \left(\int_a^b f(x,y)\, dx \right) dy \right|.$$

and justify carefully why each term can be made less than $\varepsilon/3$ by choosing the partition appropriately.

CHAPTER 5

Sequences of Functions

In this chapter we study the convergence of sequences of functions. Two kinds of convergence, pointwise and uniform, are introduced in Section 5.1, and in Section 5.2 several limit theorems, which depend on uniform convergence, are proved . Section 5.4 (Integral Equations) and Section 5.5 (The Calculus of Variations) show why these limit theorems are useful and important. In Section 5.3, the point of view shifts from individual sequences of functions to sets of functions. The supremum norm is introduced, and the related notions of Cauchy sequence and completeness are studied. These ideas lead naturally to the definition of metric space in Section 5.6 and the study of completeness in Section 5.7. An important class of metric spaces that have linear structure, normed linear spaces, is introduced in Section 5.8.

5.1   Pointwise and Uniform Convergence

Sequences of real numbers have played a central role in this text so far. For example, sequences are used in the definitions of continuous and differentiable functions. The Riemann integral is the limit of a sequence of Riemann sums. The invariant probabilities of a two-state Markov process were found by taking the limits of sequences. Newton's method shows how to find the roots of a polynomial by constructing sequences of approximations. On the other hand, many of the most important objects which one studies in analysis are functions. The solution of a differential or integral equation is a function. The solution of a partial differential equation is a function of several variables. Often one determines these solutions by constructing a sequence of functions, $f_n(x)$, that gets closer and closer to the solution $f(x)$ as $n \to \infty$. This is the technique we shall use when we study integral equations (Section 5.4) and ordinary differential equations (Section 7.1). But what do we mean by "closer and closer"? What does it mean for a sequence of *functions* to get closer and closer to a limiting *function*? As we shall see, there are many different notions of convergence for sequences of functions. In this section we introduce two of the simplest, pointwise and uniform convergence. Other kinds of convergence are discussed in Section 5.3 and later sections.

Definition.   A sequence of functions, f_n, defined on the same set E is said to **converge pointwise** to a limiting function, f on E, if

$$\lim_{n\to\infty} f_n(x) \;=\; f(x), \qquad \text{for every } x \in E. \tag{1}$$

That is, given $\varepsilon > 0$ and $x \in E$, there exists an N so that $n \geq N$ implies

$$|f_n(x) - f(x)| \;\leq\; \varepsilon. \tag{2}$$

Example 1   Consider the functions $f_n(x) = x^2 - \frac{1}{2n}x + 1$ defined on the whole real line $\mathbb{R}$. Set $f(x) = x^2 + 1$. The graphs of the first few f_n and f are shown in Figure 5.1.1. For each fixed x

$$x^2 - \frac{1}{2n}x + 1 \;\to\; x^2 + 1$$

as $n \to \infty$. That is, $f_n(x) \to f(x)$ for each x as $n \to \infty$, so by definition, the sequence of functions f_n converges pointwise to the function f.

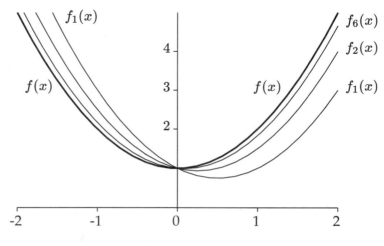

Figure 5.1.1

Example 2   Consider the sequence of functions $f_n(x) = x^n$. What happens to the sequence of numbers $f_n(x)$ for different choices of x?   If $|x| > 1$, the sequence of numbers x^n diverges. If $|x| < 1$, the sequence of numbers x^n converges to zero. If $x = 1$, the sequence of numbers x^n converges to 1. If $x = -1$, the sequence of numbers x^n does not converge because it oscillates between 1 and -1. Thus, if E is the whole real line $\mathbb{R}$, the sequence of functions $\{f_n\}$ does *not* converge pointwise since the sequence of numbers $\{f_n(x)\}$ does not converge for every $x \in E$. If E is the interval $[-1, 1]$, the sequence of functions $\{f_n\}$ does not converge pointwise since the sequence of numbers $\{f_n(x)\}$ does not converge for $x = -1$. If E is the interval $(-1, 1]$, the sequence of functions $\{f_n\}$ does converge pointwise since the sequence of numbers $\{f_n(x)\}$ converges for each $x \in E$. Thus we see that whether or not a sequence of functions converges pointwise may depend on the set on which the functions are defined.

Example 3   Let's consider the same sequence of functions as in Example 2, $f_n(x) = x^n$, this time defined on the interval $[0, 1]$.

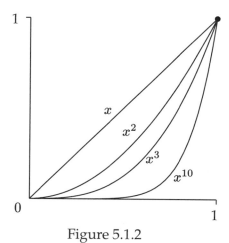

Figure 5.1.2

The sequence of functions $\{f_n\}$ converges pointwise to the function

$$f(x) = \begin{cases} 0 & \text{if } 0 \le x < 1 \\ 1 & \text{if } x = 1. \end{cases}$$

Thus, a sequence of continuous functions on a finite interval may converge pointwise to a limiting function that is not continuous.

Example 4   Consider the sequence of functions $\{f_n\}$ on the interval $[0, 1]$, whose graphs are shown in Figure 5.1.3. On the interval $[0, 2^{-n}]$, the graph of f_n consists of the straight line from $(0, 0)$ to $(2^{-1}2^{-n}, 2^n)$ and the straight line from $(2^{-1}2^{-n}, 2^n)$ to $(2^{-n}, 0)$. For all $x \geq 2^{-n}$, $f_n(x) \equiv 0$.

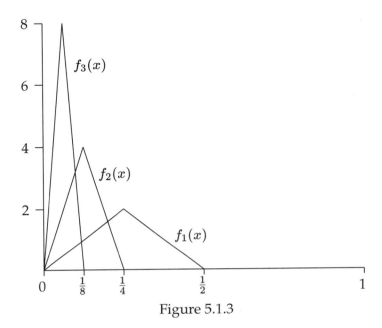

Figure 5.1.3

For any fixed $x > 0$, the sequence of numbers $\{f_n(x)\}$ converges to zero as $n \to \infty$. In fact, if we choose N so that $2^{-N} \leq x$, then $f_n(x) = 0$ for all $n \geq N$. If $x = 0$, then $f_n(0) = 0$ for all n, so $f_n(0) \to 0$. Thus the sequence of functions f_n converges pointwise to the zero function, $f(x) \equiv 0$, on the interval $[0, 1]$. Notice, however, that for each n, $\int_0^1 f_n(x)dx = 2^{-1}$ since the area under each graph is just one half the base (2^{-n}) of the triangle times the height (2^n). Therefore,

$$\lim_{n \to \infty} \int_0^1 f_n(x)\,dx = 2^{-1} \neq 0 = \int_0^1 \lim_{n \to \infty} f_n(x)\,dx.$$

Pointwise convergence is an extremely natural notion. It makes sense to say that a sequence of functions converges to a limiting function if the values of the functions converge at each x. But Examples 3 and 4 show that pointwise convergence may not be very useful. In Example 3 we saw that the pointwise limit of a sequence of continuous functions might not be continuous. In Example 4 we saw that pointwise convergence is not enough to guarantee that the limit of the integrals of the f_n is the integral of the limiting function f. We therefore want to define a stronger notion of convergence that gives us some control over the properties of the limiting function.

Definition. Suppose that the sequence of functions $\{f_n\}$ converges pointwise to a function f on the set E. Then, the sequence $\{f_n\}$ is said to **converge uniformly** to f if, given $\varepsilon > 0$, there exists an N so that $n \geq N$ implies

$$|f_n(x) - f(x)| \leq \varepsilon \quad \text{for all } x \in E. \tag{3}$$

There is a very important difference between (2) and (3). In pointwise convergence, one might have to choose a different N for each different x. In uniform convergence, there is an N which works for *all* x in the set E. This is a much stronger requirement, and we will see in Section 5.2 that this is just the condition we need. The name "uniform convergence" comes from the fact that, given $\varepsilon > 0$, we can choose an N so that the graphs of all the f_n, for $n \geq N$, lie in an ε-band about the graph of the limiting function f; that is, they are uniformly close to f. Note that, by definition, uniform convergence implies pointwise convergence. Let's examine the sequences of functions in the above examples to see whether they converge uniformly.

Example 1 (revisited) Suppose that $\varepsilon > 0$ is given. Notice that

$$|f_n(x) - f(x)| = \frac{|x|}{2n}.$$

In order to make the right-hand side $\leq \varepsilon$, we will have to choose n large, but how large depends on x. The bigger x is, the larger n will have to be chosen so that $|f_n(x) - f(x)| \leq \varepsilon$. In particular, given $\varepsilon > 0$ and any n, there is an x so that $|f_n(x) - f(x)| \geq \varepsilon$. Thus f_n does not converge uniformly to f on $\mathbb{R}$. However, the convergence *is* uniform on each finite interval $[a, b]$. To see this, let $\varepsilon > 0$ be given. Choose $N \geq \max\{|a|, |b|\}/2\varepsilon$. Then, for *all* $x \in [a, b]$, $n \geq N$ implies

$$|f_n(x) - f(x)| \leq \frac{|x|}{2N} \leq \frac{\max\{|a|, |b|\}}{2N} \leq \varepsilon.$$

Thus f_n converges uniformly to f on $[a, b]$.

Example 3 (revisited)   The functions $f_n(x) = x^n$ converge pointwise but do not converge uniformly on $[0, 1]$. Let $\varepsilon > 0$ be given. The problem is not at $x = 1$, where all the functions have value 1, but for x near 1. For each $x < 1$ in $[0, 1]$, the sequence of numbers x^n converges to zero as $n \to \infty$, but the closer x is to 1 the longer it will take until x^n is less than ε. In particular, given any small $\varepsilon > 0$ and any N, there is an $x \in [0, 1]$ so that $x^N \geq \varepsilon$. Thus, there is no N so that (3) holds for all x in $[0, 1]$. On the other hand, the functions f_n do converge uniformly to the zero function on $[0, \mu]$ if $\mu < 1$.

Example 4 (revisited)   Looking at the graphs of the functions f_n in Figure 5.1.3, we can see why they do not converge to the zero function uniformly. The closer x is to zero, the longer it takes before $|f_n(x)| \leq \varepsilon$ for any given ε. In particular, given any $0 < \varepsilon \leq 1$ and any n, there is an x (corresponding, say, to the peak) so that $f_n(x) \geq 1$. Thus, (3) cannot hold for any choice of N, so $\{f_n\}$ does not converge uniformly to the zero function on $[0, 1]$. Since uniform convergence implies pointwise convergence and $\{f_n\}$ converges pointwise to the zero function, $\{f_n\}$ can not converge uniformly to any other function. The sequence $\{f_n\}$ does converge uniformly to the zero function on any interval of the form $[\mu, 1]$ if $0 < \mu \leq 1$.

We will often use the notation $f_n \to f$ to mean that the sequence of functions converges to the limiting function f. The arrow itself doesn't indicate whether the convergence is pointwise or uniform, so we will always say $f_n \to f$ pointwise or $f_n \to f$ uniformly. In Section 5.3 we discuss other notions of convergence.

Problems

1. Prove that the sequence $f_n(x) = x^n$ converges uniformly to zero on any interval of the form $[0, \mu]$ if $\mu < 1$.

2. Suppose that g is a continuous function on $[a, b]$.

   (a) Prove that if $f_n \to f$ pointwise, then $gf_n \to gf$ pointwise.

   (b) Prove that if $f_n \to f$ uniformly, then $gf_n \to gf$ uniformly.

3. Let E be a subset of $\mathbb{R}$. Suppose that $f_n \to f$ uniformly on E and $g_n \to g$ uniformly on E. Prove that $f_n + g_n \to f + g$ uniformly on E.

4. Suppose that g is a continuous function on $[a, b]$ such that $g(x) > 0$ for each $x \in [a, b]$. Prove that if $f_n \to f$ uniformly, then $\frac{f_n}{g} \to \frac{f}{g}$ uniformly.

5. Prove that $f_n(x) = (x - \frac{1}{n})^2$ converges uniformly on any finite interval.

6. Prove that $\sin(x + \frac{1}{n}) \to \sin x$ uniformly on $\mathbb{R}$. Hint: use the Mean Value Theorem.

7. Let $f_n(x) = \frac{nx}{1+n^2x^2}$. Prove that $f_n \to 0$ pointwise but not uniformly on $[0, 1]$.

8. Let $f_n(x) = \frac{x}{n}e^{-\frac{x}{n}}$. Prove that $f_n \to 0$ pointwise but not uniformly on $[0, \infty)$.

9. Suppose that g is a continuous function on $[0, 1]$ and that $g(1) = 0$. Define $f_n(x) = g(x)x^n$. Prove that $f_n \to 0$ uniformly.

10. Let f be a uniformly continuous function on $\mathbb{R}$ and define $f_n(x) \equiv f(x + \frac{1}{n})$. Prove that $f_n \to f$ uniformly.

11. Let $g(x) = e^{-x^2}$ and define $f_n(x) = g(x - n)$. Prove that $f_n \to 0$ pointwise but not uniformly on $\mathbb{R}$.

12. Let $\{f_n\}$ be a sequence of continuous functions such that $f_n \to f$ uniformly on $\mathbb{R}$. Suppose that $x_n \to x_o$. Prove that $\lim_{n \to \infty} f_n(x_n) = f(x_o)$.

13. Let f be a continuous function on $[0, 1]$. Under what conditions on f will the sequence $f_n(x) \equiv f(x)^n$ converge pointwise? Uniformly?

14. (a) Prove that $(1 + \frac{x}{n})^n \to e^x$ for all x. Hint: see problem 11 of Section 4.3.

    (b) Prove that $(1 + \frac{x}{n})^n \to e^x$ uniformly on any finite interval $[a, b]$.

    (c) Prove that the convergence is not uniform on $\mathbb{R}$.

5.2 Limit Theorems

In this section we prove several theorems which guarantee that limiting functions have nice properties if the convergence is uniform.

❑ **Theorem 5.2.1**   Let $\{f_n\}$ be a sequence of continuous functions on a set $E \subseteq \mathbb{R}$. If $f_n \to f$ uniformly on E as $n \to \infty$, then f is continuous on E.

Proof.   Let $x_o \in E$ and $\varepsilon > 0$ be given. Since $f_n \to f$ uniformly, we can choose an N so that $n \geq N$ implies that

$$|f_n(x) - f(x)| \leq \varepsilon/3 \qquad \text{for all } x \in E. \tag{4}$$

Further, since f_N is continuous, we can find a δ so that

$$|f_N(x) - f_N(x_o)| \;\leq\; \varepsilon/3 \qquad \text{if } |x - x_o| \leq \delta. \tag{5}$$

Using (4) and (5), we see that

$$
\begin{aligned}
|f(x) - f(x_o)| \;&\leq\; |f(x) - f_N(x)| + |f_N(x) - f_N(x_o)| + |f_N(x_o) - f(x_o)| \\
&\leq\; \varepsilon/3 \;+\; \varepsilon/3 \;+\; \varepsilon/3 \\
&=\; \varepsilon
\end{aligned}
$$

for all x such that $|x - x_o| \leq \delta$. By Theorem 3.1.3, this proves that f is continuous at x_o. Since x_o was arbitrary, f is continuous on E.   ❏

Example 3 of Section 5.1 shows that the limiting function may not be continuous if the convergence is not uniform.

❏ **Theorem 5.2.2**   Let $\{f_n\}$ be a sequence of continuous functions on a finite interval $[a, b]$. If $f_n \to f$ uniformly as $n \to \infty$, then

$$\lim_{n \to \infty} \int_a^b f_n(x) \, dx \;=\; \int_a^b f(x) \, dx.$$

Proof.   Let $\varepsilon > 0$ be given. Since $f_n \to f$ uniformly, we can choose an N so that $n \geq N$ implies that

$$|f_n(x) - f(x)| \;\leq\; \frac{\varepsilon}{b - a} \qquad \text{for all } x \, \epsilon \, [a, b].$$

Thus,

$$
\begin{aligned}
\left| \int_a^b f_n(x) \, dx - \int_a^b f(x) \, dx \right| \;&=\; \left| \int_a^b (f_n(x) - f(x)) \, dx \right| \\
&\leq\; \int_a^b |f_n(x) - f(x)| \, dx \\
&\leq\; \frac{\varepsilon}{b - a} \int_a^b dx \\
&=\; \varepsilon
\end{aligned}
$$

if $n \geq N$. Thus $\int_a^b f_n(x) \, dx \to \int_a^b f(x) \, dx$. In the next to last step we used the estimate in Corollary 3.3.6.   ❏

Example 4 in Section 5.1 shows that the conclusion of Theorem 5.2.2 may not hold if the convergence is not uniform.

Example 1 To see that the hypothesis that $[a, b]$ is a finite interval is needed in Theorem 5.2.2, consider the following sequence of continuous functions on $[0, \infty)$. Let $f_n(x) = 1/n$ on the interval $[0, n)$; let f_n be the straight line from the point $(n, 1/n)$ to the point $(n + 1, 0)$ on the interval $[n, n + 1]$; let $f_n(x) = 0$ for $x > n + 1$. Since $0 \le f_n(x) \le 1/n$ for all x, it is easy to see that $f_n \to 0$ uniformly as $n \to \infty$. However,

$$\int_0^\infty f_n(x)\, dx = 1 + \frac{1}{2n}.$$

Thus, $\lim_{n \to \infty} \int_0^\infty f_n(x)\, dx = 1$, while the integral of the limiting function is 0.

❏ **Theorem 5.2.3** Let $\{f_n\}$ be a sequence of continuously differentiable functions on a finite interval $[a, b]$. Suppose that the sequence of derivatives $\{f_n'\}$ converges uniformly to a function g on $[a, b]$ and that for one point, x_o, the sequence of real numbers $\{f_n(x_o)\}$ converges. Then, $\{f_n\}$ converges uniformly to a continuous function f on $[a, b]$, f is continuously differentiable, and $f' = g$.

Proof. Since f' is continuous, the Fundamental Theorem of Calculus guarantees that

$$f_n(x) = f_n(x_o) + \int_{x_o}^x f_n'(t)\, dt \tag{6}$$

for each $x \epsilon [a, b]$. Now, $f_n' \to g$ uniformly, so by Theorem 5.2.2,

$$\lim_{n \to \infty} \int_{x_o}^x f_n'(t)\, dt = \int_{x_o}^x g(t)\, dt.$$

In addition, $\{f_n(x_o)\}$ converges by hypothesis, so the right hand side of (6) converges as $n \to \infty$. Therefore, for each x, the left-hand side of (6) converges, too. If we define $f(x) \equiv \lim_{n \to \infty} f_n(x)$ and $c \equiv \lim_{n \to \infty} f_n(x_o)$, then

$$f(x) = c + \int_{x_o}^x g(t)dt \tag{7}$$

for each $x \epsilon [a, b]$. Since g is continuous, we know by the Fundamental Theorem of Calculus that f is continuously differentiable and $f' = g$. It remains to show that the convergence $f_n \to f$ is uniform. Let $\varepsilon > 0$ be given. Choose N_1 so that $n \geq N_1$ implies $|c - f_n(x_o)| \leq \varepsilon/2$. Choose $N_2 > N_1$ so that $n \geq N_2$ implies that $|f_n'(t) - g(t)| \leq \frac{\varepsilon}{2(b-a)}$. Then, if $n \geq N_2$,

$$
\begin{aligned}
|f_n(x) - f(x)| &\leq |f_n(x_o) - c| + \left| \int_{x_o}^x (f_n'(t) - g(t))dt \right| \\
&\leq \varepsilon/2 + \int_{x_o}^x |f_n'(t) - g(t)|dt \\
&\leq \varepsilon/2 + \frac{\varepsilon}{2(b-a)} \int_{x_o}^x dt \\
&\leq \varepsilon/2 + \varepsilon/2
\end{aligned}
$$

for all $x \in [a, b]$. Thus, $f_n \to f$ uniformly.   ❏

Finally, we prove that one can differentiate under the integral sign with respect to a parameter. This theorem plays an important role in our derivation of the Euler equation in Section 5.5.

❏ **Theorem 5.2.4**   Let f be a continuous function on the rectangle $Q \equiv [a, b] \times [c, d]$. Suppose that for each x the function $f(x, \cdot)$ is differentiable on $[c, d]$ and f_y is continuous on Q. Define a function F on $[c, d]$ by

$$
F(y) \equiv \int_a^b f(x, y)\, dx.
$$

Then F is continuously differentiable and

$$
F'(y) = \int_a^b f_y(x, y)\, dx. \tag{8}
$$

Proof.   To show that $F(y)$ is differentiable, suppose that $h_n \to 0$ and $h_n \neq 0$. We want to show that the limit of the difference quotient

$$
\lim_{n \to \infty} \frac{F(y + h_n) - F(y)}{h_n} = \lim_{n \to \infty} \int_a^b \frac{f(x, y + h_n) - f(x, y)}{h_n}\, dx \tag{9}
$$

exists and equals the right hand side of (8). Since f is continuously differentiable in the second variable,

$$
\frac{f(x, y + h_n) - f(x, y)}{h_n} \to f_y(x, y) \tag{10}
$$

as $n \to \infty$ for each fixed x. What we need to know is that we can bring the limit on the right-hand side of (9) inside the integral. This is exactly the question treated in Theorem 5.2.2. According to the hypothesis of that theorem, we need to know that the convergence in (10) is uniform in x in order to interchange the limit and the integral. By the Mean Value Theorem, for each fixed x,

$$\frac{f(x, y + h_n) - f(x, y)}{h_n} = f_y(x, \xi)$$

for some $\xi \in [y, y + h_n]$. By hypothesis, f_y is continuous on the rectangle. This implies, by Theorem 4.6.1(c), that f_y is uniformly continuous on the rectangle. Thus, given $\varepsilon > 0$, we can choose a δ so that for all $x \in [a, b]$, we have $|f_y(x, z) - f_y(x, y)| \leq \varepsilon$ if $|z - y| \leq \delta$. Therefore, choosing n large enough so that $|h_n| \leq \delta$ guarantees that

$$\left| \frac{f(x, y + h_n) - f(x, y)}{h_n} - f_y(x, y) \right| \leq |f_y(x, \xi) - f_y(x, y)|$$
$$\leq \varepsilon,$$

so the convergence in (10) is uniform for x in $[c, d]$. Thus, by Theorem 5.2.2, the limit in (9) exists and can be computed by taking the limit inside the integral. This proves (8). The proof that F' is continuous is left to the reader as problem 10.   ❏

Problems

1. Let $f_n(x) = e^{-nx}$ on the interval $[0, 1]$. Explain why the sequence of functions $\{f_n\}$ converges pointwise on $[0, 1]$. What is the limiting function? Is it continuous? Is the convergence uniform?

2. Compute the following limits:

   (a) $\lim_{n \to \infty} \int_0^1 (x + \frac{1}{n})^2 \, dx$.

   (b) $\lim_{n \to \infty} \int_1^2 e^{-nx} \, dx$.

   (c) $\lim_{n \to \infty} \int_0^{\frac{\pi}{2}} \sin(x + \frac{1}{n}) \, dx$.

   (d) $\lim_{n \to \infty} \int_0^1 (1 + \frac{x}{n})^n \, dx$.

3. Give an example of a sequence of continuously differentiable functions $\{f_n\}$ on $[0, 1]$ so that $f_n \to f$ uniformly but f is not differentiable at all points of $[0, 1]$. Hint: draw graphs first.

4. Let $\{f_n\}$ be a sequence of continuous functions on a finite interval $[a, b]$ that converges uniformly to f. Show that for all continuous functions g on $[a, b]$,

$$\lim_{n \to \infty} \int_a^b f_n(t)g(t)dt = \int_a^b f(t)g(t)dt.$$

5. Suppose that f is continuous and $|f(x)| < 1$ for all $x \in [a, b]$. Prove that

$$\lim_{n \to \infty} \int_a^b (f(t))^n dt = 0.$$

6. Let $\{f_n\}$ be a sequence of continuous functions that converges uniformly on $[0, 1]$. Show that there is an M so that $|f_n(x)| \leq M$ for all n and all $x \in [0, 1]$.

7. Suppose that $\{f_n\}$ is a sequence of continuous functions on an open interval (a, b) that converges uniformly to f on (a, b). Suppose that each f_n is uniformly continuous on (a, b). Prove that f is uniformly continuous on (a, b).

8. Show by example that the hypothesis of Theorem 5.2.3, that $\{f_n(x)\}$ converges for at least one x, is needed to obtain the conclusion.

9. Let $\{f_n\}$ be a sequence of continuous functions defined on $\mathbb{R}$, and suppose that $f_n \to f$ uniformly on every finite interval $[a, b]$. Prove that f is a continuous function on $\mathbb{R}$. If each of the functions f_n is bounded, is it necessarily true that f is bounded?

10. Complete the proof of Theorem 5.2.4 by showing that F' is continuous.

11. Let f be a continuous function on $[0, \infty)$ which equals zero outside the interval $[a, b]$. For each $\lambda > 0$, define

$$F(\lambda) \equiv \int_0^\infty e^{-\lambda x} f(x)dx.$$

By using Theorem 5.2.4, prove that F is infinitely often continuously differentiable on $(0, \infty)$. Remark: F is called the **Laplace transform** of f.

12. Let f_n be a sequence of continuous functions defined on $\mathbb{R}$, and suppose that $f_n \to f$ uniformly on every finite interval $[a, b]$. Suppose that there is a nonnegative continuous function g on $\mathbb{R}$ such that $|f_n(x)| \leq g(x)$ for all n and all $x \in \mathbb{R}$ and suppose that the improper Riemann integral $\int_{-\infty}^\infty g(x)\, dx < \infty$ exists.

   (a) Prove that

$$\lim_{n \to \infty} \int_{-\infty}^\infty f_n(x)\, dx = \int_{-\infty}^\infty f(x)\, dx.$$

   (b) Find a counterexample that shows that if the hypothesis $f_n(x) \leq g(x)$ is omitted, then the conclusion may not hold.

13. Compute $\lim_{n\to\infty} \int_{-\infty}^{\infty} \frac{1}{n+x^2}\, dx$.

14. Suppose that f is bounded and continuous on $[0, \infty]$. Prove that the Laplace transform of f is infinitely often continuously differentiable. Hint: use problem 12 and the idea in the proof of Theorem 5.2.4.

15. Suppose that $\{f_n\}$ is a sequence of continuous functions on a finite interval $[a, b]$. Suppose that $f_n \to 0$ pointwise on $[a, b]$ and that for each $x \in [a, b]$ the sequence of numbers $\{f_n(x)\}$ is nonincreasing. Prove that $f_n \to 0$ uniformly. Hint: if not, there is an $\varepsilon > 0$ and points x_n so that $f_n(x_n) \geq \varepsilon$. Use the Bolzano-Weierstrass theorem to get a contradiction.

5.3   The Supremum Norm

Recall the reason for the definition of Cauchy sequence that we introduced in Section 2.4. We wanted a criterion for convergence of a sequence $\{a_n\}$ of real numbers that depended on the sequence itself, not on the limit. The reason was simple. In many situations we construct a sequence $\{a_n\}$ by some procedure and then need to know that it converges. The limit is not given to us; its existence is, in fact, the goal of our calculations. The Cauchy criterion allows us to assert the existence of the limit, given estimates on the sequence itself. We used this idea in the Bolzano-Weierstrass theorem, in the definition of the Riemann integral and in many proofs. Since the definitions of "continuous" and "differentiable" involve limits, the Cauchy criterion for sequences of real numbers is involved, either explicitly or implicitly, in almost every theorem that we have proved.

Many of the functions which are "solutions" of problems in analysis (for example, integral equations or differential equations) are constructed as the limit of a sequence of functions. Thus, we need a Cauchy criterion for sequences of functions analogous to the Cauchy criterion for sequences of real numbers. First, we need a way of saying when two functions are "close," and second, in order to do that, we need a notion of "size" for functions.

Definition.   Let f be a bounded function on a set $E \subseteq \mathbb{R}$. We define

$$\|f\|_\infty \equiv \sup_{x \in E} |f(x)|.$$

The number $\|f\|_\infty$ is called the **supremum norm** or the *sup* norm of the function f.

Thus, $\|f\|_\infty$ is just the supremum of the values of $|f(x)|$ on the set E. Usually the set E will be an interval, finite or infinite. Notice that $\|f\|_\infty$ depends on the set E, but we will not usually indicate that in the symbol $\|f\|_\infty$. If f is a continuous function on a finite interval $[a, b]$, then $|f|$ is also continuous on $[a, b]$, so $\|f\|_\infty$ is just the maximum of $|f(x)|$ on $[a, b]$.

Proposition 5.3.1   Let f and g be functions defined on a set $E \subseteq \mathbb{R}$. Then,

(a) $\|f\|_\infty \geq 0$ and $\|f\|_\infty = 0$ if and only if f is the zero function on E.

(b) For every $\alpha \in \mathbb{R}$, we have $\|\alpha f\|_\infty = |\alpha|\, \|f\|_\infty$.

(c) $\|f + g\|_\infty \leq \|f\|_\infty + \|g\|_\infty$ (the triangle inequality).

Proof.   The proofs follow quite easily from the properties of *sup* proven in Section 2.5. Since $|f(x)| \geq 0$ for each $x \in E$, the supremum over all such $|f(x)|$ must be nonnegative. Further, it is clear that $\|f\|_\infty = 0$ if and only if $|f(x)| = 0$ for all $x \in E$, in which case f is the zero function on E by definition. This proves (a).

To prove (b), note that

$$
\begin{aligned}
\|\alpha f\|_\infty &= \sup\{|\alpha f(x)| \mid x \in E\} \\
&= \sup\{|\alpha|\, |f(x)| \mid x \in E\} \\
&= |\alpha|\, \sup\{|f(x)| \mid x \in E\} \\
&= |\alpha|\, \|f\|_\infty,
\end{aligned}
$$

where we used problem 8 of Section 2.5 in the next to last step.

Finally, for each $x \in E$, we know that

$$
|f(x) + g(x)| \leq |f(x)| + |g(x)| \leq \|f\|_\infty + \|g\|_\infty.
$$

Thus, $\|f\|_\infty + \|g\|_\infty$ is an upper bound for $\{|f(x) + g(x)| \mid x \in E\}$, and therefore,

$$
\begin{aligned}
\|f + g\|_\infty &= \sup\{|f(x) + g(x)| \mid x \in E\} \\
&\leq \|f\|_\infty + \|g\|_\infty
\end{aligned}
$$

which proves (c).   ❏

Note that the properties of $\|f\|_\infty$ that we have just proven are analogous to the properties of absolute value in measuring the size of real numbers: (a) $|x| \geq 0$ and $|x| = 0$ if and only if $x = 0$; (b) $|ax| = |a|\,|x|$ for all x; (c) for all x and y, $|x + y| \leq |x| + |y|$. The sup norm is not the only way one can measure the size of functions. Other notions of size are discussed below.

Using the sup norm, we can define convergence.

Definition. A sequence of functions $\{f_n\}$ on a set $E \subseteq \mathbb{R}$ is said to **converge in the sup norm** to a function f if, given $\varepsilon > 0$, there is an N such that

$$\|f_n - f\|_\infty \leq \varepsilon \qquad \text{for } n \geq N.$$

In other words, $\{f_n\}$ converges to f in the sup norm if and only if

$$\lim_{n \to \infty} \|f_n - f\|_\infty = 0.$$

Proposition 5.3.2 Let $\{f_n\}$ be a sequence of functions defined on a set $E \subseteq \mathbb{R}$. Let f be a function on E. Then, f_n converges to f uniformly on E if and only if f_n converges to f in the *sup* norm on E.

Proof. Suppose that f_n converges to f uniformly and let $\varepsilon > 0$ be given. We can choose an N so that

$$|f_n(x) - f(x)| \ \leq \ \varepsilon \qquad \text{for all } n \geq N \text{ and all } x \in E. \tag{11}$$

Therefore,

$$\|f_n - f\|_\infty \ \leq \ \varepsilon \qquad \text{for } n \geq N. \tag{12}$$

Thus, by definition, f_n converges to f in the *sup* norm.

Conversely, if f_n converges to f in the *sup* norm, then (12) holds, which implies (11). Therefore, f_n converges to f uniformly. ❏

We have reformulated uniform convergence in terms of the sup norm so that we can make the following definition, which is the analogue of the definition of Cauchy sequence in Section 2.4.

Definition. A sequence of functions $\{f_n\}$ on a set $E \subseteq \mathbb{R}$ is said to be a **Cauchy sequence in the sup norm** if, given any $\varepsilon > 0$, there is an N such that

$$\|f_n - f_m\|_\infty \ \leq \ \varepsilon \qquad \text{for all } n \geq N \text{ and } m \geq N. \tag{13}$$

If f_n converges to f in the *sup* norm, then it is not hard to see that the sequence f_n is a Cauchy sequence (problem 1). The more difficult question is this. Let $\mathcal{S}$ be a set of functions. Does a Cauchy sequence of functions in $\mathcal{S}$ necessarily have a limit in $\mathcal{S}$? If $\mathcal{S}$ is the set of continuous functions on a set E, the answer is yes.

❏ **Theorem 5.3.3** Let $\{f_n\}$ be a sequence of continuous functions on $E \subseteq \mathbb{R}$. Suppose that $\{f_n\}$ is a Cauchy sequence in the *sup* norm. Then, there exists a continuous function f on E such that

$$\lim_{n \to \infty} \|f_n - f\|_\infty = 0.$$

Proof. Since $\{f_n\}$ is a Cauchy sequence in the sup norm, given $\varepsilon > 0$, we can choose N so that (13) holds. For each $x \in \mathbb{R}$, $|f_n(x) - f_m(x)| \le \|f_n - f_m\|_\infty$, so

$$|f_n(x) - f_m(x)| \;\le\; \varepsilon \qquad \text{for all } n \ge N \text{ and } m \ge N. \qquad (14)$$

Thus, for each x, the sequence of real numbers $\{f_n(x)\}$ is a Cauchy sequence and therefore converges because the real numbers are complete. Define $f(x) \equiv \lim_{n \to \infty} f_n(x)$. Since the absolute value is a continuous function,

$$|f(x) - f_m(x)| \;=\; \lim_{n \to \infty} |f_n(x) - f_m(x)|,$$

so (14) implies that

$$|f(x) - f_m(x)| \;\le\; \varepsilon \qquad \text{for all } m \ge N \text{ and all } x \in E.$$

This shows that $f_m \to f$ uniformly. Therefore, f is a continuous function (by Theorem 5.2.1), and f_n converges to f in the sup norm (by Proposition 5.3.2). ❏

Because of Theorem 5.3.3, we say that $C[a, b]$, the set of continuous functions on a finite interval $[a, b]$, is **complete in the sup norm**. That is, every sequence of continuous functions which is a Cauchy sequence in the *sup* norm on $[a, b]$ converges to a limiting continuous function. This fact will play as important a role in the subsequent chapters as did the convergence of Cauchy sequences of real numbers in the previous chapters.

If we use the *sup* norm, the "size" of a continuous function on $[a, b]$ is just the maximum value of $|f(x)|$. There are other reasonable notions of size. For example, we could define

$$\|f\|_1 = \int_a^b |f(x)|\, dx,$$

in which case we are taking the size of f to be the area under the graph of $|f(x)|$. More generally, we define for $1 \le p < \infty$,

$$\|f\|_p = \left(\int_a^b |f(x)|^p\, dx \right)^{\frac{1}{p}},$$

although these sizes do not have simple geometric interpretations if $p > 1$. Each of these is called a "norm" on $C[a, b]$ because it satisfies the three properties, (a), (b), and (c), of Proposition 5.3.1. This is easy to see in case $p = 1$ (problem 4), but harder to see for $p > 1$. These different norms are useful in different circumstances. For example, we will see that the L^2 norm $\|f\|_2$ arises naturally in the study of Fourier series (Chapter 9). We give a formal definition of "norm" in Section 5.8. The point of introducing norms here is to point out that there are other reasonable ways to measure the size of functions and to raise the question of whether $C[a, b]$ is complete if we use one of these other norms.

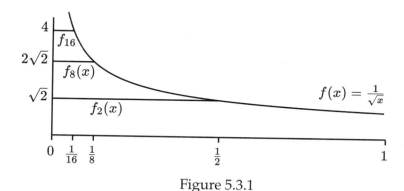

Figure 5.3.1

Example 1   Consider the functions f_n on $[0, 1]$ depicted in Figure 5.3.1 and defined by

$$f_n(x) = \begin{cases} \sqrt{n} & \text{if } 0 \le x \le \frac{1}{n} \\ \frac{1}{\sqrt{x}} & \text{if } \frac{1}{n} < x \le 1 \end{cases}$$

Each f_n is a positive continuous function on $[0, 1]$, and if $m \geq n$ we have $f_m(x) \geq f_n(x)$ for each x, so

$$
\begin{aligned}
\|f_m - f_n\|_1 &= \int_0^1 |f_m(x) - f_n(x)|\, dx \\
&= \int_0^1 (f_m(x) - f_n(x))\, dx \\
&= \left(2 - \frac{1}{\sqrt{m}}\right) - \left(2 - \frac{1}{\sqrt{n}}\right) \\
&= \frac{1}{\sqrt{n}} - \frac{1}{\sqrt{m}}.
\end{aligned}
$$

Thus, given $\varepsilon > 0$, we can choose N so that

$$
\|f_m - f_n\|_1 \leq \varepsilon \qquad \text{for all } n \geq N \text{ and } m \geq N.
$$

Therefore, $\{f_n\}$ is a sequence of continuous functions that is a Cauchy sequence in the norm $\| \cdot \|_1$. Can there exist a continuous function f so that $\|f - f_n\|_1 \to 0$ as $n \to \infty$? Evidently not, because any such f would be bounded by Theorem 3.2.1, and for large enough n the function f_n will have more and more area over the bound of f; thus, it would be impossible for $\|f_n - f\|_1$ to converge to zero. Thus, $\{f_n\}$ is a Cauchy sequence of functions in $C[0, 1]$ in the norm $\| \cdot \|_1$ that does not have a limit in $C[0, 1]$. Thus $C[0, 1]$ is *not* complete in the norm $\| \cdot \|_1$. In fact, $C[0, 1]$ is not complete in any of the norms $\| \cdot \|_p$ if $1 \leq p < \infty$.

This leads us to a very important point about the Riemann integral which was developed by A. Cauchy and B. Riemann in the 19th century. We have seen that the Riemann integral is naturally and easily defined on continuous functions and can be extended to functions which have only finitely many discontinuities. There is a theorem (which we have not proven) which shows that the Riemann integral exists, that is, the infimum of the upper sums is equal to the supremum of the lower sums for all functions that are "almost" continuous in the sense that their singularities occur on a "small" set. The above discussion shows why a more general theory of integration is necessary. A variety of norms, like the $\| \cdot \|_p$ norms, arise naturally in different analytical situations. To use these norms effectively one must use spaces of functions that are complete in these norms so that one knows that Cauchy sequences of functions have limits. Since spaces of continuous functions, or almost continuous functions, are not complete in these norms, one needs to use larger sets of functions which are complete. These sets of functions must contain at least the functions that one obtains as limits of Cauchy sequences of continuous functions. Since these limiting functions will, in general, have large sets on which they are not continuous, we need a theory of how to integrate such functions. This theory, which was developed in the early 20th century by Lebesgue and others, is covered in more advanced textbooks. Our point here is that the motivation for the Lebesgue theory is not that one wants to integrate singular functions per se but that one needs to work with sets of functions that are complete.

Finally, we remark that for a continuous function f on $[0,1]$, it is not too hard to show that $\|f\|_p$ converges to $\|f\|_\infty$ as $p \to \infty$. See project 1. This is why the *sup* norm is denoted by the symbol $\| \cdot \|_\infty$.

Problems

1. Let f_n be a sequence of bounded functions on a set E, and suppose that f is a bounded function such that $\|f_n - f\|_\infty \to 0$ as $n \to \infty$. Prove that $\{f_n\}$ is a Cauchy sequence in the *sup* norm.

2. Let f and g be continuous functions on $[a,b]$.

    (a) Use the triangle inequality to prove that

    $$\big| \, \|f\|_\infty - \|g\|_\infty \, \big| \le \|f - g\|_\infty.$$

    (b) Suppose that $f_n \to f$ in the sup norm. Prove that $\|f_n\|_\infty \to \|f\|_\infty$.

3. Let $C_b(\mathbb{R})$ denote the set of bounded continuous functions on $\mathbb{R}$. Prove that $C_b(\mathbb{R})$ is complete in the sup norm. Hint: use problem 2 to show that a Cauchy sequence of functions in the sup norm must have a uniform bound.

4. For $f \in C[a,b]$, define $\|f\|_1 = \int_a^b |f(x)| \, dx$. Show that $\| \cdot \|_1$ satisfies the three properties, (a), (b), and (c), of Proposition 5.3.1.

5. Let the functions f_n be defined on $[0,1]$ by

    $$f_n(x) = \begin{cases} 1 & 0 \le x \le \frac{1}{2} \\ 1 - n(x - \frac{1}{2}) & \frac{1}{2} < x \le \frac{1}{2} + \frac{1}{n} \\ 0 & \frac{1}{2} + \frac{1}{n} < x \le 1 \end{cases}.$$

    and define

    $$f(x) = \begin{cases} 1 & 0 \le x \le \frac{1}{2} \\ 0 & \frac{1}{2} < x \le 1. \end{cases}$$

(a) Prove that $f_n \to f$ pointwise on $[0, 1]$. Hint: draw the graph of f_n.

(b) Prove that $\|f - f_n\|_\infty = 1$ for each n so that f_n does not converge to f in the *sup* norm.

(c) Explain how you could have predicted the result of part (b) simply by using Theorem 5.2.1.

(d) Prove that $\|f - f_n\|_1 \to 0$ as $n \to \infty$.

6. Let f_n be the functions in Example 1 and define $f(x) \equiv \frac{1}{\sqrt{x}}$ on $(0, 1]$. Prove that each of the functions $|f - f_n|$ has an improper Riemann integral on $[0, 1]$ and

$$\lim_{n \to \infty} \|f - f_n\|_1 \; = \; \lim_{n \to \infty} \int_0^1 |f(x) - f_n(x)| dx \; = \; 0.$$

7. Let $C^{(1)}[a, b]$ denote the set of continuously differentiable functions on a finite interval $[a, b]$. For each f in $C^{(1)}[a, b]$, define $\|f\| \equiv \|f\|_\infty + \|f'\|_\infty$.

(a) Show that $\|f\|$ has the properties (a), (b), and (c) of Proposition 5.3.1.

(b) Prove that $C^{(1)}[a, b]$ is complete in the norm $\|f\|$. Hint: follow the proof of Theorem 5.3.3 and use Theorems 5.2.1 and 5.2.3.

8. Let $C([a, b]; \mathbb{R}^2)$ denote the set of pairs, $(f(x), g(x))$, of continuous functions on $[a, b]$. Define a norm on $C([a, b]; \mathbb{R}^2)$ by

$$\|(f, g)\|_\infty \; \equiv \; \|f\|_\infty + \|g\|_\infty.$$

(a) Prove that this norm satisfies the properties (a), (b), and (c) of Proposition 5.3.1.

(b) Prove that $C([a, b]; \mathbb{R}^2)$ is complete.

9. Let Q be a closed, finite, rectangle in the plane, and let $C(Q)$ denote the set of real-valued continuous functions on Q. For f in $C(Q)$, define

$$\|f\|_\infty \; \equiv \; \sup_{(x,y) \in Q} |f(x, y)|.$$

(a) Prove that this norm satisfies the properties (a), (b), and (c) of Proposition 5.3.1.

(b) Prove that $C(Q)$ is complete.

10. Let $C_o(\mathbb{R})$ denote the set of continuous functions on $\mathbb{R}$ such that $\lim_{x \to \infty} f(x) = 0$ and $\lim_{x \to -\infty} f(x) = 0$. Prove that $C_o(\mathbb{R})$ is complete in the $\|\cdot\|_\infty$ norm. Hint: since $C_o(\mathbb{R}) \subseteq C_b(\mathbb{R})$ the limit of a Cauchy sequence certainly exists and, by problem 3, is in $C_b(\mathbb{R})$.

5.4   Integral Equations

An equation for an unknown function $\psi(x)$ that involves the integral of $\psi(x)$ or the integral of a function of $\psi(x)$ is called an integral equation. A simple example is

$$\psi(x) \;=\; f(x) \;+\; \lambda \int_a^b K(x,y)\psi(y)\,dy. \tag{15}$$

Here $K(x,y)$ is a given function of two variables, f is a given function of one variable and λ is a parameter. We can't solve directly for $\psi(x)$ since $\psi(x)$ occurs in the integral as well as on the left-hand side. This is a *linear* integral equation because the integral term depends linearly on $\psi(x)$. A nonlinear integral equation is discussed in project 2. Integral equations often arise in the study of ordinary and partial differential equations. Examples of applications to ordinary differential equations can be found in Project 3 and Section 7.1. Integral equations can also occur over infinite intervals and can have variables in the limits of integration; see problems 10 and 11. For more information and the further development of the subject, see [22].

We will prove that (15) has solutions in the case where λ is small, f is continuous on $[a,b]$, and K is a continuous function on the square

$$Q \;\equiv\; [a,b] \times [a,b].$$

The continuity of K implies that K is bounded and uniformly continuous on Q (Theorem 4.6.1).

❑ **Theorem 5.4.1**   Let f be a continuous function on the finite interval $[a,b]$ and let K be continuous on Q. Then, if λ is small enough, there exists a unique continuous function $\psi(x)$ on $[a,b]$ that solves (15).

Proof.   Let $\psi_0(x)$ be any continuous function on $[a,b]$. Define functions $\psi_1(x), \psi_2(x), \ldots$, recursively by the formula

$$\psi_n(x) \;=\; f(x) \;+\; \lambda \int_a^b K(x,y)\psi_{n-1}(y)\,dy. \tag{16}$$

We begin by showing that the functions ψ_n are continuous. Suppose that ψ_n is continuous for some n. For each fixed x, $K(x,y)$ is a continuous function of y. By Theorem 3.1.1(c), $K(x,y)\psi_n(y)$ is a continuous function of y, so the integral in (16) makes sense. Now let x_1 and x_2 be points in $[a, b]$. Then,

$$|\psi_{n+1}(x_1)-\psi_{n+1}(x_2)|$$

$$\leq \quad |f(x_1) - f(x_2)| \;+\; \left| \lambda \int_a^b (K(x_1, y) - K(x_2, y)) \psi_n(y)\, dy \right| \qquad (17)$$

$$\leq \quad |f(x_1) - f(x_2)| \;+\; |\lambda| \int_a^b |K(x_1, y) - K(x_2, y)|\, |\psi_n(y)|\, dy. \qquad (18)$$

Let $\varepsilon > 0$ be given. Since f is continuous, we can choose a δ_1 such that $|f(x_1) - f(x_2)| \leq \frac{\varepsilon}{2}$ if $|x_1 - x_2| \leq \delta_1$. To handle the second term in (18), note that, because ψ_n is continuous, there is a constant C_n so that $|\psi_n(y)| \leq C_n$ for all $y \in [a, b]$. Since K is uniformly continuous on the square, we can choose a δ_2 so that $|x_1 - x_2| \leq \delta_2$ implies

$$|K(x_1, y) - K(x_2, y)| \;\leq\; \frac{\varepsilon/2}{C_n(b - a)} \qquad \text{for all } y \in [a, b]. \qquad (19)$$

Using (19) to estimate the integral in (18), we find

$$|\psi_{n+1}(x_1) - \psi_{n+1}(x_2)| \;\leq\; \varepsilon/2 \;+\; \varepsilon/2$$

if $|x_1 - x_2| \leq \min\{\delta_1, \delta_2\}$, which shows that ψ_{n+1} is continuous. By assumption, $\psi_0(x)$ is continuous. So, by induction, all the ψ_n are continuous functions on $[a, b]$.

Next, we will show that $\{\psi_n\}$ is a Cauchy sequence in the *sup* norm if λ is sufficiently small. Let M be an upper bound for K on the square. Then, for $n \geq 1$,

$$|\psi_{n+1}(x) - \psi_n(x)| \quad \leq \quad \left| \lambda \int_a^b K(x, y)(\psi_n(y) - \psi_{n-1}(y))\, dy \right|$$

$$\leq \quad |\lambda| \int_a^b |K(x, y)|\, |\psi_n(y) - \psi_{n-1}(y)|\, dy$$

$$\leq \quad |\lambda| \int_a^b M\|\psi_n - \psi_{n-1}\|_\infty\, dy$$

$$= \quad |\lambda| M(b - a)\|\psi_n - \psi_{n-1}\|_\infty.$$

We have repeatedly used the fact that if one replaces an integrand by something larger, the integral gets bigger (Theorem 3.3.4). For simplicity, set

$$\alpha \equiv |\lambda| M(b - a)$$

and choose λ small enough so that $\alpha < 1$. We showed above that

$$|\psi_{n+1}(x) - \psi_n(x)| \leq \alpha \|\psi_n - \psi_{n-1}\|_\infty$$

for all $x \in [a, b]$. Thus, taking the *sup* of the left-hand side, we find

$$\|\psi_{n+1} - \psi_n\|_\infty \leq \alpha \|\psi_n - \psi_{n-1}\|_\infty.$$

Iterating this inequality gives

$$\|\psi_{n+1} - \psi_n\|_\infty \leq (\alpha)^n \|\psi_1 - \psi_0\|_\infty. \tag{20}$$

Suppose that $n \geq N$ and $m \geq N$, and let $m \geq n$ be the larger of the two integers. Then, by the triangle inequality (Proposition 5.3.1(c)),

$$
\begin{aligned}
\|\psi_m - \psi_n\|_\infty &= \left\| \sum_{j=n+1}^{m} (\psi_j - \psi_{j-1}) \right\|_\infty \\
&\leq \sum_{j=n+1}^{m} \|\psi_j - \psi_{j-1}\|_\infty \\
&\leq \sum_{j=n+1}^{m} \alpha^{j-1} \|\psi_1 - \psi_0\|_\infty \\
&\leq \left(\sum_{j=n}^{m-1} \alpha^j \right) \|\psi_1 - \psi_0\|_\infty.
\end{aligned}
$$

The sum in parentheses is part of the tail of the geometric series in α. Using the estimate in part (b) of problem 1, we find that

$$\sum_{j=n}^{m-1} \alpha^j = \alpha^n \sum_{j=0}^{m-n-1} \alpha^j \leq \frac{\alpha^n}{1-\alpha} \leq \frac{\alpha^N}{1-\alpha}$$

since $n \geq N$ and $0 < \alpha < 1$. Because $\|\psi_1 - \psi_0\|_\infty$ is a fixed number and $\alpha < 1$, it is clear that we can choose N large enough so that

$$\|\psi_m - \psi_n\|_\infty \leq \varepsilon$$

if $n \geq N$ and $m \geq N$. This proves that the sequence $\{\psi_n\}$ is a Cauchy sequence in the *sup* norm on $[a, b]$. By Theorem 5.3.3, there exists a continuous function ψ on $[a, b]$ such that $\|\psi_n - \psi\|_\infty \to 0$. By Proposition 5.3.2, $\psi_n \to \psi$ uniformly.

We are not done since we must still show that ψ satisfies (15). Consider equation (16). For each x, the left-hand side converges to $\psi(x)$. On the other hand, for each fixed x, $K(x, y)$ is a continuous function of y; thus, since $\psi_n \to \psi$ uniformly, we conclude (problem 4 of Section 5.2) that

$$\int_a^b K(x, y)\psi_{n-1}(y)\, dy \longrightarrow \int_a^b K(x, y)\psi(y)\, dy$$

as $n \to \infty$. Therefore, $\psi(x)$ satisfies (15). Suppose that $\phi(x)$ is another continuous function that satisfies (15). If we subtract the $\phi(x)$ equation from the $\psi(x)$ equation and estimate as above, we find

$$\|\psi - \phi\|_\infty \leq \alpha \|\psi - \phi\|_\infty.$$

Since $\alpha < 1$, this equation can hold only if $\|\psi - \phi\|_\infty = 0$, which implies that $\phi(x) = \psi(x)$ for all x. Thus, the solution $\psi(x)$ is unique. ❑

Notice that the proof not only guarantees the existence of a $\psi(x)$ satisfying (15) but also shows that, if we start with any $\psi_0(x)$ and iterate the relation (16), we can approximate $\psi(x)$ by $\psi_n(x)$ after n steps, and the estimates show how good our approximation will be.

Problems

1. Let $S_n \equiv \sum_{j=0}^n \alpha^j$.

    (a) Show that $\alpha S_n + 1 = S_n + \alpha^{n+1}$.

    (b) Suppose $0 \leq \alpha < 1$. Show that $S_n \leq \frac{1}{1-\alpha}$.

2. Consider equation (15) with $\lambda = 1$. By carefully considering the steps in the proof of Theorem 5.4.1, explain why (15) will have a unique continuous solution if $(b - a)$ is small enough or if $\|K\|_\infty$ is small enough.

3. Explain why there is a unique continuous function on $[0, 3]$ that satisfies

$$\psi(x) \;=\; \sin x \;+\; \frac{1}{4}\int_0^3 e^{-x-y^2}\psi(y)\, dy.$$

4. Explain why there is a unique continuous function on $[0, 2]$ that satisfies

$$\psi(x) \;=\; e^{x^2} \;+\; \int_0^2 xye^{-xy}\psi(y)\, dy.$$

5. Suppose that $K(x, y) = xy$, $f(x) = 1$, $a = 0$, $b = 1$, and $|\lambda| < 3$ in equation (15). Start with $\psi_0(x) \equiv 1$ and find ψ. Show that there is no solution if $\lambda = 3$. Hint: notice that any solution must be of the form $\psi(x) = 1 + cx$ for some c.

6. Suppose that $K(x, y) = g(x)h(y)$ and that $\int_a^b g(x)h(x)\, dx = 0$. Let $\psi_0(x) = f(x)$. Show that all iterates equal the first iterate and find a simple formula for the solution.

7. Suppose that $K(x, y) = (2 + x^2) \sin(x^2 + y^2)$, $a = 0$, $b = 1$, and $\lambda = \frac{1}{5}$. Let $f(x) \equiv 1$. Choose a $\psi_0(x)$ and estimate how close the nth iterate will be to the solution. Hint: estimate $\|\psi_m - \psi_n\|_\infty$ in terms of $\|\psi_1 - \psi_0\|_\infty$ and let $m \to \infty$. Then estimate $\|\psi_1 - \psi_0\|_\infty$. Note that your estimate will depend on your choice of $\psi_0(x)$.

8. Suppose that there is a nonzero continuous function $h(x)$ so that

$$h(x) = \lambda \int_a^b K(x, y)h(y)\, dy.$$

Show that if $\psi(x)$ is a solution of (15), then $\psi(x) + h(x)$ is also a solution. Let $a = 0$, $b = \pi$, and $K(x, y) = \sin x \sin y$. Show that there is a choice of λ so that such an $h(x)$ exists. Why doesn't this nonuniqueness violate Theorem 5.4.1?

9. Prove that there is at most one continuous function on the interval $[0, 2]$ that satisfies

$$\psi(x) = f(x) + \int_0^2 e^{-(x-y)^2} \cos(.3\psi(y))\, dy.$$

Hint: estimate and use the Mean Value Theorem.

10. (a) Show that there is a unique, bounded, continuous function ψ on the interval $[0, \infty)$ that solves

$$\psi(x) = e^{-2x} + \int_0^\infty e^{-2x-2y} \sin(x - y)\psi(y)\, dy.$$

Hint: follow the proof of Theorem 5.4.1.

(b) Prove that $\|\psi\|_\infty \le 2$. Hint: iterate $\|\psi\|_\infty \le 1 + \frac{1}{2}\|\psi\|_\infty$.

11. Let K be a continuous function on the square $S \equiv [0, 1] \times [0, 1]$ and let f be a continuous function on $[0, 1]$. We want to solve the **Volterra integral equation**:

$$\psi(x) = f(x) + \int_0^x K(x, y)\psi(y)\, dy. \tag{21}$$

Let $\psi_0(x) \equiv 1$ and define $\psi_n(x)$ recursively by

$$\psi_{n+1}(x) = f(x) + \int_0^x K(x, y)\psi_n(y)\, dy.$$

(a) Prove that $\psi_n(x)$ is continuous on $[0, 1]$ for each n.

(b) Let $M \equiv \sup_S |K|$ and $C \equiv \|\psi_1 - \psi_0\|_\infty$. Prove that

$$|\psi_2(x) - \psi_1(x)| \leq MCx \quad \text{for all } x \in [0, 1].$$

(c) Prove that for each $n \geq 2$,

$$|\psi_{n+1}(x) - \psi_n(x)| \leq \frac{M^n C x^n}{n!} \quad \text{for all } x \in [0, 1].$$

(d) Prove that ψ_n converges uniformly to a solution of (21). Hint: the series $\sum \frac{M^n}{n!}$ converges.

5.5   The Calculus of Variations

One of the familiar uses of calculus is to find the maxima and minima of functions of a real variable. However, in many applications the quantities which one wants to maximize or minimize depend on functions rather than points. For example, the flight path of an airplane flying between two cities is given by specifying the three functions, $x(t)$, $y(t)$, and $z(t)$, which give the coordinates of the plane as a function of time. If we are given the wind speeds at each point of space, we might want to know the flight path that minimizes the amount of gasoline used. The rate of growth of a yeast colony in a tank depends in a complicated way on the amount of glucose, $g(t)$, available at each time t. How should we choose $g(t)$ so that the growth is the greatest with the minimum expenditure of glucose. The branch of mathematics in which one studies methods for maximizing and minimizing quantities that depend on functions is called the Calculus of Variations. It has a long and rich history which greatly influenced the development of analysis in the 18th and 19th centuries. See the discussion in [28]. In this section we illustrate some the simplest ideas by examining two classical questions from geometry and physics. Further developments can by found in [14] and [18].

Example 1   Let (x_1, y_1) and (x_2, y_2) be two points in the plane. What is the curve between them that has the shortest length? We all know that the answer is the straight line that connects the points, and that certainly seems reasonable geometrically. But can we prove it analytically? To keep the analysis simple, we will only consider curves between the points that are the graphs of functions. Let $y(x)$ be such a function, that is, $y(x_1) = y_1$ and $y(x_2) = y_2$, and suppose that $y(x)$ is continuously differentiable. See Figure 5.5.1(a). Then the arc length of the graph of $y(x)$ between the points (x_1, y_1) and (x_2, y_2) is

$$J(y') \equiv \int_{x_1}^{x_2} \sqrt{1 + (y'(x))^2}\,dx.$$

Project 2(a) of Chapter 3 outlines the derivation of this formula. J is called a **functional** because its domain consists of a set of functions (namely the continuously differentiable functions whose graphs go through the two points) and its value is a real number. We write $J(y')$ because J depends explicitly on y' and depends on x and $y(x)$ only through y'. We want to know which function $y(x)$ minimizes J.

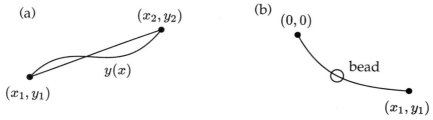

Figure 5.5.1

Example 2   In 1696, John Bernoulli posed the following problem. Let (x_1, y_1) be a point in the plane with $x_1 \neq 0$ and $y_1 < 0$. A wire, whose graph is given by the function $y(x)$ connects the origin $(0, 0)$ to the point (x_1, y_1). At time $t = 0$, a bead on the wire at $(0, 0)$ is released and slides down the wire to (x_1, y_1) under the influence of gravity which we assume points in the negative y direction. If we neglect friction and air resistance, what is the correct shape of the wire so that the transit time is the shortest? This is known as the brachistochrone (shortest time) problem. See Figure 5.5.1(b).

Let $y(x)$ be the function whose graph has the shape of the wire. Let $v(x)$ be the speed of the bead when the x-coordinate is x. Let $s(x)$ denote the length already traversed and $t(x)$ the time already spent when the x-coordinate is x. Then,

$$v(x) = \frac{ds}{dt} = \frac{ds}{dx}\frac{dx}{dt}$$

or

$$\frac{dx}{dt} = \frac{v(x)}{\frac{ds}{dx}}.$$

By Theorem 4.5.2,

$$\frac{dt}{dx} = \frac{\frac{ds}{dx}}{v(x)}.$$

From project 2(a) in Chapter 3 we know that the rate of change of arc length with respect to x is given by $\frac{ds}{dx} = \sqrt{1 + (y'(x))^2}$. To figure out how $v(x)$ and $y(x)$ are related, we use the fact that the total energy (kinetic plus potential),

$$E \equiv \frac{1}{2}mv(x)^2 + mgy(x),$$

is conserved since we are neglecting friction. Here m is the mass of the bead and g is the acceleration of gravity. Since the bead starts at rest at the origin, we can take the energy to be zero at $t = 0$. Solving, we find that $v(x) = \sqrt{-2gy(x)}$ for all x. Therefore, by the Fundamental Theorem of Calculus,

$$
\begin{aligned}
\text{Total time elapsed} \quad &= \quad t(x_1) - t(0) \\[2mm]
&= \quad \int_0^{x_1} \frac{dt}{dx}\, dx \\[2mm]
&= \quad \int_0^{x_1} \frac{\sqrt{1 + (y'(x))^2}}{\sqrt{-2gy(x)}}\, dx.
\end{aligned}
$$

Thus, we want to find how to choose $y(x)$ so that the functional

$$J(y, y') \equiv \int_0^{x_1} \frac{\sqrt{1 + (y'(x))^2}}{\sqrt{-2gy(x)}}\, dx$$

is minimized. We write $J(y, y')$ because J depends explicitly on both $y(x)$ and $y'(x)$.

These two examples suggest a very general question. Let f be a function of three independent variables. How do we find the functions $y(x)$, whose graphs pass through the points (x_1, y_1) and (x_2, y_2), so that the the functional

$$J(x, y(x), y'(x)) \equiv \int_{x_1}^{x_2} f(x, y(x), y'(x))\, dx \tag{22}$$

is maximized or minimized. We have allowed J to depend explicitly on x though that was not the case in the above two examples.

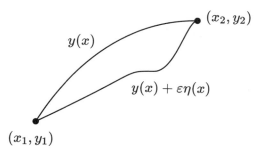

Figure 5.5.2

Here is the idea. Suppose that $y(x)$ minimizes J, and let $\eta(x)$ be a continuously differentiable function such that $\eta(x_1) = 0 = \eta(x_2)$. For every ε, the graph of $y(x) + \varepsilon\eta(x)$ passes through the points (x_1, y_1) and (x_2, y_2). See Figure 5.5.2. Note that ε is allowed to take negative values. Define a real-valued function I on $\mathbb{R}$ by

$$I(\varepsilon) \equiv J(x, y(x) + \varepsilon\eta(x), y'(x) + \varepsilon\eta'(x)).$$

If $y(x)$ minimizes J, then the function $I(\varepsilon)$ should reach its minimum at $\varepsilon = 0$, and this should be true for every choice of $\eta(x)$ that satisfies $\eta(x_1) = 0 = \eta(x_2)$. If $I(\varepsilon)$ is continuously differentiable, then, according to Theorem 4.2.1,

$$I'(0) = 0, \tag{23}$$

and this should be true for all choices of $\eta(x)$. If y is a twice continuously differentiable function such that (23) holds for all such $\eta(x)$, then y is called an **extremal** for the functional J. We want to determine what condition this puts on the function $y(x)$.

❏ **Theorem 5.5.1** Suppose that f is a twice continuously differentiable function of three variables and that the functional J is defined by (22). Suppose that $y(x)$ is an extremal for J. Then, $y(x)$ satisfies the differential equation

$$f_y(x, y(x), y'(x)) - \frac{d}{dx} f_{y'}(x, y(x), y'(x)) = 0. \tag{24}$$

Note. We follow tradition in the Calculus of Variations and denote the partial derivatives of f with respect to the second and third variables by f_y and $f_{y'}$ respectively. Equation (24) is known as the **Euler equation** after Leonhard Euler(1707 − 1783).

Proof. Let $\eta(x)$ be a twice continuously differentiable function satisfying $\eta(x_1) = 0 = \eta(x_2)$. If $I(\varepsilon)$ is defined by (22), then $I'(0) = 0$ since y is an extremal. Using Theorem 5.2.4, we can compute $I'(\varepsilon)$ by differentiating under the integral sign and using the chain rule:

$$I'(\varepsilon) \;=\; \int_{x_1}^{x_2} f_y(x, y(x) + \varepsilon\eta(x)y'(x) + \varepsilon\eta'(x))\eta(x)\,dx$$

$$+ \int_{x_1}^{x_2} f_{y'}(x, y(x) + \varepsilon\eta(x), y'(x) + \varepsilon\eta'(x))\eta'(x)\,dx.$$

Thus,

$$I'(0) \;=\; \int_{x_1}^{x_2} f_y(x, y(x), y'(x))\eta(x) + f_{y'}(x, y(x), y'(x))\eta'(x)\,dx \qquad (25)$$

$$= \int_{x_1}^{x_2} \{f_y(x, y(x), y'(x)) - \frac{d}{dx} f_{y'}(x, y(x), y'(x))\}\eta(x)\,dx, \qquad (26)$$

where we integrated by parts in the second term and used the assumption that η vanishes at the endpoints. Therefore, if $y(x)$ is an extremal,

$$\int_{x_1}^{x_2} \{f_y(x, y(x), y'(x)) - \frac{d}{dx} f_{y'}(x, y(x), y'(x))\}\eta(x)\,dx \;=\; 0$$

for all twice continuously differentiable functions $\eta(x)$ vanishing at the endpoints. Since $y(x)$ is twice continuously differentiable by assumption, the expression in brackets is continuous and it follows easily (problem 1) that the expression in brackets must be identically zero.   ❏

Example 1 (revisited)   In Example 1, $f(x, y, y') = \sqrt{1 + (y')^2}$. Calculating the derivatives in Euler's equation and rearranging, we find

$$f_y(x, y(x), y'(x)) - \frac{d}{dx} f_{y'}(x, y(x), y'(x)) \;=\; \frac{d}{dx} \frac{y'}{(1 + (y')^2)^{\frac{1}{2}}}$$

$$= \frac{y''}{(1 + (y')^2)^{\frac{3}{2}}}.$$

Thus, if $y(x)$ is an extremal, then $y''(x) = 0$ for all $x \in [x_1, x_2]$. Therefore, $y(x)$ is a straight line, and since it must pass through (x_1, y_1) and (x_2, y_2) the function $y(x)$ is determined. So far so good. The only extremal is the straight line between the points. But how do we know whether the extremal is a minimum, a maximum, or neither? Even if we can show that it is a local minimum (it minimizes J over all curves through the points that are close to $y(x)$), how do we know it is a global minimum? In general, such questions are very difficult, but in this case we can answer them directly because of the simplicity of the functional. Let κ be the slope of the extremal $y(x)$. Then

$$I(\varepsilon) = \int_{x_1}^{x_2} \sqrt{1 + (\kappa + \varepsilon\eta'(x))^2} \, dx.$$

A short calculation (problem 2) shows that

$$I''(\varepsilon) = \int_{x_1}^{x_2} \frac{(\eta'(x))^2}{(1 + (\kappa + \varepsilon\eta'(x))^2)^{\frac{3}{2}}} \, dx,$$

and so $I''(\varepsilon)$ is strictly positive for all ε if $\eta(x)$ is not the zero function on $[x_1, x_2]$. It follows from the Fundamental Theorem of Calculus that $I'(\varepsilon) > 0$ if $\varepsilon > 0$ and $I'(\varepsilon) < 0$ if $\varepsilon < 0$. From this, it follows that $I(\varepsilon) > I(0)$ if $\varepsilon \neq 0$. Since every function whose graph goes through (x_1, y_1) and (x_2, y_2) can be written in the form $y(x) + \eta(x)$ for an $\eta(x)$ that vanishes at x_1 and x_2, we conclude that the straight line is the absolute global minimum of the length functional J over the whole class of twice differentiable functions whose graphs go through (x_1, y_1) and (x_2, y_2).

The Euler equation is normally a second-order differential equation. However, if the integrand f of the functional J does not depend on x explicitly (as in Examples 1 and 2), then we can find a first-order differential equation satisfied y.

$$\frac{d}{dx}(y'f_{y'} - f) = y''f_{y'} + f_{y'y}(y')^2 + f_{y'y'}y'y'' - f_y y' - y''f_{y'}$$

$$= (-y')(f_y - \frac{d}{dx}f_{y'})$$

$$= 0.$$

Thus

$$f_{y'}y' - f = C \tag{27}$$

for some constant C. Since (27) is a first-order differential equation, it is usually easier to solve.

Example 2 (revisited)   For the Brachistochrone,

$$f(y, y') = \frac{\sqrt{1 + (y'(x))^2}}{\sqrt{-2gy(x)}}.$$

So, using (27) and carrying out the differentiations, we compute that

$$\frac{(y')^2}{\sqrt{1 + (y'(x))^2}\sqrt{-2gy(x)}} - \frac{\sqrt{1 + (y'(x))^2}}{\sqrt{-2gy(x)}} = C.$$

Rearranging and squaring both sides, we find that any extremal $y(x)$ must satisfy the differential equation

$$y(x)(1 + (y'(x))^2) = -\frac{1}{2gC^2}.$$

The method for solving this differential equation, which requires a special change of variables, is outlined in project 4.

Problems

1. Suppose that $H(x)$ is a continuous function on the interval $[x_1, x_2]$ and that

$$\int_{x_1}^{x_2} H(x)\eta(x)\, dx = 0 \tag{28}$$

for all twice continuously differentiable functions $\eta(x)$ that vanish at the endpoints. Prove that $H(x) \equiv 0$ in the interval as follows:

   (a) Let $[a, b]$ be any finite interval. Show how to construct a twice continuously differentiable function on $\mathbb{R}$ which is strictly positive on the open interval (a, b) and identically zero everywhere else. Hint: use pieces of polynomials.

   (b) If x_o is a point of $[x_1, x_2]$ such that $H(x_o) \neq 0$, show how to choose $\eta(x)$ so that hypothesis (28) is violated.

2. Carry out the second derivative calculation in Example 1, revisited.

3. Find a curve passing through $(1, 2)$ and $(2, 4)$ that is an extremal for the functional

$$J(x, y') = \int_1^2 xy'(x) + (y'(x))^2\, dx.$$

4. Find a curve passing through $(1, 1)$ and $(2, 2)$ that is an extremal for the functional

$$J(x, y') = \int_1^2 \frac{(y')^2}{x^3} \, dx.$$

5. Find a curve passing through $(0, 0)$ and $(1, 1)$ that is an extremal for the functional

$$J(x, y, y') = \int_0^1 (y'(x))^2 + 12xy(x) \, dx.$$

6. Find a curve passing through $(0, 0)$ and $(1, 1)$ that is an extremal for the functional

$$J(y, y') = \int_0^1 (y'(x))^2 + 2y(x) \, dx.$$

7. Find a curve passing through $(0, 0)$ and $(\frac{\pi}{2}, 1)$ that is an extremal for the functional

$$J(y, y') = \int_0^{\frac{\pi}{2}} (y'(x))^2 - (y(x))^2 \, dx.$$

8. Let $y(x)$ be a twice differentiable function whose graph passes through the points (x_1, y_1) and (x_2, y_2) in the plane, where $y_1 > 0$ and $y_2 > 0$.

   (a) Show that the surface area generated when the curve is revolved around the x-axis is given by (project 2(d) of Chapter 3)

   $$J(y, y') = 2\pi \int_{x_1}^{x_2} y(x)\sqrt{1 + (y'(x))^2} \, dx.$$

   (b) Show that if $y(x)$ is an extremal, then $y(x)$ satisfies

   $$\frac{y(x)}{\sqrt{1 + (y'(x))^2}} = C_1.$$

   (c) Show that the functions $y(x) = C_2 \cosh\left(\frac{x - C_1}{C_2}\right)$ satisfy the differential equation for all choices of C_1 and C_2.

   (d) Show that C_1 and C_2 can be chosen so that the graph of y passes through (x_1, y_1) and (x_2, y_2).

9. Explain why you would not expect the functional

   $$J(y) = \int_{x_1}^{x_2} (y(x))^2 \, dx$$

   to have a minimum in the class of twice continuously differentiable functions whose graphs go through the points (x_1, y_1) and (x_2, y_2).

5.6   Metric Spaces

There is a very strong similarity between the way we studied convergent sequences of numbers in Chapter 2 and the way we studied convergent sequences of functions in Sections 5.1 - 5.3. For real numbers $x \in \mathbb{R}$, we have a notion of "size," $|x|$. This allowed us to define the distance between two numbers x and y as $|x - y|$. Using this notion of distance, we then defined what it means for a sequence $\{a_n\}$ to converge to a limit a. Namely, given $\varepsilon > 0$, there is an N so that $n \geq N$ implies that $|a_n - a| \leq \varepsilon$. We then introduced the idea of Cauchy sequence and the idea of completeness, and the rest, so to speak, is history.

For continuous functions on a finite interval $[a, b]$, we also have a notion of "size," $\|f\|_\infty = \sup_x |f(x)|$. The sup norm allows us to define the distance between two functions f and g as $\|f - g\|_\infty$. In Section 5.3 we saw that a sequence of functions $\{f_n\}$ converges uniformly to a limit function f if and only if, given $\varepsilon > 0$, there is an N so that $n \geq N$ implies that $\|f - f_n\|_\infty \leq \varepsilon$. We then introduced the idea of Cauchy sequence and the idea of completeness, and we showed that the set of continuous functions, $C[a, b]$, is complete.

Although the two sets of objects, $\mathbb{R}$ and $C[a, b]$, are very different, the progression of ideas is very similar. In both cases, the definition of "convergence" depends on having a way to measure the "distance" between two elements of the set. This suggests that there is a general idea here which is worth studying. We begin by saying what properties distance functions, called metrics, should have.

Definition.   If $\mathcal{M}$ is a set, a function ρ from $\mathcal{M} \times \mathcal{M}$ to $[0, \infty)$ is called a **metric** if

(a) for all x and y in $\mathcal{M}$, $\rho(x, y) \geq 0$ and $\rho(x, y) = 0$ if and only if $x = y$.

(b) $\rho(x, y) = \rho(y, x)$

(c) for any three points x, y, and z in $\mathcal{M}$, $\rho(x, z) \leq \rho(x, y) + \rho(y, z)$.

The pair $(\mathcal{M}, \rho)$ is called a **metric space**.

The three conditions are very intuitive. The first statement says that the "distance" between distinct points is always positive. The second says that the distance from x to y is always the same as the distance from y to x. The third says that the distance from x to z is less than or equal to the distance from x to y plus the distance from y to z. This idea is very familiar from the usual idea of distance in the plane. See Figure 5.6.1. In the spaces that we have previously considered, we shall see that property (c) is really just the triangle inequality .

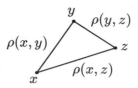

Figure 5.6.1

Example 1 Let $M = \mathbb{R}$ and define $\rho(x, y) \equiv |x - y|$. Then properties (a) and (b) are immediate since the absolute value function is even and the absolute values of all numbers except zero are positive. Since

$$
\begin{aligned}
\rho(x, z) = |x - z| &= |(x - y) + (y - z)| \\
&\leq |x - y| + |y - z| \\
&= \rho(x, y) + \rho(y, z),
\end{aligned}
$$

we see that property (c) holds too. Thus ρ is a metric on $\mathbb{R}$. Note that the crucial step in proving (c) was the triangle inequality.

Example 2   Let $M = C[a, b]$ and define $\rho_\infty(f, g) \equiv \|f - g\|_\infty$. If $\rho_\infty(f, g) = 0$, then $|f(x) - g(x)| = 0$ for all $x \in [a, b]$. Thus the functions f and g are identical. As is Example 1, property (b) holds because absolute value is an even function. Let f, g, and h be any three functions in $C[a, b]$. Then,

$$
\begin{aligned}
\rho_\infty(f, g) = \|f - g\|_\infty &= \|(f - h) + (h - g)\|_\infty \\
&\leq \|f - h\|_\infty + \|h - g\|_\infty \\
&= \rho_\infty(f, h) + \rho_\infty(h, g),
\end{aligned}
$$

so property (c) holds too. Thus ρ_∞ is a metric on $C[a, b]$. The crucial step in proving (c) was the triangle inequality (Proposition 5.3.1(c)).

Example 3   Let $x = (x_1, x_2, ..., x_n)$ and $y = (y_1, y_2, ..., y_n)$ be points in $\mathbb{R}^n$. We define the **Euclidean metric** on $\mathbb{R}^n$ by

$$
\rho_2(x, y) \equiv \left(\sum_{i=1}^n |x_i - y_i|^2 \right)^{\frac{1}{2}}.
$$

In the case $n = 1$, this is just the metric in Example 1. As in the above examples, properties (a) and (b) are easy to verify. In the case $n = 2$, property (c) follows easily from the triangle inequality proved in problem 10 of Section 2.2. The proof in the general case, which is somewhat harder, is given in Example 3 of Section 5.8. For the moment, we restrict our attention to $\mathbb{R}^2$, where

$$\rho_2((x_1, y_1), (x_2, y_2)) \equiv \sqrt{|x_1 - x_2|^2 + |y_1 - y_2|^2}.$$

The same set can have different metrics. In problem 1, the student is asked to verify that both

$$\rho_1((x_1, y_1), (x_2, y_2)) \equiv |x_1 - x_2| + |y_1 - y_2|$$

and

$$\rho_{\max}((x_1, y_1), (x_2, y_2)) \equiv \max\{|x_1 - x_2|, |y_1 - y_2|\}$$

are metrics on $\mathbb{R}^2$. Let us now denote a general point in $\mathbb{R}^2$ by (x, y) and ask what is the set of points (x, y) that are a distance ≤ 1 from the origin $(0, 0)$. Since the three metrics, ρ_2, ρ_1, and $\rho_{\max}$ measure distance differently, it is not surprising that the three sets,

$$\{(x, y) \in \mathbb{R}^2 \mid \rho_2((x, y), (0, 0)) \leq 1\} = \{(x, y) \in \mathbb{R}^2 \mid x^2 + y^2 \leq 1\}$$
$$\{(x, y) \in \mathbb{R}^2 \mid \rho_1((x, y), (0, 0)) \leq 1\} = \{(x, y) \in \mathbb{R}^2 \mid |x| + |y| \leq 1\}$$
$$\{(x, y) \in \mathbb{R}^2 \mid \rho_{\max}((x, y), (0, 0)) \leq 1\} = \{(x, y) \in \mathbb{R}^2 \mid \max\{|x|, |y|\} \leq 1\},$$

are different, as shown in Figure 5.6.2. Thus different metrics correspond to different geometries, which is why metrics play an important role in geometry; see, for example, [11] or [40].

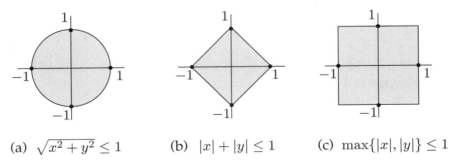

(a) $\sqrt{x^2 + y^2} \leq 1$ (b) $|x| + |y| \leq 1$ (c) $\max\{|x|, |y|\} \leq 1$

Figure 5.6.2

Notice that metric spaces are not required to have any linear structure; that is, they need not be vector spaces. No notion of addition or scalar multiplication occurs in the definition. Thus for any metric space $(\mathcal{M}, \rho)$ and any subset $\mathcal{M}_1 \subseteq \mathcal{M}$, $(\mathcal{M}_1, \rho)$ is also a metric space. For example, the set of points (x, y) in $\mathbb{R}^2$ such that $x^2 = y$ is a metric space under any of the three metrics on $\mathbb{R}^2$ discussed in Example 3. If $\mathcal{M}$ is a vector space and the vector space has a norm $\| \cdot \|$, then $\rho(x, y) \equiv \|x - y\|$ defines a metric on $\mathcal{M}$ because (c) follows from the triangle inequality for norms. Vector spaces with norms are discussed in Section 5.8.

Example 4   Let S_2 denote the set of points in $\mathbb{R}^3$ that are a distance 1 from the origin in the Euclidean metric. That is,

$$S_2 \equiv \{(x, y, z) \in \mathbb{R}^3 \mid x^2 + y^2 + z^2 = 1\}.$$

Since S_2 is a subset of $\mathbb{R}^3$, the Euclidean metric, ρ_2, is a metric on S_2. Here is a different metric. For any two points α and β in S_2, let $\rho_g(\alpha, \beta)$ be the length of the shortest curve between α and β which remains on the surface S_2. Here "length" means the Euclidean length of the curve as a curve in $\mathbb{R}^3$. These shortest curves that remain on the surface are called **geodesics**, and ρ_g is called the geodesic metric. The great circles are the circles on S_2 whose centers are $(0, 0, 0)$. Unless α and β are antipodal, there is a unique great circle through α and β. It can be shown that the geodesic between α and β is the shorter piece of the great circle through α and β. If α and β are antipodal, then there are infinitely many great circles through α and β. This does not create a problem since they all give the same geodesic distance, π. It is clear that ρ_g satisfies properties (a) and (b) in the definition of metric. Property (c) is harder to prove though it looks "obvious" if the three points are close together.

Example 5   Let $\mathcal{Q}$ be an alphabet of symbols, finite or infinite, and let N be a positive integer. Define $\mathcal{M}$ to be the set of strings of symbols of length N which can be made from the symbols in $\mathcal{Q}$. Let δ be the function on $\mathcal{Q} \times \mathcal{Q}$ such that $\delta(q, q) = 0$ and $\delta(q, p) = 1$ if $q \neq p$ for all symbols q and p in $\mathcal{Q}$. If $x = q_1 q_2 \ldots q_N$ and $y = p_1 p_2 \ldots p_N$ are two strings of symbols in $\mathcal{M}$, we define

$$\rho(x, y) \equiv \sum_{i=1}^{N} \delta(q_i, p_i). \tag{29}$$

It is not hard to show that ρ is a metric on $\mathcal{M}$ (problem 2). Metrics like ρ are useful in coding theory, where one wants to know how "close" two messages are; see Section 10.2. Here we will briefly describe why such metrics are useful in molecular biology.

Deoxyribonucleic acid (DNA) is a two-stranded polymer, each strand consisting of a sequence of four building blocks joined together linearly along a sugar-phosphate backbone. The four building blocks are the nucleotides adenine (A), thymine (T), cytosine (C), and guanine (G). Since A only binds to T and C to G, the linear sequence of A's, T's, C's and G's along one strand determines the other strand and therefore the whole DNA molecule. The length of the sequence ranges from 4.6 million for an *E. coli* bacterium to about 3 billion in human DNA. Short segments of the DNA molecule code for the production of specific proteins. Every protein consists of a linear chain of amino acids (typically 50 to 1500) selected from a fixed list of 20. Experimental techniques allow one to determine the sequence of relatively small segments of DNA molecules and the entire amino acid sequence of some proteins.

There are several reasons why one wants to compare two different linear sequences in order to say how "similar" they are. If the DNAs of two different animals are similar, then the animals are probably close to each other on the evolutionary tree. Very similar short segments of DNA probably code for similar proteins. If the sequences of amino acids in two proteins are similar, their three-dimensional structures may be similar and their functions may also be similar. The three-dimensional structures and the functions of proteins are both difficult to determine. Thus, astute comparisons of the sequence of a new protein with the sequences of well-known proteins in data bases may suggest reasonable hypotheses about structure and function.

In the case of DNA, several biological facts make the situation more complicated. First, the sequences that we wish to compare may have different lengths. We can handle this by counting as a mismatch a symbol matched with nothing. Second, DNA mutates by substitutions (a symbol replaced by another symbol), additions (a symbol placed between two there already), and deletions (a symbol removed). In the case of a deletion, we can hold the place of the deleted symbol with a new symbol "-". So, given two sequences, a natural question is to find the relative placement so that the distance between them is minimal. Or one can ask the same question but permit additions or deletions or substitutions. Even for short sequences, difficult combinatorial questions are involved. For moderately long sequences, efficient algorithms for machine computation must be devised. See project 5. The notion of "similar" and all the rest of our discussion depends of course on the metric that is chosen. Metrics with different weights, for example, setting $\delta(x, -) = \frac{1}{2}$, have also been used. For more information about the use of metrics in molecular biology, see [30].

Once one has a metric, there is a natural notion of convergence.

Definition.   A sequence of points $\{x_n\}$ in a metric space $(\mathcal{M}, \rho)$ is said to **converge** to $x \in \mathcal{M}$ if $\rho(x_n, x) \to 0$ as $n \to \infty$. The point x is called the **limit** of the sequence $\{x_n\}$, and we also write $\lim_{n \to \infty} x_n = x$ or simply $x_n \to x$.

In problems $7, 8,$ and 9, the student is asked to show that convergent sequences in metric spaces have many of the same properties as convergent sequences of real numbers. Cauchy sequences and completeness are discussed in the next section.

Does it matter what metric one uses on a set? As we indicated in Section 5.3, where we discussed different norms on $C[a, b]$, the answer is yes, in general. On the other hand, it wouldn't make much difference if we used the metric $3\rho_2$ instead of the Euclidean metric ρ_2 on $\mathbb{R}^n$. To make this idea precise, we make a definition.

Definition.   Two metrics, ρ and σ, on a set $\mathcal{M}$ are said to be **equivalent** if, for every $x \in \mathcal{M}$ and $\varepsilon > 0$, there is a $\delta > 0$ such that for all $y \in \mathcal{M}$

$$\rho(x, y) < \delta \qquad \text{implies} \qquad \sigma(x, y) < \varepsilon \tag{30}$$

and

$$\sigma(x, y) < \delta \qquad \text{implies} \qquad \rho(x, y) < \varepsilon. \tag{31}$$

Examples of equivalent and nonequivalent metrics are developed in the problems. Equivalent metrics have the same convergent sequences.

Proposition 5.6.1   Let ρ and σ be two equivalent metrics on a set $\mathcal{M}$ and suppose that $\{x_n\}$ is a sequence in $\mathcal{M}$. Then $x_n \to x$ in the metric ρ if and only if $x_n \to x$ in the metric σ.

Proof.   Suppose that $x_n \to x$ in the metric ρ. Let $\varepsilon > 0$ be given, and choose a $\delta > 0$ which satisfies (30) for this x and ε. Since $\rho(x_n, x) \to 0$, we can choose an N so that $n \geq N$ implies that $\rho(x_n, x) < \delta$. By (30),

$n \geq N$ therefore implies that $\sigma(x_n, x) < \varepsilon$. Thus, $x_n \to x$ in the metric σ. The proof of the converse is the same. ❏

Problems

1. Prove that the functions ρ_1 and ρ_{max} defined in Example 3 are indeed metrics on $\mathbb{R}^2$.

2. Prove that the function ρ defined in Example 5 is a metric.

3. Prove that $\rho(f, g) \equiv \int_a^b |f(x) - g(x)|\, dx$ is a metric on $C[a, b]$.

4. Suppose that ρ is a metric on $\mathcal{M}$. Prove that the following are also metrics:

    (a) $\rho_1 \equiv 5\rho$.

    (b) $\rho_2 \equiv \min\{1, \rho\}$.

5. Suppose that ρ_1 and ρ_2 are metrics on $\mathcal{M}$. Prove that the following are also metrics:

    (a) $\rho \equiv \rho_1 + \rho_2$.

    (b) $\rho_2 \equiv \max\{\rho_1, \rho_2\}$.

6. Prove that $\rho(x, y) \equiv \left| \frac{1}{x} - \frac{1}{y} \right|$ is a metric on $(0, \infty)$.

7. Let $(\mathcal{M}, \rho)$ be a metric space and suppose that $\{x_n\}$ is a sequence in $(\mathcal{M}, \rho)$ so that $\lim_{n \to \infty} x_n = x$ and $\lim_{n \to \infty} x_n = y$. Prove that $x = y$.

8. Suppose that $x_n \to x$ and $y_n \to y$ in metric space $(\mathcal{M}, \rho)$. Prove that $\lim_{n \to \infty} \rho(x_n, y_n) = \rho(x, y)$.

9. Let $(\mathcal{M}, \rho)$ be a metric space. A point $d \in \mathcal{M}$ is called a **limit point** of a sequence $\{x_n\}$ if for every $\varepsilon > 0$ there is an n so that $\rho(d, x_n) \leq \varepsilon$. Prove that d is a limit point of $\{x_n\}$ if and only if $\{x_n\}$ has a subsequence which converges to d.

10. A metric space $(\mathcal{M}, \rho)$ is said to be **discrete** if for every $x \in \mathcal{M}$ there is an $\varepsilon > 0$ so that $\rho(x, y) < \varepsilon$ implies $y = x$.

    (a) Define a function δ on $\mathbb{R}$ by $\delta(x, x) = 0$ and $\delta(x, y) = 1$ if $x \neq y$. Prove that $(\mathbb{R}, \delta)$ is discrete.

    (b) Prove that $(\mathbb{R}, \rho_2)$ is not discrete.

    (c) Which of the metrics in Examples $1 - 5$ are discrete?

    (d) In a discrete metric space $(\mathcal{M}, \rho)$, what are the convergent sequences?

11. Two metrics, ρ and σ, on a set $\mathcal{M}$ are said to be **uniformly equivalent** if there exist positive constants c_1 and c_2 such that

$$c_1\rho(x,y) \leq \sigma(x,y) \leq c_2\rho(x,y)$$

for all x and y in $\mathcal{M}$. Prove that if ρ and σ are uniformly equivalent, then they are equivalent.

12. Prove that the metrics ρ_1, $\rho_{\max}$, and ρ_2 defined in Example 3 are uniformly equivalent.

13. Are the metrics ρ_∞ of Example 2 and ρ of problem 3 equivalent?

14. Let ρ be the function defined on $\mathbb{R} \times \mathbb{R}$ by

$$\rho(x,y) = \frac{|x-y|}{1+|x-y|}.$$

    (a) Prove that ρ is a metric.

    (b) Prove that ρ is equivalent to the Euclidean metric ρ_2.

    (c) Prove that ρ is not uniformly equivalent to ρ_2.

15. Let ψ be a continuous function on $[a,b]$. Define a function ρ_ψ on $C[a,b] \times C[a,b]$ by

$$\rho_\psi(f,g) \equiv \int_a^b \psi(x)|f(x)-g(x)|\,dx.$$

    (a) Prove that ρ_ψ is a metric on $C[a,b]$ if $\psi(x) > 0$ for all $x \in [a,b]$.

    (b) Explain why the condition $\psi(x) \geq 0$ is not enough to guarantee that ρ_ψ is a metric on $C[a,b]$.

    (c) Suppose that ψ and ϕ are continuous functions on $[a,b]$ which satisfy $\psi(x) > 0$ and $\phi(x) > 0$ for all x. Prove that the metrics ρ_ψ and ρ_ϕ are uniformly equivalent.

5.7 The Contraction Mapping Principle

If a set has a metric, then the notions of Cauchy sequence and completeness can be defined analogously to their definitions on $\mathbb{R}$.

Definition. A sequence of points $\{x_n\}$ in a metric space $(\mathcal{M},\rho)$ is said to be a **Cauchy sequence** if, given $\varepsilon > 0$, there is an N so that $\rho(x_n,x_m) \leq \varepsilon$ for all $n \geq N$ and $m \geq N$.

Definition.   If every Cauchy sequence in a metric space $(\mathcal{M}, \rho)$ has a limit in $(\mathcal{M}, \rho)$, we say that $\mathcal{M}$ is **complete** in the metric ρ and refer to the pair $(\mathcal{M}, \rho)$ as a **complete metric space**.

As we have seen in many special cases, it is easy to prove that a convergent sequence is automatically a Cauchy sequence (problem 1). It is the converse, completeness, which is crucial for the constructions of analysis. The main objects of calculus, derivatives, integrals, and series, are defined by limiting operations and therefore depend heavily on the completeness of $\mathbb{R}$. The proof of the Bolzano-Weierstrass theorem, which is the main tool for proving the properties of continuous functions in Section 3.2, also depends on the completeness of $\mathbb{R}$. We proved in Section 5.3 that $C[a, b]$ is complete in the sup norm, and this fact played a central role in the construction of solutions of integral equations in Section 5.4. The completeness of $C[a, b]$ will also be used in the construction of local solutions of differential equations in Section 7.1.

Example 1   Let $\mathcal{M} \equiv \{(x, y) \in \mathbb{R}^2 \mid y = x^2\}$ with the Euclidean metric, ρ_2. Since ρ_2 is a metric on $\mathbb{R}^2$, it is automatically a metric on $\mathcal{M}$ because $\mathcal{M}$ is a subset of $\mathbb{R}^2$. Let $\{p_n\}$ be a Cauchy sequence in $(\mathcal{M}, \rho_2)$. Then, $\{p_n\}$ is a Cauchy sequence in $\mathbb{R}^2$ and therefore has a limit $p = (x, y)$ in $\mathbb{R}^2$ since $\mathbb{R}^2$ is complete (problem 8 of Section 2.4). If $p_n = (x_n, y_n)$, then, by problem 10(c) of Section 2.2, we know that $x_n \to x$ and $y_n \to y$. Thus, by Theorem 2.2.5,

$$y = \lim_{n \to \infty} y_n = \lim_{n \to \infty} x_n^2 = \left(\lim_{n \to \infty} x_n\right)^2 = x^2.$$

Therefore, the limit point (x, y) is in $\mathcal{M}$, so $(\mathcal{M}, \rho_2)$ is complete.

Example 2   Let $\mathcal{M} \equiv \{f \in C[a, b] \mid f(x) \geq 0 \text{ for } x \in [a, b]\}$ with the metric ρ_∞, which comes from the sup norm. Since ρ_∞ is a metric on $C[a, b]$, it is a metric on $\mathcal{M}$. Let $\{f_n\}$ be a Cauchy sequence in $\mathcal{M}$. Since $\{f_n\}$ is a Cauchy sequence in $C[a, b]$ and $C[a, b]$ is complete (Theorem 5.3.2), there is a continuous function f on $[a, b]$ such that $f_n \to f$ uniformly. For each x, $f_n(x) \geq 0$ for all n, and this implies that $f(x) = \lim_{n \to \infty} f_n(x) \geq 0$. Therefore $f \in \mathcal{M}$, so $(\mathcal{M}, \rho_\infty)$ is complete.

Many other examples of complete and incomplete metric spaces are given in the problems. We turn now to an important theorem whose proof uses completeness.

A function T from a metric space $(\mathcal{M}, \rho)$ to itself is called a **contraction** if there is an α which satisfies $0 \le \alpha < 1$ so that

$$\rho(T(x), T(y)) \le \alpha \rho(x, y) \qquad \text{for all } x, y \in \mathcal{M}.$$

Thus a contraction shrinks the distance between points. The importance of this concept is that on complete metric spaces contractions have **fixed points**, that is, points x such that $T(x) = x$.

❏ **Theorem 5.7.1 (The Contraction Mapping Principle)** Let T be a contraction on a complete metric space $(\mathcal{M}, \rho)$. Then there is a unique point $x \in \mathcal{M}$ such that $T(x) = x$. Furthermore, if x_0 is any point in $\mathcal{M}$ and we define $x_{n+1} = T(x_n)$, then $x_n \to x$ as $n \to \infty$.

Proof. Uniqueness is easy, for suppose that there are two fixed points, x and $\bar{x}$. Then,

$$\rho(x, \bar{x}) = \rho(T(x), T(\bar{x})) \le \alpha \rho(x, \bar{x}),$$

which implies that $\rho(x, \bar{x}) = 0$ since $\alpha < 1$. Thus, by the definition of metric, $x = \bar{x}$. To prove existence, we first show that for any $x_0 \in \mathcal{M}$, the sequence $\{x_n\}$ defined by $x_{n+1} = T(x_n)$ is a Cauchy sequence. Since

$$\rho(x_{n+1}, x_n) = \rho(T(x_n), T(x_{n-1})) \le \alpha \rho(x_n, x_{n-1}),$$

iteration shows that $\rho(x_{n+1}, x_n) \le \alpha^n \rho(x_1, x_0)$. Therefore, by the triangle inequality, if $m > n$,

$$\begin{aligned}
\rho(x_m, x_n) &\le \rho(x_m, x_{m-1}) + \rho(x_{m-1}, x_{m-2}) + \ldots + \rho(x_{n+1}, x_n) \\
&\le \left(\sum_{j=n}^{m-1} \alpha^j \right) \rho(x_1, x_0).
\end{aligned}$$

Since $\alpha < 1$, the geometric series $\sum \alpha^j$ converges. Now, if $\rho(x_1, x_0) = 0$, then x_0 is a fixed point and we are done. Otherwise, $\rho(x_1, x_0) > 0$, and, given $\varepsilon > 0$, we can choose an N so that $\sum_{j=n}^{m-1} \alpha^j \le \varepsilon \rho(x_1, x_0)^{-1}$ if $n \ge N$ and $m \ge N$. For such n and m, $\rho(x_m, x_n) \le \varepsilon$, which proves that $\{x_n\}$ is a Cauchy sequence. Since $\mathcal{M}$ is complete, there is an $x \in \mathcal{M}$ so that $x_n \to x$. Finally, by the triangle inequality,

$$\begin{aligned}
\rho(T(x), x) &\le \rho(T(x), T(x_n)) + \rho(T(x_n), x_n) + \rho(x_n, x) \\
&\le \alpha \rho(x, x_n) + \rho(x_{n+1}, x_n) + \rho(x_n, x) \\
&\le \alpha \rho(x, x_n) + \rho(x_{n+1}, x) + \rho(x, x_n) + \rho(x_n, x).
\end{aligned}$$

Since $x_n \to x$, all the terms on the right converge to zero. Therefore, we conclude that $\rho(T(x), x) = 0$, and so, by the definition of metric, $T(x) = x$. ❏

Example 3   The contraction mapping principle can be used to shorten the proof of the existence of solutions of integral equations in Section 5.4. We take $C[a, b]$ with the metric ρ_∞ as our metric space and define a function T on $C[a, b]$ by

$$T(\psi)(x) \;=\; f(x) \;+\; \int_a^b K(x, y)\psi(y)dy. \tag{32}$$

As in the proof of Theorem 5.4.1, we prove that $T(\psi)(x)$ is continuous on $[a, b]$ if ψ is. This shows that T is a function from $\mathcal{M}$ to $\mathcal{M}$. Again, as in that proof, we estimate

$$|T(\varphi)(x) - T(\psi)(x)| \;\leq\; |\lambda| \int_a^b |K(x, y)||\varphi(y) - \psi(y)|dy$$

$$\leq\; |\lambda| M(b - a)\|\varphi - \psi\|_\infty.$$

Taking the supremum of the left-hand side, we see that

$$\|T(\varphi) - T(\psi)\|_\infty \;\leq\; \alpha\|\varphi - \psi\|_\infty,$$

where $\alpha = |\lambda| M(b-a)$. If $\alpha < 1$, the contraction mapping principle guarantees that there is a unique continuous function ψ such that $T(\psi) = \psi$. Replacing $T(\psi)$ by ψ in (32) shows that ψ solves the integral equation. Thus, the contraction mapping principle eliminates the need for the convergence part of the proof of Theorem 5.4.1.

Let f be a real-valued function on $\mathbb{R}$ and consider the iteration

$$x_{n+1} \;=\; f(x_n),$$

which starts with a given point x_0. A point $\bar{x}$ is called a **fixed point** of the iteration (or of f) if $f(\bar{x}) = \bar{x}$. A fixed point is called **stable** if there is a $\delta > 0$ so that $x_0 \in [\bar{x} - \delta, \bar{x} + \delta]$ implies that $x_n \to \bar{x}$. In Section 2.7, we investigated an iteration called the quadratic map and proved that some of its fixed points are stable.

❏ **Theorem 5.7.2**   Suppose that f is continuously differentiable in an open interval about a fixed point $\bar{x}$. If $|f'(\bar{x})| < 1$, then $\bar{x}$ is a stable fixed point.

Proof.   Since f' is continuous, there is a $\beta < 1$ and a $\delta > 0$ such that $|f'(x)| < \beta < 1$ for all $x \in [\bar{x} - \delta, \bar{x} + \delta]$. Let $\mathcal{M} = [\bar{x} - \delta, \bar{x} + \delta]$ with the Euclidean metric. If $x \in \mathcal{M}$,

$$
\begin{aligned}
|f(x) - \bar{x}| &= |f(x) - f(\bar{x})| \\
&= |f'(c)||x - \bar{x}| \\
&\leq \beta|x - \bar{x}|
\end{aligned}
$$

by the Mean Value Theorem. This proves that $f(x)$ is closer to $\bar{x}$ than x, so f takes $\mathcal{M}$ into itself. Using the Mean Value Theorem again, we see that if x and y are in $\mathcal{M}$, then

$$
|f(x) - f(y)| \leq \beta|x - y|,
$$

which shows that f is a contraction on $\mathcal{M}$ since $\beta < 1$. It follows from the contraction mapping principle that $\bar{x}$ is the only fixed point in $\mathcal{M}$ and whatever x_0 we choose in $\mathcal{M}$, the sequence of points defined by the iteration $x_{n+1} = f(x_n)$ will converge to $\bar{x}$. Thus $\bar{x}$ is a stable fixed point. ❏

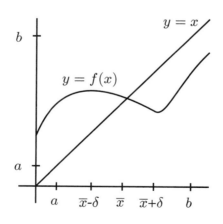

Figure 5.7.1

Example 4   Let's see what information Theorem 5.7.2 gives us about the quadratic map,

$$
f(x) \equiv ax(1 - x),
$$

in the case $1 < a < 3$. It is easy to check that the only fixed points in the interval $[0, 1]$ are the points 0 and $\bar{x} = \frac{a-1}{a}$. After calculating that $f'(x) = a(1 - 2x)$, we substitute $\bar{x}$ and find $f'(\bar{x}) = 2 - a$. Since $1 < a < 3$, we see that $|f(\bar{x})| < 1$, so the hypothesis of Theorem 5.7.2 is satisfied. Therefore,

$\bar{x}$ is stable, and if we start close enough to $\bar{x}$, the iteration will converge to $\bar{x}$. Notice that this gives us some information quite easily. Theorem 2.7.2, whose proof is quite long and difficult, shows much more. For *any* $x_0 \in (0, 1)$, the sequence $\{x_n\}$ converges to $\bar{x}$.

The contraction mapping principle shows the importance of completeness. But what if we are working in a space $\mathcal{M}$ that is not complete, for example, $C[a, b]$ with the L^1 or L^2 norm? There is a theorem which says that any metric space can be enlarged to become complete. To say exactly what this means, we need two definitions. First, a set $\mathcal{D} \subseteq \mathcal{M}$ is said to be **dense** in $\mathcal{M}$ if, given any $\varepsilon > 0$ and any $x \in \mathcal{M}$, there is a $y \in \mathcal{D}$ such that $\rho(x, y) \leq \varepsilon$. Second, a function T from one metric space $(\mathcal{M}_1, \rho_1)$ into another $(\mathcal{M}_2, \rho_2)$ is called an **isometry** if $\rho_2(T(x), T(y)) = \rho_1(x, y)$ for all $x, y \in \mathcal{M}_1$. If the range of T is all of $\mathcal{M}_2$ then $\mathcal{M}_1$ and $\mathcal{M}_2$ are said to be **isometric**. Isometric metric spaces are identical as far as their metric space properties are concerned.

❏ **Theorem 5.7.3**   Let $(\mathcal{M}_1, \rho_1)$ be a metric space. Then, there is a complete metric space $(\mathcal{M}_2, \rho_2)$ and an isometry from $\mathcal{M}_1$ into $\mathcal{M}_2$ such that the range of T is dense in $\mathcal{M}_2$.

Theorem 5.7.3 says that in terms of metric space properties $(\mathcal{M}_1, \rho_1)$ can be identified with a dense subset of a complete metric space $(\mathcal{M}_2, \rho_2)$. Though Theorem 5.7.3 says that every metric space can be "enlarged" to become complete, it is not very useful since in practice one wants to be able to characterize the added points. For example, what are the added points if one completes $C[a, b]$ in the L^2 norm? Problem 12 gives an example of an incomplete space for which one can characterize the completion.

Problems

1. Show that a convergent sequence in a metric space is a Cauchy sequence.

2. Let $(\mathcal{M}, \rho)$ be a metric space and suppose that $\{x_n\}$ and $\{y_n\}$ are Cauchy sequences in $(\mathcal{M}, \rho)$. Prove that $\lim_{n \to \infty} \rho(x_n, y_n)$ exists.

3. Which of the following subsets of $\mathbb{R}$ are complete metric spaces with the Euclidean metric? (a) $[-1, 6]$, (b) $[0, \infty)$, (c) $(0, \infty)$, (d) $\mathbb{Q}$, (e) $\mathbb{N}$.

4. Which of the following subsets of $\mathbb{R}^2$ are complete metric spaces with the Euclidean metric?

(a) $\{(x, y) \in \mathbb{R}^2 \mid x^2 + y^2 < 1\}$.

(b) $\{(x, y) \in \mathbb{R}^2 \mid x \geq 1 \text{ and } y \leq -2\}$.

(c) $\{(x, y) \in \mathbb{R}^2 \mid y \in \mathbb{N}\}$.

(d) $\{(x, y) \in \mathbb{R}^2 \mid f(x, y) = 0\}$, where f is continuous on $\mathbb{R}^2$.

5. Which of the following subsets of $C[a, b]$ are complete metric spaces with the metric ρ_∞?

   (a) $\{f \in C[a, b] \mid f(x) > 0 \text{ for } x \in [a, b]\}$.

   (b) $\{f \in C[a, b] \mid f(a) = 0\}$.

   (c) $\{f \in C[a, b] \mid f(x) = 0 \text{ for } a < c \leq x \leq d < b\}$.

   (d) $\{f \in C[a, b] \mid |f(x)| \leq 2 + f(x)^2 \text{ for } x \in [a, b]\}$.

6. Give an example to show that a discrete metric space may not be complete.

7. Consider the integral equation (15), where $f \in C[a, b]$ and K is continuous on the square $[a, b] \times [a, b]$. Suppose that f and K are non-negative. Prove that the solution ψ is nonnegative. Hint: use the metric space in Example 2.

8. Show how to use the contraction mapping principle to provide an easier proof that the nonlinear integral equation in project 2 has a unique solution.

9. For $b > 0$, show that $(0, b)$ is complete in the metric $\rho(x, y) = \left| \frac{1}{x} - \frac{1}{y} \right|$.

10. (a) Prove that if ρ and σ are uniformly equivalent metrics on $\mathcal{M}$, then $(\mathcal{M}, \rho)$ is complete if and only if $(\mathcal{M}, \sigma)$ is complete.

    (b) Suppose that ρ and σ are equivalent metrics on $\mathcal{M}$. Show by example that it is possible that $(\mathcal{M}, \rho)$ is complete but $(\mathcal{M}, \sigma)$ is not complete. Hint: see problem 9.

11. Let $(\mathcal{M}_i, \rho_i)$, $i = 1, \ldots, N$, be a finite collection of complete metric spaces. Let $\mathcal{M}$ be the product of the spaces $\mathcal{M}_i$; that is, $\mathcal{M}$ consists of the N-tuples $(x_1, x_2, \ldots, x_N)$ with $x_i \in \mathcal{M}_i$ for each i. For two such N-tuples, $x = (x_1, x_2, \ldots, x_N)$ and $y = (y_1, y_2, \ldots, y_N)$, define

$$\rho(x, y) \equiv \sum_{i=1}^{N} \rho_i(x_i, y_i).$$

Prove that $(\mathcal{M}, \rho)$ is a complete metric space.

12. Let $\mathcal{M}$ be the set of continuous functions on $\mathbb{R}$ which vanish outside a finite interval (the interval may depend on the function).

    (a) Show that $\mathcal{M}$ is a metric space in the sup norm.

    (b) Show that $\mathcal{M}$ is not complete.

(c) Show that $C_o(\mathbb{R})$, the continuous functions which go to zero at ∞, is complete in the sup norm (problem 10 of Section 5.3).

(d) Prove that $\mathcal{M}$ is dense in $C_o(\mathbb{R})$.

13. (a) Let $\mathcal{M}$ be the circle of radius 1 with the center at the origin in $\mathbb{R}^2$. Let ρ_2 be the Euclidean metric, and let ρ be the metric which assigns to two points the arc length along the circle between them (going the shorter way). Prove that these metrics are uniformly equivalent.

   (b) Assume that the geodesic metric, ρ_g, of Example 4 of Section 5.6 assigns to the points α and β on S_2 the shorter of the two great circle arcs between them. Use part (a) to show that ρ_g is uniformly equivalent to ρ_2 on S_2.

   (c) Prove that (S_2, ρ_2) is complete.

   (d) Conclude that (S_2, ρ_g) is complete.

14. In studying lateral inhibition in the retina, one is led to the following kind of model. We imagine a line of cells indexed by j for $-\infty < j < \infty$, and denote by e_j a nonnegative number representing the stimulation of the jth cell. If r_j represents the response of the jth cell, we would like to solve the family of equations

$$r_j = e_j - \frac{1}{4}(r_{j-1} + r_{j+1}).$$

   (a) Prove that for every bounded sequence $\{e_j\}$, there is a unique bounded sequence $\{r_j\}$ so that these equations hold. Hint: you will need to use the fact that ℓ_∞ is a Banach space; see Section 5.8.

   (b) How can you compute the sequence $\{r_j\}$?

5.8   Normed Linear Spaces

Throughout this text we have used many sets which have natural notions of addition and scalar multiplication. We can add real numbers; we can add pairs of real numbers. If we add two continuous functions, the result is another continuous function. We have sometimes referred to these sets informally as "vector spaces." We now make this concept precise by giving a formal definition.

Definition.   A **vector space** over the real numbers is a set V, whose elements are called **vectors**, together with operations of addition and scalar multiplication that satisfy the following rules:

1. For v and w in V, the vector $v + w$ is in V.

2. For v and w in V, $v + w = w + v$.

3. For all v, w, and u in V, $(u + v) + w = u + (v + w)$.

4. There exists a unique vector 0 in V so that $v + 0 = v$ for all $v \in V$.

5. For each $v \in V$ there exists another vector in V, denoted $-v$, so that $v + (-v) = 0$.

6. For each α in $\mathbb{R}$ and v in V, αv is in V.

7. For α and β in $\mathbb{R}$ and v in V, $(\alpha + \beta)v = \alpha v + \beta v$.

8. For α and β in $\mathbb{R}$ and v in V, $\alpha(\beta v) = (\alpha \beta)v$.

9. For α in $\mathbb{R}$ and v and w in V, $\alpha(v + w) = \alpha v + \alpha w$.

Example 1   The set $\mathbb{R}^n$, consisting of all n-tuples of real numbers, $(x_1, x_2, ..., x_n)$, is a vector space which is familiar from linear algebra. Addition and scalar multiplication are defined by

$$(x_1, x_2, \ldots, x_n) + (y_1, y_2, \ldots, y_n) = (x_1 + y_1, x_2 + y_2, \ldots, x_n + y_n)$$
$$\alpha(x_1, x_2, \ldots, x_n) = (\alpha x_1, \alpha x_2, \ldots, \alpha x_n).$$

Example 2   The set of continuous functions on an interval, $C[a, b]$, is a vector space. Suppose that W is a subset of $C[a, b]$ which contains the zero function. Then properties 2, 3, 7, 8, and 9 hold automatically because every vector in W is in $C[a, b]$ and these properties hold there. To prove properties 1, 5, and 6, one needs to verify that for all v and w in W and all α and β in $\mathbb{R}$, it is true that $\alpha v + \beta w$ is in W too. So, for example, the set of continuous functions on $[a, b]$ that vanish at a particular point x_o is a vector space since this property is preserved by linear combinations. On the other hand, the set of continuous functions on $[0, 2]$ such that $f(1) = 5$ is not a vector space.

Throughout the text we have used the absolute value $|x|$ to measure the size of a real numbers and in Section 5.3 we introduced the sup norm $\|f\|_\infty$ to measure the size of bounded functions. Notice that the properties of the absolute value in Proposition 1.1.2 and the properties of $\| \cdot \|_\infty$ in Proposition 5.3.1 are the same. Other properties of $| \cdot |$ and $\| \cdot \|_\infty$ are also very similar; compare, for example, problem 10 in Section 1.1 with problem 2(a) in Section 5.3. This suggests that the absolute value and the sup norm are special cases of a more general idea.

Definition. Let V be a vector space over the real numbers. A function, $\|\cdot\|$, from V to $\mathbb{R}$ is called a **norm** if it satisfies the following conditions:

  (a) $\|v\| \geq 0$ and $\|v\| = 0$ if and only if v is the zero vector in V.

  (b) For every $\alpha \in \mathbb{R}$, we have $\|\alpha v\| = |\alpha|\, \|v\|$.

  (c) For all v and w in V, $\|v + w\| \leq \|v\| + \|w\|$ (the triangle inequality).

A vector space with a norm is called a **normed linear space**.

We have already seen several examples of norms besides the absolute value and the sup norm. The Euclidean norm on $\mathbb{R}^2$ was defined in problem 10 of Section 2.2, and the L_1 norm on $C[a, b]$ was defined in Section 5.3.

Example 3   It is not hard to show that

$$\|(x_1, x_2, ..., x_n)\|_1 \equiv \sum_{j=1}^{n} |x_j|$$

is a norm on $\mathbb{R}^n$. To prove that the **Euclidean norm**

$$\|(x_1, x_2, ..., x_n)\|_2 \equiv \left(\sum_{j=1}^{n} |x_j|^2 \right)^{\frac{1}{2}}$$

satisfies the triangle inequality is a little more difficult. The case $n = 2$ was outlined in problem 10 of Section 2.2. The general case requires the Cauchy-Schwarz inequality (problem 10). Suppose that $x = (x_1, x_2, ..., x_n)$ and $y = (y_1, y_2, ..., y_n)$ are in $\mathbb{R}^n$. Then,

$$\|x + y\|_2^2 = \sum_{i=1}^{n} |x_i + y_i|^2 \tag{33}$$

$$\leq \sum_{i=1}^{n} (|x_i| + |y_i|)^2 \tag{34}$$

$$= \sum_{i=1}^{n} |x_i|^2 + 2 \sum_{i=1}^{n} |x_i||y_i| + \sum_{i=1}^{n} |y_i|^2 \tag{35}$$

$$\leq \sum_{i=1}^{n} |x_i|^2 + 2 \left(\sum_{i=1}^{n} |x_i|^2 \right)^{\frac{1}{2}} \left(\sum_{i=1}^{n} |y_i|^2 \right)^{\frac{1}{2}} + \sum_{i=1}^{n} |y_i|^2 \tag{36}$$

$$\leq \left(\left(\sum_{i=1}^{n} |x_i|^2 \right)^{\frac{1}{2}} + \left(\sum_{i=1}^{n} |y_i|^2 \right)^{\frac{1}{2}} \right)^2 \tag{37}$$

$$= (\|x\|_2 + \|y\|_2)^2. \tag{38}$$

Taking the square root of both sides proves the triangle inequality for the Euclidean norm. The Cauchy-Schwarz inequality was used in going from (35) to (36).

If $\| \cdot \|$ is a norm on a vector space V, then it is easy to see (problem 3) that

$$\rho(x, y) \equiv \|x - y\|$$

is a metric on V. Thus, the concepts of convergence, Cauchy sequence, and completeness, which we introduced in the last two sections, have meaning on V.

Definition. Let V be a vector space with a norm $\| \cdot \|$. If every Cauchy sequence in V converges to a limit in V, then V is said to be **complete** in the norm $\| \cdot \|$. A complete normed linear space is called a **Banach space** after Stephan Banach (1892 – 1945).

We already know that $\mathbb{R}$ is complete in the absolute value norm and that $C[a, b]$ is complete in the sup norm. We know from Example 1 in Section 5.3 that $C[a, b]$ is *not* complete in the L_1 norm. The proof that $\mathbb{R}^n$ is complete in the Euclidean norm is outlined in problem 6.. Here we will investigate a new space.

Let ℓ_∞ denote the set of bounded sequences, $\{a_j\}$, of real numbers. We define addition of sequences and scalar multiplication by

$$\{a_j\} + \{b_j\} \equiv \{a_j + b_j\}$$
$$\alpha\{a_j\} \equiv \{\alpha a_j\}.$$

The sequence which is all zeros is the 0 of the vector space. We define a norm on ℓ_∞ by

$$\|\{a_j\}\|_\infty \equiv \sup_j |a_j|.$$

Since $|a_j| \geq 0$ for each j, it is clear that $\|\{a_j\}\|_\infty \geq 0$ and $\|\{a_j\}\|_\infty = 0$ if and only if $a_j = 0$ for each j. Furthermore, if $\alpha \in \mathbb{R}$,

$$\|\alpha\{a_j\}\|_\infty = \|\{\alpha a_j\}\|_\infty = \sup_j \{|\alpha a_j|\} = |\alpha| \sup_j \{|a_j|\} = |\alpha| \|\{a_j\}\|_\infty.$$

Thus, properties (a) and (b) in the definition of norm hold.

If α and β are in $\mathbb{R}$ and $\{a_j\}$ and $\{b_j\}$ are in ℓ_∞, then

$$
\begin{aligned}
\|\alpha\{a_j\} + \beta\{b_j\}\|_\infty &= \|\{\alpha a_j + \beta b_j\}\|_\infty \\
&= \sup_j |\alpha a_j + \beta b_j| \\
&\leq \sup_j |\alpha||a_j| + \sup_j |\beta||b_j| \\
&= |\alpha| \sup_j |a_j| + |\beta| \sup_j |b_j| \\
&= |\alpha| \|\{a_j\}\|_\infty + |\beta| \|\{b_j\}\|_\infty. \\
&< \infty
\end{aligned}
$$

This proves that linear combinations of sequences in ℓ_∞ are again in ℓ_∞. Furthermore, if $\alpha = 1 = \beta$, this is just the triangle inequality. Thus, ℓ_∞ is a normed linear space.

❏ **Theorem 5.8.1** ℓ_∞ is a Banach space.

Proof.   Let $\{a^{(n)}\}$ be a sequence of elements of ℓ_∞. Each $a^{(n)}$ is a sequence $\{a_j^{(n)}\}_{j=1}^\infty$. Suppose that $\{a^{(n)}\}$ is a Cauchy sequence. That is, given $\varepsilon > 0$, there is an N so that

$$
\|a^{(n)} - a^{(m)}\|_\infty \leq \varepsilon \tag{39}
$$

for all $n \geq N, m \geq N$. Since

$$
\|a^{(n)} - a^{(m)}\|_\infty = \|\{a_j^{(n)} - a_j^{(m)}\}_{j=1}^\infty\|_\infty = \sup_j |a_j^{(n)} - a_j^{(m)}|,
$$

we see that

$$
|a_j^{(n)} - a_j^{(m)}| \leq \varepsilon \tag{40}
$$

if $n \geq N, m \geq N$, for each fixed j. Thus, each of the sequences of components $\{a_j^{(n)}\}_{n=1}^\infty$ is a Cauchy sequence of real numbers and therefore converges to a real number a_j. By problem 4(a),

$$
|\, \|a^{(n)}\|_\infty - \|a^{(n)}\|_\infty \,| \leq \|a^{(n)} - a^{(m)}\|_\infty,
$$

so because of (39), $\{\|a^{(n)}\|_\infty\}$ is a Cauchy sequence of real numbers and is therefore bounded. That is, there is an M so that $\sup_j |a_j^{(n)}| \leq M$ for all n. Thus $|a_j^{(n)}| \leq M$ for all n and all j. Since $a_j^{(n)} \to a_j$ as $n \to \infty$, this proves that $|a_j| \le M$ for all j. Therefore, the sequence $a \equiv \{a_j\}$ is bounded and is thus an element of ℓ_∞.

Finally, letting $n \to \infty$ in (40) shows that $|a_j - a_j^{(m)}| \le \varepsilon$ for each j if $m \ge N$. Thus,

$$\|a - a^{(m)}\|_\infty = \sup_j |a_j - a_j^{(m)}| \le \varepsilon$$

if $m \ge N$. This proves that $a^{(m)} \to a$ in ℓ_∞ as $m \to \infty$. Thus, ℓ_∞ is complete. ❏

If normed linear spaces are a special case of metric spaces, why do we treat them separately? The reason is that the vector space structure allows us to define the important concept of linear transformation.

Definition. A function T from a normed linear space V to a normed linear space W is called a **linear transformation** if

$$T(\alpha u + \beta v) = \alpha T(u) + \beta T(v) \tag{41}$$

for all u and v in V and α and β in $\mathbb{R}$.

Example 4 To define a function T from $\mathbb{R}^n$ to $\mathbb{R}^m$, one must specify the m components of $y = T(x)$ for each $x \in \mathbb{R}^n$. If for each i, the component y_i is a linear combination of the components x_j of x, that is,

$$y_i = a_{i1}x_1 + a_{i2}x_2 + \ldots + a_{in}x_n,$$

then T is a linear transformation. It is represented (in the standard basis) by the matrix $\{a_{ij}\}$.

Example 5 If K is a continuous function on the square $[a, b] \times [a, b]$, we can define

$$S(\psi)(x) = \int_a^b K(x, y)\psi(y)\, dy.$$

We showed in Section 5.4 that if ψ is a continuous function on $[a, b]$, then $S(\psi)$ is a continuous function on $[a, b]$. That is, S is a function from $C[a, b]$ to itself. Notice that S is a linear transformation because

$$S(\alpha\psi + \beta\phi)(x) = \int_a^b K(x, y)(\alpha\psi(y) + \beta\phi(y))\, dy$$

$$= \alpha \int_a^b K(x, y)\psi(y)\, dy + \beta \int_a^b K(x, y)\phi(y)\, dy$$

$$= \alpha S(\psi)(x) + \beta S(\phi)(x).$$

Example 6   Consider the operation of differentiation $\frac{d}{dx}$. Let $C^{(1)}[a, b]$ denote the set of continuously differentiable functions on $[a, b]$. We know from problem 7 of Section 5.3 that $C^{(1)}[a, b]$ is a Banach space with the norm $\|f\| = \|f\|_\infty + \|f'\|_\infty$. If we differentiate a function in $C^{(1)}[a, b]$ we get a continuous function. That is, $\frac{d}{dx}$ is a function from $C^{(1)}[a, b]$ to $C[a, b]$. In fact, $\frac{d}{dx}$ is a linear transformation since

$$\frac{d}{dx}(\alpha f(x) + \beta g(x)) \;=\; \alpha \frac{d}{dx} f(x) \;+\; \beta \frac{d}{dx} g(x).$$

These examples show that some of the most important objects that one wants to study are linear transformations on Banach spaces. If the Banach spaces are finite dimensional as in Example 4, the study of linear transformations is called linear algebra. If the underlying Banach spaces are infinite dimensional, as in Examples 5 and 6, the study of linear transformations is part of a branch of mathematics called functional analysis.

Problems

1. Which of the following subsets of $C[a, b]$ are vector spaces?

   (a) The continuous functions, f, which satisfy $f(a) = 1$.

   (b) $C^{(1)}[a, b]$.

   (c) The continuous functions, f, which satisfy $\int_a^b f(x)dx = 0$.

   (d) The functions $f \in C^{(2)}[a, b]$ that satisfy

   $$f''(x) \;+\; (2x^2 + 1)f'(x) \;+\; (\sin x)f(x) \;=\; 0.$$

2. Which of the following subsets of $C(\mathbb{R})$ are vector spaces?

   (a) $C_b(\mathbb{R})$, the bounded continuous functions on $\mathbb{R}$.

   (b) $C_o(\mathbb{R})$, the continuous functions that go to zero at $\pm\infty$.

   (c) The continuous functions that go to 1 at $\pm\infty$.

   (d) $C^{(1)}(\mathbb{R})$, the continuously differentiable functions on $\mathbb{R}$.

   (e) The continuous functions on $\mathbb{R}$ that vanish at $x = 5$.

   (f) The continuous functions on $\mathbb{R}$ that satisfy $|f(x)| \leq ce^{x^2}$ for some $c \in \mathbb{R}$ which can depend on f.

3. Let $\| \cdot \|$ be a norm on a vector space V. Prove that $\rho(x, y) \equiv \|x - y\|$ is a metric on V.

4. Let V be a vector space with a norm $\| \cdot \|$.

   (a) Prove that for all $v \in V$ and $w \in V$,

   $$| \|v\| - \|w\| | \leq \|v - w\|.$$

   (b) Suppose that $v_n \to v$ in V. Prove that $\|v_n\| \to \|v\|$.

5. Show that every convergent sequence in a normed linear space is a Cauchy sequence.

6. Prove that $\mathbb{R}^n$ is complete in the Euclidean norm. Hint: show that a sequence is Cauchy in $\mathbb{R}^n$ if and only if each of the sequences of components is Cauchy in $\mathbb{R}$.

7. Show that
   $$\|f\| \equiv \int_{-\infty}^{\infty} |f(x)| e^{-x^2} dx$$

   is a norm on the space of bounded continuous functions on $\mathbb{R}$.

8. Two norms, $\| \cdot \|_1$ and $\| \cdot \|_2$, are called **equivalent** if there are positive constants, c and d, so that

   $$c\|v\|_1 \leq \|v\|_2 \leq d\|v\|_1$$

   for all $v \in V$.

   (a) Prove that if $\| \cdot \|_1$ and $\| \cdot \|_2$ are equivalent, then V is complete in $\| \cdot \|_1$ if and only if V is complete in $\| \cdot \|_2$.

   (b) Prove that the $\|\cdot\|_1$ norm and the Euclidean norm $\|\cdot\|_2$ are equivalent on $\mathbb{R}^n$ by showing that

   $$\|x\|_2^2 \leq \|x\|_1^2 \leq n\|x\|_2^2.$$

   (c) Prove that the sup norm and the L_1 norm are not equivalent on $C[a, b]$.

9. Recall from linear algebra that a set of vectors $\{v_i\}_{i=1}^m$ is said to be **linearly independent** if no linear combination $\alpha_1 v_1 + \alpha_2 v_2 + \ldots + \alpha_j v_n$ is the zero vector unless $\alpha_i = 0$ for all i. A vector space V is said to have **dimension** N if every set of N independent vectors $\{v_i\}_{i=1}^N$ spans V; that is, every vector in V can be written as a linear combination of the v_i. If V has dimension N for some N, V is said to be **finite dimensional**.

   (a) Show that $\mathbb{R}^n$ has dimension n.

   (b) Show that $\{1, x, x^2, \ldots, x^n\}$ is an independent set of vectors in $C[a, b]$ for each n.

(c) Prove that $C[a, b]$ is not finite dimensional.

10. Let $\{x_i\}_{i=1}^N$ and $\{y_i\}_{i=1}^N$ be real numbers not all zero. Define a quadratic, $p(\lambda)$, by

$$p(\lambda) = \sum_{i=1}^N (x_i + \lambda y_i)^2.$$

Explain why $p(\lambda)$ has either two complex roots or a double real root. Use this fact to prove the **Cauchy-Schwarz inequality**

$$\left| \sum_{i=1}^N x_i y_i \right| \leq \left(\sum_{i=1}^N x_i^2 \right)^{\frac{1}{2}} \left(\sum_{i=1}^N y_i^2 \right)^{\frac{1}{2}}. \qquad (42)$$

Under what circumstances does one get equality?

11. Let c_o denote the set of sequences, $\{a_j\}$, of real numbers such that $a_j \to 0$ as $j \to \infty$. Define

$$\|\{a_j\}\|_\infty \equiv \sup_j |a_j|.$$

(a) Explain why c_o is a normed linear space with the norm $\| \cdot \|_\infty$.

(b) Prove that c_o is complete. Hint: since $c_o \subseteq \ell_\infty$, we know that any Cauchy sequence has a limit in ℓ_∞.

(c) Show that the set of sequences which are zero after finitely many terms is dense in c_o.

(d) Show that c_o is not dense in ℓ_∞.

12. Let T be a linear transformation from a Banach space to itself and suppose that T is a contraction. What fixed points can T have?

13. Let $\mathcal{A}_n$ be the set of $n \times n$ matrices $A = \{a_{ij}\}$. Define

$$\|A\| \equiv \sup_{ij} |a_{ij}|.$$

(a) Explain why $\mathcal{A}_n$ is a vector space.

(b) Prove that $\| \cdot \|$ is a norm.

(c) Prove that $\mathcal{A}_n$ is complete in the norm $\| \cdot \|$.

(d) Let B be a fixed element of $\mathcal{A}_n$ and define $T(A) \equiv BA$ where BA denotes matrix multiplication. Show that T is a linear transformation on $\mathcal{A}_n$.

14. For $f \in C^{(2)}[a, b]$, define $\|f\| \equiv \|f\|_\infty + \|f'\|_\infty + \|f''\|_\infty$.

(a) Explain why $C^{(2)}[a, b]$ is a vector space.

(b) Prove that $\| \cdot \|$ is a norm on $C^{(2)}[a, b]$.

(c) Prove that $C^{(2)}[a, b]$ is complete in the norm $\| \cdot \|$.

(d) Prove that $\frac{d^2}{dx^2}$ is a linear transformation from $C^{(2)}[a,b]$ to $C[a,b]$.

(e) Prove that $\{f \in C^{(2)}[a,b] \mid f''(x) \equiv 0\}$ is a vector space. It is called the **kernel** of $\frac{d^2}{dx^2}$.

(f) Identify the functions in the kernel.

(g) Is the linear transformation $\frac{d^2}{dx^2}$ one-to-one?

Projects

1. The purpose of this project is to prove that if f is a continuous function on $[0,1]$, then $\|f\|_p \to \|f\|_\infty$ as $p \to \infty$.

   (a) Show that for each p, $\|f\|_p \leq \|f\|_\infty$.

   (b) Explain why $|f(x)|$ is continuous on $[0,1]$. Let x_o be the point where $|f(x)|$ achieves its maximum. Explain why $|f(x_o)| = \|f\|_\infty$.

   (c) Assume that x_o is not one of the endpoints and let $\varepsilon > 0$ be given. Explain why you can choose a $\mu > 0$ so that $|f(x)| \geq \|f\|_\infty - \varepsilon$ for all $x \in [x_o - \mu, x_o + \mu]$.

   (d) Prove that $\int_0^1 |f(x)|^p dx \geq 2\mu\,(\|f\|_\infty - \varepsilon)^p$ for all p.

   (e) Show that for p large enough, $\|f\|_\infty \geq \|f\|_p \geq \|f\|_\infty - 2\varepsilon$ and conclude that $\lim_{p\to\infty} \|f\|_p = \|f\|_\infty$.

2. The purpose of this project is to show by example that the method outlined in Section 5.4 can sometimes be used to solve nonlinear integral equations. Consider the following integral equation on the whole line $\mathbb{R}$:

$$\psi(x) \;=\; \frac{1}{2}\cos x \;+\; \frac{1}{2}\int_{x-\frac{1}{4}}^{x+\frac{1}{4}} \sin(x-y)(\psi(y))^2\, dy. \tag{43}$$

As in Section 5.4, we will try to solve the equation iteratively by choosing a $\psi_0(x)$ and then defining

$$\psi_{n+1}(x) \;=\; \frac{1}{2}\cos x \;+\; \frac{1}{2}\int_{x-\frac{1}{4}}^{x+\frac{1}{4}} \sin(x-y)(\psi_n(y))^2\, dy.$$

   (a) Recall that $C_b(\mathbb{R})$ denotes the space of bounded continuous functions on $\mathbb{R}$ and that $C_b(\mathbb{R})$ is complete (problem 3 in Section 5.3). Prove that if $\psi_n \in C_b(\mathbb{R})$, then $\psi_{n+1} \in C_b(\mathbb{R})$. Argue inductively that $\psi_n \in C_b(\mathbb{R})$ for all n if $\psi_0 \in C_b(\mathbb{R})$.

   (b) Suppose $\psi_0 \in C_b(\mathbb{R})$ and $\|\psi_0\|_\infty \leq 1$. Prove that $\|\psi_n\|_\infty \leq 1$.

(c) Use estimates similar to those in the proof of Theorem 5.4.1 to show that

$$\|\psi_{n+1} - \psi_n\|_\infty \leq \frac{1}{2}\|\psi_n - \psi_{n-1}\|_\infty.$$

Explain why the proof in Theorem 5.4.1 allows us to conclude that ψ_n converges uniformly to a function $\psi \in C_b(\mathbb{R})$.

(d) Prove that $\|\psi\|_\infty \leq 1$ and ψ satisfies (43).

(e) Prove that ψ is the unique function in $C_b(\mathbb{R})$ satisfying $\|\psi\|_\infty \leq 1$ which solves (43).

(f) Give a simpler proof using the contraction mapping principle.

3. The purpose of this project is to show that integral equations can sometimes be used to solve boundary value problems for differential equations. Let f be a continuous function on $[0, 1]$ and define

$$K(x, y) = \begin{cases} y(1 - x) & \text{if } y \leq x \\ x(1 - y) & \text{if } y \geq x. \end{cases}$$

(a) Show that K is a continuous function on $[0, 1] \times [0, 1]$.

(b) Suppose that ψ is a continuous function on $[0, 1]$ that satisfies

$$\psi(x) = -\int_0^1 K(x, y)f(y)\, dy + \lambda \int_0^1 K(x, y)\psi(y)\, dy.$$

Prove that ψ is twice continuously differentiable on $[0, 1]$ and that ψ satisfies

$$\psi''(x) + \lambda\psi(x) = f(x)$$

and the boundary conditions $\psi(0) = 0 = \psi(1)$. Hint: divide the region of integration into two parts $0 \leq y \leq x$ and $x \leq y \leq 1$ and use the Fundamental Theorem of Calculus repeatedly.

4. The purpose of this project is to solve the differential equation satisfied by the extremal function for the Brachistochrone problem. The extremal $y(x)$ satisfies $y(x)(1 + (y'(x))^2) = C_1$ for some constant C_1.

(a) Explain why $C_1 < 0$ and why $y'(x)$ must blow up as $x \searrow 0$. What does it mean geometrically that $y'(x)$ blows up?

(b) Prove that $\frac{dx}{dy} = \sqrt{\frac{y}{C_1 - y}}$.

(c) We introduce a new variable θ as a parameter and try to find both x and y in terms of θ. Let y and θ be related by the equation $y = C_1(\sin\theta)^2$. Use the chain rule to prove that

$$\frac{dx}{d\theta} = C_1(1 - \cos 2\theta).$$

(d) Find $x(\theta)$ in terms of C_1 and a new constant C_2.

(e) Show that the constants C_1 and C_2 can be chosen so that the curve $(x(\theta), y(\theta))$ passes through the points $(0, 0)$ and (x_1, y_1).

(f) Generate the graph of the curve $(x(\theta), y(\theta))$. Why do you think that the curve which gives the shortest time of descent is so steep near the origin?

5. The purpose of this project is to introduce some of the computational issues involved in DNA sequence comparison.

(a) Consider the sequences AGGCTC and AGCTCG drawn from the DNA alphabet. We use the discrete metric in which a letter opposite a deletion symbol and a letter opposite nothing are counted as full mismatches. Show that the minimum distance between the sequences is 3 if we do not allow deletion symbols to be inserted. Show that the minimum distance is 2 if we do allow deletion symbols.

(b) Design and implement a computational algorithm for finding the minimal distance (with no deletion symbols allowed) between two sequences of length 10 and length 8 constructed from the DNA alphabet.

(c) Design an algorithm which produces random DNA sequences of length 8 and 10.

(d) Conduct an experiment in which you determine the minimum distance of 1000 randomly chosen pairs of lengths 10 and 8, respectively. What fraction has distance 2, distance 3, and so forth? How likely is it that two randomly chosen pairs have a distance ≤ 3?

(e) If the lengths of the sequences are N and M instead of 10 and 8, estimate how the number of computational steps involved in finding the minimal distance grows as N and M get large.

CHAPTER 6

Series of Functions

Before starting our discussion of series, we introduce in Section 6.1 a new tool for analyzing sequences, the limit superior and the limit inferior. In Section 6.2 we define what it means for a series to converge and we prove various tests for convergence. Series of functions are introduced in Section 6.3, and power series, an important special case, are treated in Section 6.4. In Section 6.5 we review the basic properties of complex numbers and extend many of the ideas which we have developed to complex sequences and series. Finally, in Section 6.6 we give criteria that guarantee that infinite products converge and use the results to investigate the distribution of prime numbers.

6.1 Lim sup and Lim inf

Let $\{a_n\}$ be a sequence of real numbers and define for each N

$$\bar{s}_N \equiv \sup\{\, a_n \mid n \geq N\}.$$

As N gets larger, the *sup* is taken over a smaller set so the sequence of numbers $\{\bar{s}_N\}$ is monotone decreasing; that is, $\bar{s}_N \geq \bar{s}_{N+1}$. If $\{\bar{s}_N\}$ is bounded then, by Theorem 2.4.3, $\{\bar{s}_N\}$ converges to a finite number $\bar{s}$. We define $\bar{s}$ to be the **limit superior** of the sequence $\{a_n\}$ and write

$$\limsup_{n\to\infty} a_n \equiv \bar{s} \equiv \lim_{N\to\infty} \bar{s}_N.$$

We shall usually write $\limsup a_n$, omitting the subscript $n \to \infty$. If $\{\bar{s}_N\}$ is not bounded, there are only two possibilities since it is monotone decreasing. Either $\bar{s}_N = \infty$ for all N, in which case we say that $\limsup a_n \equiv \infty$, or $\bar{s}_N \to -\infty$, in which case we say that $\limsup a_n \equiv -\infty$. Similarly, we define

$$\underline{s}_N \equiv \inf\{\, a_n \mid n \geq N\}$$

and the **limit inferior** of the sequence $\{a_n\}$ by

$$\liminf_{n\to\infty} a_n \equiv \underline{s} \equiv \lim_{N\to\infty} \underline{s}_N.$$

A sequence $\{a_n\}$ may or may not have a limit, but it always has a lim sup and lim inf, though they may equal $\pm\infty$. Notice that $\underline{s}_N \leq \overline{s}_N$ for all N, so, by problem 3 in Section 2.4, we always have $\underline{s} \leq \overline{s}$.

Example 1   Let $a_n = (-1)^n$. The sequence $\{a_n\}$ certainly does not converge. However, for each N, $\overline{s}_N = 1$ and $\underline{s}_N = -1$. Thus, $\limsup a_n = 1$ and $\liminf a_n = -1$.

Example 2   Consider the sequence $a_n = 2 + (-1)^n(1 + \frac{1}{n})$. The values are depicted in Figure 6.1.1. For each N, $\overline{s}_N = sup_{n\geq N}\{a_n\}$ is a little above 3, but the amount above decreases to 0 as $N \to \infty$. Thus, we see that $\overline{s} = \lim \overline{s}_N = 3$. Similarly, for each N, $\underline{s}_N = inf_{n\geq N}\{a_n\}$ is a little below 1, but the amount below 1 decreases to 0 as $N \to \infty$. Therefore, $\underline{s} = \lim \underline{s}_N = 1$.

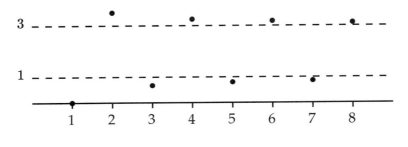

Figure 6.1.1

Example 3   Let $\{a_n\}$ be the sequence

$$1, 0, 1, -1, 1, -2, 1, -3, 1, -4, 1, -5, \ldots$$

For each N, $\overline{s}_N = 1$ and $\underline{s}_N = -\infty$. Therefore, $\limsup a_n = 1$ and $\liminf a_n = -\infty$.

We will see later that lim sup and lim inf are very useful. For the moment we prove a theorem that gives a practical and intuitive characterization of $\overline{s}$ and $\underline{s}$.

❑ **Theorem 6.1.1**   Let $\{a_n\}$ be a sequence of real numbers.

(a) If $\bar{s}$ is finite and $\varepsilon > 0$ is given, there exists an N so that $a_n \leq \bar{s} + \varepsilon$ for all $n \geq N$, and for each N there exists an $n \geq N$ so that $a_n \geq \bar{s} - \varepsilon$. Conversely, if s is a number satisfying these properties, then $s = \bar{s}$.

(b If $\underline{s}$ is finite and $\varepsilon > 0$ is given, there exists an N so that $a_n \geq \underline{s} - \varepsilon$ for all $n \geq N$, and for each N there exists an $n \geq N$ so that $a_n \leq \bar{s} + \varepsilon$. Conversely, if s is a number satisfying these properties, then $s = \underline{s}$.

Proof.   We will prove (a); the proof of (b) is similar. Let $\varepsilon > 0$ be given. Since $\bar{s}_N \to \bar{s}$ and $\bar{s}$ is finite, we can choose N so that $\bar{s}_N - \bar{s} \leq \varepsilon$. That is, $\sup\{a_n \,|\, n \geq N\} \leq \bar{s} + \varepsilon$ so $a_n \leq \bar{s} + \varepsilon$ for all $n \geq N$. Given N, suppose that there were no $n \geq N$ such that $a_n \geq \bar{s} - \varepsilon$. Then $\bar{s}_N = \sup\{a_n \,|\, n \geq N\} \leq \bar{s} - \varepsilon$, which is impossible since $\bar{s}_N$ decreases to $\bar{s}$.

Conversely, suppose that s satisfies the stated properties. Then for each $\varepsilon > 0$ there is an N so that $\bar{s}_N \leq s + \varepsilon$; thus, since $\bar{s}_N$ is monotone decreasing, $\bar{s} \leq s + \varepsilon$. Since ε is arbitrary, we conclude that $\bar{s} \leq s$. On the other hand, given any N, there is an $n \geq N$ such that $a_n \geq s - \varepsilon$. This implies that $\bar{s}_N \geq s - \varepsilon$. Thus $\bar{s} \geq s - \varepsilon$, and since ε is arbitrary we conclude that $\bar{s} \geq s$. Therefore, $\bar{s} = s$.   ❑

Corollary 6.1.2   A sequence $\{a_n\}$ of real numbers converges to a finite limit a if and only if

$$-\infty < \;\limsup a_n \;\leq\; \liminf a_n \;< \infty \tag{1}$$

in which case

$$\limsup a_n \;=\; a \;=\; \liminf a_n.$$

Proof.   Suppose that $a_n \to a$. Given $\varepsilon > 0$, there is an N so that $a - \varepsilon \leq a_n \leq a + \varepsilon$ for all $n \geq N$. Thus, for $n \geq N$,

$$a - \varepsilon \leq \bar{s}_n \leq a + \varepsilon \quad \text{and} \quad a - \varepsilon \leq \underline{s}_n \leq a + \varepsilon,$$

so

$$a - \varepsilon \leq \bar{s} \leq a + \varepsilon \quad \text{and} \quad a - \varepsilon \leq \underline{s} \leq a + \varepsilon.$$

Because ε is arbitrary, it follows that $\underline{s} = a = \bar{s}$, which implies (1). Conversely, suppose that (1) holds. Then $\underline{s}$ and $\bar{s}$ are finite and $\underline{s} \geq \bar{s}$. Since we always have $\underline{s} \leq \bar{s}$, we conclude that $\underline{s} = \bar{s}$. Define $a \equiv \underline{s} = \bar{s}$, and let $\varepsilon > 0$ be given. Then, by part (a) of Theorem 6.1.1, there is an

N_1 so that $a_n \leq a + \varepsilon$ if $n \geq N_1$, and by part (b), there is an N_2 so that $a_n \geq a - \varepsilon$ if $n \geq N_2$. Choosing $N_3 = \max\{N_1, N_2\}$, we see that $|a_n - a| \leq \varepsilon$ if $n \geq N_3$, which proves that $a_n \to a$.   ❏

Theorem 6.1.1 gives some intuition about $\limsup$ and $\liminf$. The terms of the sequence eventually get below any number that is bigger than $\bar{s}$ and keep coming back above any number that is less than $\bar{s}$. Similarly, the terms of the sequence are eventually above any number below $\underline{s}$ and keep coming back below any number above $\underline{s}$. We emphasize that a sequence $\{a_n\}$ may or may not have a limit, but $\limsup a_n$ and $\liminf a_n$ always exist.

The following technical theorem will be used when we consider infinite series, and its proof illustrates the concepts that we have defined.

❏ **Theorem 6.1.3**   Let $\{a_n\}$ be a sequence of positive numbers. Then

$$\liminf \frac{a_{n+1}}{a_n} \leq \liminf \sqrt[n]{a_n} \leq \limsup \sqrt[n]{a_n} \leq \limsup \frac{a_{n+1}}{a_n}. \qquad (2)$$

Proof.   The middle inequality holds because the $\liminf$ of any sequence is less than or equal to the $\limsup$. We will prove the inequality on the right. The proof of the one on the left is similar. Define

$$\bar{a} = \limsup_{n \to \infty} \frac{a_{n+1}}{a_n}.$$

If $\bar{a} = \infty$, we have nothing to prove. Otherwise, let $\varepsilon > 0$ be given. Then, by part (a) of Theorem 6.1.1, there is an N such that

$$\frac{a_{n+1}}{a_n} \leq \bar{a} + \varepsilon$$

for all $n \geq N$. We can rewrite this as $a_{n+1} \leq a_n(\bar{a} + \varepsilon)$. Iterating this inequality, starting at N, gives

$$a_n \leq a_N(\bar{a} + \varepsilon)^{n-N},$$

so

$$\sqrt[n]{a_n} \leq \sqrt[n]{a_N(\bar{a} + \varepsilon)^{-N}} \, (\bar{a} + \varepsilon). \qquad (3)$$

As $n \to \infty$ the right-hand side of (3) converges to $\bar{a} + \varepsilon$ since the term with the nth root converges to 1. Therefore, by the result in problem 7,

$\limsup \sqrt[n]{a_n} \le \bar{a} + \varepsilon$. Since ε was arbitrary, $\limsup \sqrt[n]{a_n} \le \bar{a}$, which is what we needed to prove. ❏

Problems

1. Find the lim sup and lim inf of each of the following sequences:

   (a) $a_n = 5 + (-1)^n$.

   (b) $a_n = 5 + (-2)^n$.

   (c) $a_n = 5 + \frac{1}{n} \sin n$.

   (d) $a_{n+1} = (3.2)a_n(1 - a_n)$, with $a_0 = \frac{1}{2}$.

   (e) $a_n = \sin n$. Hint: see project 5 of Chapter 2.

2. Let $a_n = 1$ if $n = 2^k$ for some positive integer k, and $a_n = \frac{1}{n!}$ otherwise.

   (a) Find $\limsup a_n$ and $\liminf a_n$.

   (b) Find $\limsup \frac{|a_{n+1}|}{|a_n|}$.

   (c) Find $\limsup |a_n|^{\frac{1}{n}}$.

3. Suppose that $\limsup a_n = c > 0$.

   (a) Prove that $\limsup (2a_n + 1) = 2c + 1$.

   (b) Prove that $\limsup (a_n^2) = c^2$.

4. Let $\{a_n\}$ be a sequence of real numbers and suppose that $\limsup a_n$ is finite. Prove that if $c > 0$, we have $\limsup ca_n = c \limsup a_n$.

5. Let $\{a_n\}$ be a sequence of real numbers and suppose that $\limsup a_n$ is finite. Let $\{c_n\}$ be another sequence and suppose that $c_n \to c$.

   (a) Prove that, if $c \ge 0$, then

   $$\limsup c_n a_n = c \limsup a_n. \tag{4}$$

   (b) Find a counterexample to (4) with $c < 0$.

6. Let $\{a_n\}$ and $\{b_n\}$ be sequences of real numbers. Prove that

   $$\limsup (a_n + b_n) \le \limsup a_n + \limsup b_n$$

   and give an example which shows that strict inequality can hold.

7. Suppose that $\{a_n\}$ and $\{b_n\}$ are sequences such that $a_n \le b_n$ for all n and $b_n \to b$. Prove that $\limsup a_n \le b$.

8. Let $\{a_n\}$ be a bounded sequence of real numbers. Prove that $\{a_n\}$ has a subsequence that converges to $\limsup a_n$.

9. Let $\{a_n\}$ be a bounded sequence of real numbers and let P be the set of limit points of $\{a_n\}$. Limit points are defined in Section 2.6. Prove that $\limsup a_n = \sup P$ and $\liminf a_n = \inf P$.

10. Consider the sequence

$$\frac{1}{2}, \frac{1}{3}, \frac{1}{2^2}, \frac{1}{3^2}, \frac{1}{2^3}, \frac{1}{3^3}, \frac{1}{2^4}, \frac{1}{3^4}, \ldots$$

Prove that all four quantities in (2) have different values.

6.2   Series of Real Constants

We begin our study of series of functions with the simplest special case, series of constants. Let $\{a_j\}$ be a sequence of real numbers. Throughout, we shall use the summation notation

$$\sum_{j=n}^{m} a_j \equiv a_n + a_{n+1} + \ldots + a_{m-1} + a_m.$$

If $m = \infty$, then infinitely many terms are being added up and the sum is called an infinite series. We want to determine conditions under which we can give a reasonable meaning to $\sum_{j=1}^{\infty} a_j$ and we want to prove theorems that allow us to manipulate infinite series. For each n, we define the nth **partial sum**, S_n, of the series $\sum_{j=1}^{\infty} a_j$ to be

$$S_n \equiv \sum_{j=1}^{n} a_j.$$

The sum defining S_n is over finitely many terms so there is no doubt about its meaning.

Definition.   If the sequence of partial sums $\{S_n\}$ converges to a finite number S as $n \to \infty$, we say that the infinite series $\sum_{j=1}^{\infty} a_j$ **converges** and define

$$\sum_{j=1}^{\infty} a_j \equiv S.$$

If the sequence of partial sums does not converge we say that the series **diverges**.

Example 1   If α is a real number, the series $\sum_{j=0}^{\infty} \alpha^j$ is called the **geometric series**. In this case, the index j starts at zero. Because of the special form of the terms of this series, we can calculate $S_n = \sum_{j=0}^{n} \alpha^j$ explicitly. If we multiply S_n by α, we see that

$$\alpha S_n + 1 = S_n + \alpha^{n+1}.$$

If $\alpha \neq 1$ we can solve for S_n obtaining

$$S_n = \frac{1 - \alpha^{n+1}}{1 - \alpha}.$$

If $|\alpha| < 1$, then $\alpha^{n+1} \to 0$ as $n \to \infty$. Therefore, by Theorem 2.2.6, S_n converges to $\frac{1}{1-\alpha}$ as $n \to \infty$. Thus, by definition, the series converges and

$$\sum_{j=0}^{\infty} \alpha^j = \frac{1}{1 - \alpha}.$$

To see that this is reasonable, consider the case $\alpha = \frac{1}{2}$. We have shown that as we take more and more terms, the sum of

$$1 + \frac{1}{2} + \frac{1}{4} + \frac{1}{8} + \frac{1}{16} + \frac{1}{32} + \cdots$$

gets closer and closer to 2. If $|\alpha| > 1$, then one can see from the explicit formula for S_n that S_n does not converge. In the case $\alpha = 1$, we are adding up 1's, so the series certainly does not converge. In the case $\alpha = -1$, the partial sums alternate between 1 and 0 and so do not converge.

For simplicity, we will often write $\sum_{j=1}^{\infty} a_j$ as $\sum a_j$, where the indices are understood.

Since a series converges if and only if the sequence of partial sums converges, we can use the theorems which we have proven about sequences to study series.

❏ **Theorem 6.2.1**

(a) A series $\sum_{j=1}^{\infty} a_j$ converges if and only if for each given $\varepsilon > 0$ there is an N such that

$$\left| \sum_{j=n}^{m} a_j \right| \leq \varepsilon \quad \text{for all } n \geq N \text{ and } m \geq N. \tag{5}$$

(b) If $\sum a_j$ converges, then $a_j \to 0$ as $j \to \infty$.

(c) If $\sum |a_j|$ converges, then $\sum a_j$ converges.

(d) If $\sum a_j$ converges and $\sum b_j$ converges and c and d are any real numbers, then $\sum(ca_j + db_j)$ converges and

$$\sum(ca_j + db_j) = c\sum a_j + d\sum b_j. \tag{6}$$

Proof.   Let $S_n = \sum_{j=1}^{n} a_j$ and suppose $\varepsilon > 0$ is given. If the series converges, then, by definition, $\{S_n\}$ is a Cauchy sequence. Thus, for n and m large enough, $|S_m - S_{n-1}| \leq \varepsilon$, so

$$\left| \sum_{j=n}^{m} a_j \right| = |S_m - S_{n-1}| \leq \varepsilon,$$

which proves (5). Conversely, if (5) holds, then $|S_m - S_{n-1}| \leq \varepsilon$ for n and m large enough, so $\{S_n\}$ is a Cauchy sequence. Thus $\lim_{n\to\infty} S_n$ exists and by definition the series converges.

To prove (b), let $\varepsilon > 0$ be given and choose N so that (5) holds. If we choose $m = n$, then

$$|a_n| = \left| \sum_{j=n}^{n} a_j \right| \leq \varepsilon$$

for $n \geq N$, which implies that $a_j \to 0$ as $j \to \infty$. Part (c) follows immediately from (a) and the fact that

$$\left| \sum_{j=n}^{m} a_j \right| \leq \sum_{j=n}^{m} |a_j|.$$

To prove (d), let $S_{a,n}$ and $S_{b,n}$ be the nth partial sums of $\sum a_j$ and $\sum b_j$ respectively. Let S_n be the nth partial sum of $\sum(ca_j + db_j)$. Then, $S_n = cS_{a,n} + dS_{b,n}$ for each n. Since $S_{a,n}$ converges and $S_{b,n}$ converges, Theorems 2.2.3 and 2.2.4 guarantee that S_n converges and

$$\lim_{n\to\infty} S_n = c \lim_{n\to\infty} S_{a,n} + d \lim_{n\to\infty} S_{b,n}.$$

Thus $\sum(ca_j + db_j)$ converges and (6) holds.   ❑

A series $\sum a_j$ for which $\sum |a_j|$ converges is called **absolutely convergent.** If $\sum a_j$ converges but $\sum |a_j|$ does not converge, the series is said to be **conditionally convergent**. Part (c) shows that if a series is absolutely convergent, then it is convergent. The converse is not true. Alternating series, which are discussed in project 1, are sometimes convergent but not absolutely convergent.

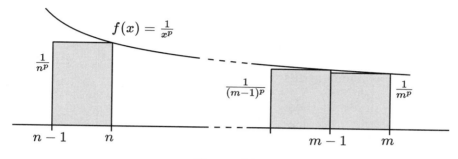

Figure 6.2.1

Example 2 We will show that the series $\sum_{j=1}^{\infty} \frac{1}{j^p}$ converges if $p > 1$. Notice that $\sum_{j=n}^{m} \frac{1}{j^p}$ is a lower sum for the integral of $f(x) = \frac{1}{x^p}$ on the interval $[n-1, m]$. See Figure 6.2.1. Thus,

$$\sum_{j=n}^{m} \frac{1}{j^p} \leq \int_{n-1}^{m} \frac{1}{x^p} \, dx$$

$$= \frac{-1}{p-1} \left(\frac{1}{m^{p-1}} - \frac{1}{(n-1)^{p-1}} \right)$$

$$\leq \frac{1}{p-1} \frac{2}{(N-1)^{p-1}}$$

if $n \geq N$ and $m \geq N$. Since $p > 1$, the expression on the right can be made as small as we like by choosing N large. Thus, by part (a) of Theorem 6.2.1, $\sum_{j=1}^{\infty} \frac{1}{j^p}$ converges.

Example 3 The series $\sum \frac{1}{j}$, which is called the **harmonic series**, does not converge. To see this, group the terms as indicated:

$$1 + \frac{1}{2} + \{\frac{1}{3} + \frac{1}{4}\} + \{\frac{1}{5} + \frac{1}{6} + \frac{1}{7} + \frac{1}{8}\} + \{\frac{1}{9} + \ldots + \frac{1}{16}\} + \ldots$$

The terms in each bracket add up to a number greater than $\frac{1}{2}$. Thus the sequence of partial sums diverges to ∞, and thus the harmonic series does not converge. Notice, however, that the *sequence* $a_j = \frac{1}{j}$ converges to zero. This shows that the converse to part (b) of Theorem 6.2.1 is false. It is *not* true that $a_j \to 0$ implies that $\sum a_j$ converges. $\sum \frac{1}{j}$ can also be shown to diverge by using an integral argument like that in Example 2 (problem 2).

❏ **Theorem 6.2.2 (The Comparison Test)**   Let $\{a_j\}$, $\{b_j\}$, and $\{c_j\}$ be sequences of nonnegative numbers such that $a_j \le b_j \le c_j$ for each j. Then,

   (a) If $\sum c_j$ converges, then $\sum b_j$ converges.

   (b) If $\sum a_j$ diverges, then $\sum b_j$ diverges.

Proof.   Suppose that $\sum c_j$ converges and let $\varepsilon > 0$ be given. Then, by part (a) of Theorem 6.2.1, there is an N such that $\sum_{j=n}^{m} c_j \le \varepsilon$ if $n \ge N$ and $m \ge N$. Since $0 \le b_j \le c_j$ for each j,

$$\left| \sum_{j=n}^{m} b_j \right| = \sum_{j=n}^{m} b_j \le \sum_{j=n}^{m} c_j \le \varepsilon.$$

Therefore, by part(a) of Theorem 6.2.1, $\sum b_j$ converges.

   To prove (b), notice that $S_n = \sum_{j=1}^{n} a_j$ is a montone increasing sequence because the a_j are nonnegative. Therefore, either S_n converges to a finite limit or $S_n \to \infty$. Since $\sum a_j$ diverges by assumption, $S_n \to \infty$. But since $a_j \le b_j$ for each j, $S_n \le \sum_{j=1}^{n} b_j$ for each n. Thus the partial sums of $\sum b_j$ diverge to ∞.   ❏

Example 4   Consider the series $\sum \frac{\sin j}{j^2 + 1}$. Since

$$\frac{|\sin j|}{j^2 + 1} \le \frac{1}{j^2}$$

and $\sum \frac{1}{j^2}$ converges, by part (a) of Theorem 6.2.2, $\sum \frac{\sin j}{j^2+1}$ is absolutely convergent. Therefore, by part (c) of Theorem 6.2.1, $\sum \frac{\sin j}{j^2+1}$ is convergent.

   A very important point to remember is that the convergence or divergence of a series is not affected by changing finitely many of its terms. This is shown by part (a) of Theorem 6.2.1 since criterion (5) is required to hold only for n and m larger than some large N. This fact is often useful when applying the comparison test because it means that the criterion $a_j \leq b_j \leq c_j$ need only hold for all j bigger than some finite number J. Though convergence depends only on the tail of the series, the sum of the series, if it converges, depends on all of the terms.

❑ **Theorem 6.2.3 (The Root Test)** Set $\alpha = \limsup |a_j|^{\frac{1}{j}}$. Then

(a) if $\alpha < 1$, the series $\sum a_j$ converges absolutely.

(b) if $\alpha > 1$, the series $\sum a_j$ diverges.

Proof. If $\alpha < 1$, choose a number β so that $\alpha < \beta < 1$. By Theorem 6.1.1 we can find a J so that $|a_j|^{\frac{1}{j}} \leq \beta$ for all $j \geq J$. Thus, $|a_j| \leq \beta^j$ for those j and $\sum a_j$ converges by comparison with the geometric series.

If $\alpha > 1$, we choose β so that $\alpha > \beta > 1$. Again, by Theorem 6.1.1, we can find infinitely many values of j so that $|a_j|^{\frac{1}{j}} \geq \beta$. Since $|a_j| > 1$ for those values of j, the sequence $\{a_j\}$ does not converge to zero. Therefore, by part (b) of Theorem 6.2.1, $\sum a_j$ cannot converge. ❑

❑ **Theorem 6.2.4 (The Ratio Test)** Let $\sum a_j$ be a series of nonzero terms. Then,

(a) if $\limsup \frac{|a_{j+1}|}{|a_j|} < 1$, the series converges absolutely.

(b) if $\liminf \frac{|a_{j+1}|}{|a_j|} > 1$, the series diverges.

Proof. Set $\alpha = \limsup |a_j|^{\frac{1}{j}}$. By Theorem 6.1.3,

$$\liminf \frac{|a_{j+1}|}{|a_j|} \leq \alpha \leq \limsup \frac{|a_{j+1}|}{|a_j|}.$$

Thus, if $\limsup \frac{|a_{j+1}|}{|a_j|} < 1$, the series converges by part (a) of Theorem 6.2.3. And, if $\liminf \frac{|a_{j+1}|}{|a_j|} > 1$, the series diverges by part (b) of Theorem 6.2.3. ❑

Neither the root test nor the ratio test give information if the respective limits equal 1. We remark that in many cases the sequence $\frac{|a_{j+1}|}{|a_j|}$ has a limit, in which case the four limits in Theorem 6.1.3 are all the same.

Example 5   Consider the series $\sum \frac{2^j}{j!}$. Since

$$\frac{|a_{j+1}|}{|a_j|} = \frac{2}{j+1} \to 0$$

as $j \to \infty$, the ratio test proves that the series converges.

We defined the sum of a series to be the limit of the sequence of partial sums if the limit exists. If we rearrange the series, that is if we add up the terms in a different order, that will certainly change the sequence of partial sums. Will the new partial sums converge? To the same limit?

❑ **Theorem 6.2.5**   Suppose that $\sum a_j$ converges absolutely and let f be a one-to-one function from $\mathbb{N}$ onto $\mathbb{N}$. Then, the series $\sum a_{f(j)}$ converges absolutely and $\sum a_{f(j)} = \sum a_j$.

Proof.   Define $S_n \equiv \sum_{j=0}^n a_j$, $S \equiv \lim_{n\to\infty} S_n$, and $T_n \equiv \sum_{j=0}^n a_{f(j)}$. Let $\varepsilon > 0$ be given. Since $\sum a_j$ converges absolutely, we can choose N_1 so that

$$\sum_{j=N_1+1}^{\infty} |a_j| \ \leq \ \varepsilon/2. \tag{7}$$

Let $J \equiv \max\{j_0, j_1, ..., j_{N_1}\}$, where j_k is the natural number such that $f(j_k) = k$. Such integers exist because f is onto. Now, choose $N = \max\{J+1, N_1\}$. By the triangle inequality,

$$|T_n - S| \ \leq \ |T_n - S_n| + |S_n - S|$$

for each n. If $n \geq N$, then $n \geq N_1$, so the second term on the right is $\leq \frac{\varepsilon}{2}$ by (7). Further, since $n \geq J+1$, the partial sum $\sum_{j=0}^n a_{f(j)}$ contains every term in the sum $\sum_{j=0}^N a_j$. Thus, the difference $T_n - S_n$ contains only terms a_j with $j \geq N+1$, and therefore $|T_n - S_n| \leq \frac{\varepsilon}{2}$, again by (7). Thus,

$$|T_n - S| \leq \varepsilon \qquad \text{for } n \geq N,$$

so $T_n \to S$; that is, $\sum a_{f(j)} = \sum a_j$. The same proof, using $|a_j|$ and $|a_{f(j)}|$ in place of a_j and $a_{f(j)}$, shows that $\sum |a_{f(j)}| = \sum |a_j|$, so $\sum a_{f(j)}$ converges absolutely.   ❑

It might seem that Theorem 6.2.5 is just a technical exercise, but it is not. It can be shown that if $\sum a_j$ is a conditionally convergent series and

a is any real number, then there is a rearrangement f so that $\sum a_{f(j)} = a$. See problem 16. In other words, one can rearrange a conditionally convergent series to get any sum one likes! More important, in some situations a sum of infinitely many terms may arise with no order specified. If the sum is absolutely convergent, then the order doesn't matter. Consider the question of whether one can multiply out two series:

$$\left(\sum_{j=0}^{\infty} a_j\right)\left(\sum_{k=0}^{\infty} b_k\right) = \sum a_j b_k. \tag{8}$$

How are we to understand the double sum on the right? We could mean

$$\sum_{j=0}^{\infty}\left(\sum_{k=0}^{\infty} a_j b_k\right) \quad \text{or} \quad \sum_{k=0}^{\infty}\left(\sum_{j=0}^{\infty} a_j b_k\right) \quad \text{or} \quad \sum_{n=0}^{\infty}\left(\sum_{k=0}^{n} a_k b_{n-k}\right)$$

Each of these expressions uses a different ordering of the pairs of nonnegative integers (j, k). Theorem 6.2.5 tells us that if we can show that $\sum a_j b_k$ is absolutely convergent in any ordering, then it is absolutely convergent in all orderings, and most important, the sum $\sum a_j b_k$ is the same in all orderings.

❏ **Theorem 6.2.6** Suppose that $\sum a_j$ and $\sum b_k$ are absolutely convergent. Then $\sum a_j b_k$ is absolutely convergent and (8) holds.

Proof. Let $A = \sum |a_j|$ and $B = \sum |b_k|$. Consider the following subsequence of the partial sums of the absolute values of the terms $a_j b_k$ in a special ordering,

$$
\begin{aligned}
S_N &\equiv \sum_{k=0}^{N}\left(\sum_{j=0}^{N} |a_j||b_k|\right) \\
&= \sum_{k=0}^{N} |b_k|\left(\sum_{j=0}^{N} |a_j|\right) \\
&= \left(\sum_{j=0}^{N} |a_j|\right)\left(\sum_{k=0}^{N} |b_k|\right) \\
&\leq AB.
\end{aligned}
$$

Thus, this subsequence of partial sums, $\{S_N\}$, is bounded and since it is monotone increasing, it converges by Theorem 2.4.3. By problem 9 of

Section 2.6, the sequence of partial sums converges. Thus, in this order-ing, $\sum a_j b_k$ is absolutely convergent. By Theorem 6.2.5, $\sum a_j b_k$ is abso-lutely convergent in all orderings and $\sum a_j b_k$ is the same in all orderings. So, using the usual rules of arithmetic,

$$\left(\sum_{j=0}^{N} a_j \right) \left(\sum_{k=0}^{N} b_k \right) = \sum_{k=0}^{N} \left(\sum_{j=0}^{N} a_j b_k \right).$$

Taking the limit of both sides as $N \to \infty$ yields (8).   ❑

One can use this theorem to show that $e^x e^y = e^{x+y}$ by multiplying out the series for the terms on the left and regrouping (problem 12 of Section 6.5). We remark that the conclusions of Theorem 6.2.5 are still true if only one of the two series $\sum a_j$ and $\sum b_k$ is absolutely convergent, but the proof is harder.

Problems

1. Determine whether the following series converge or diverge:

   (a) $\sum \frac{3^j}{j!}$ (b) $\sum \frac{\sqrt{n}}{1+n^2}$ (c) $\sum \frac{2^j}{j^2}$ (d) $\sum \frac{j^2}{j!}$

   (e) $\sum (\sqrt{j+1} - \sqrt{j})$ (f) $\sum \frac{\sqrt{j+1} - \sqrt{j}}{j}$ (g) $\sum e^{-j+\sin j}$ (h) $\sum \frac{j+\cos j}{j}$

2. Use an integral argument similar to that in Example 2 to show that the harmonic series diverges.

3. How many terms of the series $\sum \frac{1}{j^2}$ do we have to add up to be sure we are within 10^{-4} of the sum? Hint: estimate the tail of the series by using integrals, as we did in Example 2.

4. Consider the series $\sum_{j=1}^{\infty} \frac{1}{j(j+1)}$.

   (a) Prove that the series converges by the comparison test.
   (b) What information does the ratio test give?
   (c) Use the formula $\frac{1}{j(j+1)} = \frac{1}{j} - \frac{1}{j+1}$ to compute the partial sums. Show explicitly that the partial sums converge.

5. Consider the series

$$\sum_{j=0}^{\infty} 2^{(-1)^j} \left(\frac{1}{2} \right)^j.$$

Show that the ratio test gives no information but that the root test and the comparison test show convergence.

6. Show that the series

$$\sum_{j=1}^{\infty} \frac{j+1}{(j+2)10^{10}}$$

diverges. What happens if you calculate the first few partial sums on your hand calculator?

7. Compute as many terms of $\sum \frac{(-1)^{j+1}}{j}$ as necessary to provide strong numerical evidence that the series converges. See also project 1.

8. Show that the series $\sum_{j=1}^{\infty} \frac{2^j + j}{3^j - j}$ converges.

9. Show that the series $\sum_{j=1}^{\infty} \sin\left(\frac{1}{j}\right)^2$ converges. Hint: use the Mean Value Theorem.

10. Establish the convergence or divergence of $\sum_{j=1}^{\infty} \ln\left(1 + \frac{1}{j}\right)$. Hint: use the definition of $\ln x$ in Example 2 of Section 4.3.

11. Establish the convergence or divergence of $\sum_{j=1}^{\infty} \frac{(1 + \frac{1}{j})^{2j}}{e^j}$.

12. Prove that if the sequence $\{b_j\}$ is bounded and $\sum |a_j|$ converges, then $\sum a_j b_j$ converges.

13. Prove that if $a_j \geq 0$ for all j and $\sum a_j$ converges, then $\sum a_j^2$ converges.

14. Suppose that $a_j \geq 0$ and that $\sum a_j$ converges.

    (a) Show by example that it is not necessarily true that $\sum \sqrt{a_j}$ converges.

    (b) Show that $\sum \frac{\sqrt{a_j}}{j}$ converges. Hint: use the Cauchy-Schwarz inequality.

15. Suppose that $a_j \geq 0$ and that $\sum a_j$ diverges. Prove that $\sum \frac{a_j}{1 + a_j}$ diverges. Hint: first show that if it converges, then $a_j \to 0$.

16. Consider the series $\sum \frac{(-1)^{j+1}}{j}$. Since the signs alternate and the absolute values of the terms decrease and converge to zero, this series satisfies the hypotheses of the Alternating Series Theorem in project 1. Therefore, it converges. However, it is conditionally convergent since the harmonic series diverges.

    (a) Prove that the sum of the positive terms diverges to infinity. Prove that the sum of the absolute values of the negative terms diverges to infinity.

    (b) Let a be a given real number. Rearrange the series by choosing only positive terms, starting at the beginning of the series, until the sum is greater than a. Choose as many negative terms as needed to bring the sum below a. Continue in this manner and use the fact that the terms in the series converge to zero to show that this can be done in such a way that the sequence of partial sums converges to a.

6.3   The Weierstrass M-test

Suppose that we have a sequence of functions, $\{f_j(x)\}$, and we try to form the infinite sum

$$f(x) \;=\; \sum_{j=1}^{\infty} f_j(x). \tag{9}$$

In general, since each f_j depends on x, the sum will depend on x when it exists. Suppose that the interval $[a, b]$ is in the domain of all of the functions f_j. We say that the series of functions $\sum f_j$ **converges** to f on $[a, b]$ if, for each $x \in [a, b]$, the series of numbers $\sum f_j(x)$ converges to $f(x)$. By the definition of convergence, this just means that the sequence of partial sums

$$S_n(x) \;\equiv\; \sum_{j=1}^{n} f_j(x) \tag{10}$$

converges pointwise to $f(x)$ on $[a, b]$. We have already seen that pointwise convergence does not give us much control over the limiting function. Thus, we want conditions which guarantee that the partial sums $S_n(x)$ converge uniformly to $f(x)$, in which case we say that series (9) **converges uniformly**.

❏ **Theorem 6.3.1 (The Weierstrass M-test)**   Let $\{f_j(x)\}$ be a sequence of functions defined on a set $E \subseteq \mathbb{R}$. Suppose that for each j there is a constant M_j such that $|f_j(x)| \leq M_j$ for all $x \in E$ and that $\sum M_j$ converges. Then $S_n(x)$ converges uniformly to $f(x)$. If each f_j is continuous on E, then f is continuous on E.

Proof.   Let $\varepsilon > 0$ be given. Since $\sum M_j$ converges, we can choose an N so that $\sum_{j=n+1}^{m} M_j \leq \varepsilon$ if $n \geq N$ and $m \geq N$. We estimate

$$|S_m(x) - S_n(x)| \;=\; \left| \sum_{j=n+1}^{m} f_j(x) \right|$$

$$\leq\; \sum_{j=n+1}^{m} |f_j(x)|$$

$$\leq\; \sum_{j=n+1}^{m} M_j$$

$$\leq\; \varepsilon.$$

Thus, for each $x \in E$, $S_n(x)$ is a Cauchy sequence of numbers and therefore converges to a limit $f(x)$, which is, by definition, the sum of the series. Letting $m \to \infty$ in the above inequality, we find

$$|f(x) - S_n(x)| \leq \varepsilon$$

for each $x \in E$ if $n \geq N$. Thus, the series converges uniformly. Finally, suppose that each f_j is continuous on E. Then, by Theorem 3.1.1(a), S_n is continuous for each n. Since $S_n \to f$ uniformly, we conclude from Theorem 5.2.1 that f is continuous on E.   □

❑ **Theorem 6.3.2**   Let $\{f_j(x)\}$ be a sequence of continuous functions defined on a finite interval $[a, b]$. Suppose that the series $\sum_{j=1}^{\infty} f_j(x)$ converges uniformly to $f(x)$ on $[a, b]$. Then for each $x \in [a, b]$

$$\int_a^x f(t)\, dt \;=\; \sum_{j=1}^{\infty} \int_a^x f_j(t)\, dt, \tag{11}$$

and the series on the right converges uniformly in x.

Proof.   Since the series $\sum f_j(x)$ converges uniformly to $f(x)$, we can choose an N so that $|\sum_{j=1}^n f_j(x) - f(x)| \leq \varepsilon/(b-a)$ for all $x \in [a, b]$ if $n \geq N$. Thus,

$$\left| \int_a^x f(t)\, dt - \sum_{j=1}^n \int_a^x f_j(t)\, dt \right| = \left| \int_a^x f(t) - \sum_{j=1}^n f_j(t)\, dt \right|$$

$$\leq \int_a^x \left| f(t) - \sum_{j=1}^n f_j(t) \right| dt$$

$$\leq \int_a^x \frac{\varepsilon}{b-a}\, dt$$

$$\leq \varepsilon$$

for all $x \in [a, b]$. Thus the right-hand side of (11) converges uniformly to the left hand side.   □

Theorem 6.3.2 shows that if a series of continuous functions converges uniformly, then we can integrate it term by term. The following theorem shows that under reasonable hypotheses we can differentiate a series term by term.

❑ **Theorem 6.3.3** Let $\{f_j(x)\}$ be a sequence of continuously differentiable functions defined on a finite or infinite interval $[a, b]$. Suppose that the sum $\sum f_j(x)$ converges uniformly to $f(x)$ on $[a, b]$ and that $\sum f_j'(x)$ converges uniformly on $[a, b]$. Then, f is continuously differentiable on $[a, b]$ and

$$f'(x) \;=\; \sum_{j=1}^{\infty} f_j'(x). \tag{12}$$

Proof.   As above, define $S_n(x) = \sum_{j=1}^{n} f_j(x)$. Then, by hypothesis, $S_n(x) \to f(x)$ uniformly and S_n' converges uniformly. By Theorem 5.2.3, $f(x)$ is continuously differentiable and $f'(x) = \lim S_n'$. Since $S_n'(x) = \sum_{j=1}^{n} f_j'(x)$,

$$f'(x) \;=\; \lim_{n \to \infty} S_n'(x) \;=\; \sum_{j=1}^{\infty} f_j'(x).$$

❑

Although the hypotheses of Theorems 6.3.2 and 6.3.3 do not mention the Weierstrass M-test, the uniform convergence in the hypotheses is usually proven in practice by using Theorem 6.2.1 and the M-test. We note that the theorems in this section are easy to prove because we have done the hard work in Sections 5.2 and 5.3 already.

Example 1   Let $\{a_n\}$ be a sequence such that $|a_n| \le C/n^p$ where $p > 1$. Define a function $f(x)$ by

$$f(x) \;\equiv\; \sum_{n=1}^{\infty} a_n \sin nx.$$

Since $|a_n \sin nx| \le C/n^p$, the series converges uniformly by the Weierstrass M-test. Thus, f is well defined and continuous on $\mathbb{R}$ since $a_n \sin nx$ is continuous for each n. By Theorem 6.3.2, we can integrate f by integrating the series term by term. So, for example,

$$\int_0^{2\pi} f(x)\, dx \;=\; \sum_{n=1}^{\infty} a_n \int_0^{2\pi} \sin nx\, dx \;=\; 0.$$

Suppose that $p > 2$. Then, if we differentiate the series term by term, we obtain

$$\sum_{n=1}^{\infty} n a_n \cos nx.$$

Since $|na_n \cos nx| \leq C/n^{p-1}$, this series converges uniformly by the M-test because $p - 1 > 1$. Thus, by Theorem 6.3.3, f is continuously differentiable and

$$f'(x) = \sum_{n=1}^{\infty} na_n \cos nx.$$

Series like these are called Fourier series. We study Fourier series in Chapter 9.

Example 2   We will use the function g defined in problem 13 of Section 4.1 to construct a continuous function on $\mathbb{R}$ that is nowhere differentiable. Note that g is continuous on $\mathbb{R}$, bounded by 1, and satisfies

$$|g(x) - g(y)| \leq |x - y| \tag{13}$$

for all x and y in $\mathbb{R}$. Equality holds in (13) if there is no integer strictly between x and y. Define

$$f(x) \equiv \sum_{j=0}^{\infty} \left(\frac{3}{4}\right)^j g(4^j x)$$

The jth term in the sum is a continuous function which is bounded by $\left(\frac{3}{4}\right)^j$. Thus, by the Weierstrass M-test, f is a continuous function on $\mathbb{R}$.

Let $x \in \mathbb{R}$ be given. We shall show that f is not differentiable at x by exhibiting a sequence $\{h_n\}$ with $h_n \to 0$ such that

$$\frac{f(x + h_n) - f(x)}{h_n} \longrightarrow \infty.$$

Define $h_n = \pm\frac{1}{2}4^{-n}$, where we choose the plus or minus sign for each n so that there is no integer strictly between $4^n x$ and $4^n x + 4^n h_n$. Fix n. Then,

$$\frac{g(4^j(x + h_n)) - g(4^j x)}{h_n} \quad \begin{cases} = 0, & \text{if } j > n \\ = \pm 4^n, & \text{if } j = n \\ \leq 4^j, & \text{if } 0 \leq j \leq n - 1. \end{cases}$$

For $j > n$, this is true because $4^j(x + h_n)$ and $4^j x$ differ by a multiple of 2 and g is periodic with period 2. For $j \leq n - 1$, the inequality follows from (13), and if $j = n$, the quotient equals $\pm 4^n$ because equality holds in (13) if there is no integer between x and y. We can now compute that

$$\left| \frac{f(x+h_n) - f(x)}{h_n} \right| = \left| \sum_{j=0}^{\infty} \left(\frac{3}{4}\right)^j \frac{g(4^j(x+h_n)) - g(4^j x)}{h_n} \right| \tag{14}$$

$$= \left| \sum_{j=0}^{n} \left(\frac{3}{4}\right)^j \frac{g(4^j(x+h_n)) - g(4^j x)}{h_n} \right| \tag{15}$$

$$\geq 3^n - \left| \sum_{j=0}^{n-1} \left(\frac{3}{4}\right)^j \frac{g(4^j(x+h_n)) - g(4^j x)}{h_n} \right| \tag{16}$$

$$\geq 3^n - \sum_{j=0}^{n-1} 3^j \tag{17}$$

$$= \frac{1}{2}(3^n + 1). \tag{18}$$

Thus, the difference quotient for f diverges as $n \to \infty$ and $h_n \to 0$. Since x was chosen arbitrarily, f is not differentiable anywhere. In going from (15) to (16) we used the inequality $|a + b| \geq |a| - |b|$. In going from (17) to (18) we used the explicit formula for the partial sum of the geometric series.

Problems

1. Show that the series $\sum_{j=0}^{\infty} x^j$ converges uniformly in the interval $[-\beta, \beta]$ if $|\beta| < 1$.

2. (a) Show that the series $\sum_{j=0}^{\infty} e^{-jx} x^j$ converges uniformly on $[0, \infty)$. Hint: how large can xe^{-x} be for $x \geq 0$?

   (b) Compute the sum of the series.

3. (a) Prove that $\sum_{j=0}^{\infty} x^j$ is differentiable on $(-1, 1)$ and

$$\frac{d}{dx} \sum_{j=0}^{\infty} x^j = \sum_{j=0}^{\infty} (j+1)x^j.$$

   (b) Use the fact that $\sum_{j=0}^{\infty} x^j = \frac{1}{1-x}$ on $(-1, 1)$ to find a formula for $\sum_{j=0}^{\infty} (j+1)x^j$.

   (c) Use this to calculate $\sum_{j=0}^{\infty} \frac{j+1}{2^j}$ exactly.

4. Find a formula for $\sum_{j=1}^{\infty} \frac{x^j}{j}$ if x is in $(-1, 1)$.

5. (a) Show that the series

$$f(x) \equiv x + \sum_{j=1}^{\infty} x(1-x)^j$$

converges for every x in $[0, 1]$.

   (b) What is $f(x)$?

   (c) Is the convergence uniform on $[0, 1]$?

6. (a) Show that the series

$$f(x) \equiv \sum_{j=0}^{\infty} \frac{x^j}{j!}$$

converges for every x in $\mathbb{R}$.

   (b) Prove that f is differentiable and $f'(x) = f(x)$ for all x.

7. Let $f(x)$ be defined by

$$f(x) \equiv \sum_{j=1}^{\infty} \frac{1}{x^2 + j^2}.$$

   (a) Prove that f is a well-defined, continuous function on the whole real line.

   (b) Prove that f is continuously differentiable and find a series representation for f'.

8. Let N be a positive integer and suppose that $p > N$. Let $\{a_j\}$ be a sequence of numbers satisfying $|a_j| \leq C/j^p$ for some constant C. Prove that

$$f(x) = \sum_{j=1}^{\infty} a_j \sin(2\pi j x)$$

is $N - 1$ times continuously differentiable.

9. Suppose that $p > 1$ and that $\{a_j\}_{j=0}^{\infty}$ is a sequence of numbers satisfying $|a_j| \leq C/j^p$ for $j > 0$ for some constant C. Define f by

$$f(x) \equiv \sum_{j=0}^{\infty} a_j \cos(2\pi j x).$$

Compute $\int_0^1 f(x)dx.$

10. Let $\{a_j\}$ be a bounded sequence and, for $t > 0$, define

$$f(t) \equiv \sum_{j=0}^{\infty} a_j e^{-jt^2}.$$

Prove that f is infinitely often continuously differentiable for $t > 0$. Evaluate f explicitly in the case where $a_j = 1$ for all j.

11. (a) Show that the series

$$f(x) \equiv \sum \frac{1}{j2^j} \sin jx$$

converges uniformly on $[0, 2\pi]$.

 (b) Prove that f is uniformly continuous on $[0, 2\pi]$.

 (c) Let $\varepsilon = 10^{-3}$. Find a $\delta > 0$ so that $|x - y| \leq \delta$ implies that $|f(x) - f(y)| \leq \varepsilon$. Hint: use the Mean Value Theorem.

12. Prove that

$$\zeta(x) \equiv \sum_{j=1}^{\infty} \frac{1}{j^x}$$

defines a continuous function for $x > 1$. The function $\zeta(x)$ is called the **Riemann zeta function**.

13. Let Q be a rectangle $[a, b] \times [c, d]$ in the plane. Let $f_j(x, y)$ be a sequence of continuous functions on Q that satisfy $|f_j(x, y)| \leq M_j$ for all $(x, y) \in Q$. Suppose that $\sum_{j=0}^{\infty} M_j < \infty$. Prove that

$$f(x, y) = \sum_{j=0}^{\infty} f_j(x, y)$$

is a well-defined continuous function on Q.

14. Let V be a Banach space with norm $\| \cdot \|$. Let $\{x_i\}_{i=0}^{\infty}$ be a sequence of elements of V such that $\sum_{j=0}^{\infty} \|x_i\| < \infty$. Prove that $\sum_{j=0}^{\infty} x_i$ is a well-defined element of V.

15. Prove that

$$f(x) \equiv \sum_{j=1}^{\infty} \frac{e^{-(x-j)^2}}{j^2}$$

is a continuous function on $\mathbb{R}$ that goes to zero as $x \to \pm\infty$.

6.4   Power Series

A special class of series of functions arises naturally. In Section 4.3 we introduced the Taylor polynomials,

$$T^{(n)}(x, x_o) \equiv f(x_o) + f'(x_o)\,(x - x_o) + \frac{f''(x_o)}{2!}\,(x - x_o)^2 +$$

$$\ldots + \frac{f^{(n)}(x_o)}{n!}\,(x - x_o)^n,$$

as approximations to a given function $f(x)$ near a point x_o. Recall that f must be n times differentiable at x_o for $T^{(n)}(x, x_o)$ to be well defined. If f is $n+1$ times continuously differentiable near x_o, Taylor's theorem gives us an error estimate for $|f(x) - T^{(n)}(x, x_o)|$, and if f is infinitely often continuously differentiable then the Taylor polynomials exist for all n. The nth Taylor polynomial $T^{(n)}$ is the nth partial sum of the infinite series of functions

$$\sum_{j=0}^{\infty} \frac{f^{(j)}(x_o)}{j!}\,(x - x_o)^j.$$

This raises the natural question of whether the infinite series equals the function itself, that is, whether

$$f(x) \;=\; \sum_{j=0}^{\infty} \frac{f^{(j)}(x_o)}{j!}\,(x - x_o)^j.$$

The sum on the right is called the **Taylor series** of the function f. In the special case when $x_o = 0$ the series is called the **Maclaurin series** for f. There are really two separate important questions here. Where does a series of the form $\sum a_j(x - x_o)^j$ converge? Such a series is called a **power series**. And if the Taylor series of a function f converges, does it converge to $f(x)$? We begin with the first question, which has a straightforward answer.

   Let $\rho \equiv \limsup |a_j|^{\frac{1}{j}}$ and define $R = 1/\rho$ if ρ is finite and nonzero. If $\rho = 0$, we define $R = \infty$, and if $\rho = \infty$, we define $R = 0$. R is called the **radius of convergence** of the series

$$\sum_{j=0}^{\infty} a_j(x - x_o)^j, \tag{19}$$

a name which is justified by the following theorem.

❏ **Theorem 6.4.1** Let R be the radius of convergence of series (19). Then the series converges for all x in the open interval $(x_o - R, x_o + R)$ and diverges for all x outside the closed interval $[x_o - R, x_o + R]$. For each $0 \leq r < R$, the series converges uniformly on the interval $[x_o - r, x_o + r]$.

Proof. Suppose that R is finite and nonzero. Then,

$$\limsup (|a_j||x - x_o|^j)^{\frac{1}{j}} = \limsup (|x - x_o||a_j|^{\frac{1}{j}}) \tag{20}$$

$$= |x - x_o| \limsup |a_j|^{\frac{1}{j}} \tag{21}$$

$$= |x - x_o|/R. \tag{22}$$

In the second step we used problem 4 of Section 6.1. Thus, if $|x - x_o| < R$, (19) converges by the root test (Theorem 6.2.3), and if $|x - x_o| > R$, (19) diverges by the root test. If $R = \infty$, then (21) is zero, so the series converges for all x by the root test. Finally, if $R = 0$, the series diverges for $x \neq x_o$ by the root test.

It remains to show the uniform convergence. Suppose that $r < R$, and choose γ so that $\frac{r}{R} < \gamma < 1$. If $|x - x_o| \leq r$, then by (22) we have

$$\limsup (|a_j||x - x_o|^j)^{\frac{1}{j}} \leq \frac{r}{R} < \gamma.$$

By Theorem 6.1.1, we can choose a J so that

$$(|a_j||x - x_o|^j)^{\frac{1}{j}} < \gamma$$

for all $j \geq J$. Now choose $M_j = \gamma^j$ for $j \geq J$ and define

$$M_j = \sup \{|a_j(x - x_o)^j| \mid x \in [x_o - r, x_o + r]\}$$

for $j = 0, 1, 2, ..., J - 1$. Then

$$|a_j(x - x_o)^j| \leq M_j$$

for all j and all $x \in [x_o - r, x_o + r]$. Since $\gamma < 1$, the series $\sum M_j$ converges, so, by the Weierstrass M-test, (19) converges uniformly on $[x_o - r, x_o + r]$.
❏

Example 1 Consider the geometric series $\sum_{j=0}^{\infty} x^j$. Since $a_j = 1$ for all j, it is clear that $\limsup |a_j|^{\frac{1}{j}} = 1$, so $R = 1$. Thus Theorem 6.4.1 confirms what we already know about the geometric series, namely, that it converges for $|x| < 1$ and diverges for $|x| > 1$. The theorem gives no information about $x = \pm R$, but we can see in this case that the series diverges for $x = \pm 1$. We have already computed that the sum of the series is $f(x) = \frac{1}{1-x}$ for $|x| < 1$. This is a perfectly nice infinitely differentiable function everywhere on $\mathbb{R}$ except for $x = 1$, but the series $\sum_{j=0}^{\infty} x^j$ equals the function only on the interval $(-1, 1)$. To represent $f(x)$ around the point $x = 3$, we write

$$\frac{1}{1-x} = -\frac{1}{2}\frac{1}{1+\frac{x-3}{2}}$$

$$= -\frac{1}{2}\sum_{j=0}^{\infty}(-1)^j\left(\frac{x-3}{2}\right)^j,$$

which converges if $|x - 3| < 2$. This is the Taylor series for f around the point $x_o = 3$, as one can check by computing the derivatives of f at 3, and it represents the function in the interval $(1, 5)$. What we see here is a general phenomenon: the radius of convergence of the Taylor series is the distance from x_o to the nearest "singularity" of the function. The reason for this will be clarified when we study analytic functions of a complex variable in Chapter 8.

❏ **Theorem 6.4.2**   Let R be the radius of convergence of the series (19). Then the function $f(x)$ to which the series converges in $(x_o - R, x_o + R)$ is infinitely often continuously differentiable in $(x_o - R, x_o + R)$, and the derivatives can be computed by differentiating the series term by term. Furthermore, (19) is just the Taylor series of f expanded about the point x_o.

Proof.   We shall give a sketch of the proof. Let $r < R$. We know that $f(x) = \sum_{j=0}^{\infty} a_j(x - x_o)^j$ and that the powers of $x - x_o$ are continuously differentiable. Thus, Theorem 6.3.3 guarantees that f is continuously differentiable and the derivative can be computed term by term if we can show that the series of term-by term-derivatives

$$\sum_{j=1}^{\infty} j a_j(x - x_o)^{j-1} \tag{23}$$

converges uniformly. It follows from problem 3b of Section 6.1 that

$$\limsup_{j\to\infty}(j|a_j|)^{\frac{1}{j-1}} = \limsup_{j\to\infty}(j^{\frac{1}{j-1}})(|a_j|^{\frac{1}{j-1}})$$

$$= \lim_{j \to \infty} (j^{\frac{1}{j-1}}) \limsup_{j \to \infty} (|a_j|^{\frac{1}{j}})^{\frac{j}{j-1}}$$

$$= \limsup_{j \to \infty} (|a_j|^{\frac{1}{j}})$$

since $j^{\frac{1}{j-1}} \to 1$. Thus (23) has the same radius of convergence as (19). Therefore, Theorem 6.4.1 implies that (23) converges uniformly on $(x_o - r, x_o + r)$. Thus, by Theorem 6.3.3, $f(x)$ is continuously differentiable on $(x_o - r, x_o + r)$ and

$$f'(x) = \sum_{j=1}^{\infty} j a_j (x - x_o)^{j-1}. \tag{24}$$

Since r was an arbitrary number less than R, (24) holds on $(x_o - R, x_o + R)$. We now apply the same idea to show that f'' exists and that

$$f''(x) = \sum_{j=2}^{\infty} j(j-1) a_j (x - x_o)^{j-2} \tag{25}$$

on $(x_o - R, x_o + R)$. The crucial step is to show that (25) has the same radius of convergence as (19). The argument, which is similar to that above, uses the fact that $\lim_{n \to \infty} (j(j-1))^{\frac{1}{j-2}} = 1$. Continuing in this manner, we prove that

$$f^{(n)}(x) = \sum_{j=n}^{\infty} j(j-1) \ldots (j-n+1) a_j (x - x_o)^{j-n} \tag{26}$$

by showing that the series on the right-hand side has the same radius of convergence as (19). We omit the details, which are very similar to those outlined above. Notice that that if we evaluate both sides of (26) at x_o, we find that

$$f^{(n)}(x_o) = n(n-1) \ldots (1) a_n$$

since the other terms on the right vanish when $x = x_o$. Thus $a_j = \frac{f^{(j)}(x_o)}{j!}$, and so the original series (19) is just the Taylor series for f expanded about the point x_o. ❑

Example 2 (the exponential function)   We define

$$e^x = \sum_{j=0}^{\infty} \frac{x^j}{j!}. \tag{27}$$

Since $\frac{|a_{j+1}|}{|a_j|} = \frac{1}{j+1} \to 0$, Theorem 6.1.3 guarantees that $\limsup |a_j|^{\frac{1}{j}} = 0$. Thus the radius of convergence of the right-hand side of (27) is $R = \infty$. By Theorem 6.4.2, e^x is infinitely often differentiable and its derivative can be computed by differentiating term by term in (27). When one does this, one obtains the same series, so $(e^x)' = e^x$. We will show that e^x is the inverse function of $\ln x$ as defined in Example 2 of Section 4.3. If ψ is the inverse function of the natural logarithm, then ψ is defined everywhere since the range of the natural logarithm is $\mathbb{R}$, and for all positive x we have $\psi(\ln x) = x$. By Theorem 4.5.2, we know that ψ is differentiable at $\ln x$ and

$$\psi'(\ln x) = \frac{1}{(\ln x)'} = x = \psi(\ln x).$$

Thus, for all real numbers $y = \ln x$, we have $\psi'(y) = \psi(y)$. By the uniqueness of the solutions of ordinary differential equations (Theorem 7.1.1), $\psi(x) = Ce^x$ for some constant C. Since $\ln 1 = 0$, we know that $\psi(0) = 1$, so $C = 1$. Finally, problem 8 in Section 4.5 shows that the inverse function to the natural log satisfies $\psi(x + y) = \psi(x)\psi(y)$, which proves that $e^x e^y = e^{x+y}$.

Example 3 (the trigonometric functions)   We define $\sin x$ and $\cos x$ by the power series

$$\sin x = \sum_{j=0}^{\infty} (-1)^j \frac{x^{2j+1}}{(2j+1)!} \tag{28}$$

$$\cos x = \sum_{j=0}^{\infty} (-1)^j \frac{x^{2j}}{(2j)!}. \tag{29}$$

As in Example 2, the ratio test shows that the radius of convergence of both series is $R = \infty$. By Theorem 6.4.2, both functions are infinitely differentiable and, by differentiating the series term by term, we see that the usual formulas, $(\sin x)' = \cos x$ and $(\cos x)' = -\sin x$, hold. All of the other trigonometric functions are defined in terms of $\sin x$ and $\cos x$ so their differentiation formulas can be derived from these properties of $\sin x$ and $\cos x$.

We have shown that power series converge on intervals and where they converge they are the Taylor series of their limits. Unfortunately, it is not true that the Taylor series of an infinitely differentiable function

f necessarily converges. Even more surprising, the Taylor series may converge but not to the function f.

Example 4   Consider the function $f(x) = e^{-\frac{1}{x^2}}$, which we initially take to be defined for $x \neq 0$. Away from zero, f is clearly infinitely differentiable since it is the composition of two infinitely differentiable functions. We define $f(0) = 0$. Since $-\frac{1}{x^2} \to -\infty$ as $x \to 0$, we see that $f(x) \to 0$ as $x \to 0$. Thus f is continuous at zero. To see that f is differentiable at zero, consider the difference quotient

$$\frac{f(h) - f(0)}{h} = \frac{1}{he^{\frac{1}{h^2}}}.$$

Since $|he^{\frac{1}{h^2}}| \to \infty$ as $h \to 0$, we see that the difference quotient converges to zero. Thus, f is differentiable at $x = 0$ and $f'(0) = 0$. Furthermore, away from zero, $f'(x) = \frac{2}{x^3}e^{-\frac{1}{x^2}}$ by the chain rule. Since $f'(x) \to 0$ as $x \to 0$, f is continuously differentiable. Continuing in this way, one can show, by explicitly taking limits, that f is infinitely often continuously differentiable and all its derivatives are zero at zero. The proofs use only the fact that

$$|h^k e^{\frac{1}{h^2}}| \to \infty$$

as $h \to 0$ for any positive k. This can be proven directly or by using the power series for e^x. Since all the derivatives of f have value zero at zero, the Taylor series for f is identically zero. Thus the Taylor series converges but does not equal the function anywhere except at $x = 0$.

In Chapter 8, we show that the infinitely differentiable functions that are the limits of their Taylor series are the restrictions to the real axis of analytic functions in the complex plane.

Problems

1. Find the radius of convergence of the series $\sum a_j x^j$ for each of the following choices of the coefficients a_j:

    (a) 3^j (b) $\frac{3^j}{j!}$ (c) $\ln j^3$ (d) j^3

    (e) $\frac{\sin j}{j}$ (f) $2j + 1$ (g) 2^{j^2} (h) Prob. 10 of Sec. 6.1.

2. Given the following conditions on the coefficients $\{a_j\}$, what can you say about the radius of convergence of $\sum a_j x^j$?

   (a) $0 < m_1 \leq a_j \leq m_2$, for some constants m_1 and m_2.

   (b) $2^j \leq a_j \leq 3^j$.

   (c) $j^2 \leq a_j \leq j^3$.

3. What do you think is the radius of convergence of the Taylor series for $\ln x$ expanded about $x_o = 1$? Find the series and prove it. What do you think is the radius of convergence of the Taylor series for $\ln x$ expanded about 4? Find it and prove it.

4. What is the radius of convergence of a power series whose coefficients a_n are those given in problem 2 of Section 6.1?

5. Using power series, prove that $\frac{\sin x}{x} \to 1$ as $x \to 0$.

6. Using power series, prove that $\frac{\tan x - x}{x^2} \to 0$ as $x \to 0$.

7. (a) Use power series to evaluate directly the limits in problems 3(a) and 3(c) of Section 4.3.

   (b) Suppose that f and g can be represented by convergent Taylor series in an interval about x_o. Prove l'Hospital's rule (Theorem 4.3.2) and its generalization in problem 13 of Section 4.3 without using the Mean Value Theorem or Taylor's theorem.

8. In calculus, the function $f(x) = e^{-x^2}$ is always given as an example of a function that "you can't integrate" because one can't guess an elementary function F such that $F'(x) = e^{-x^2}$. Find a power series for F.

9. Is there a power series which converges to the function $f(x) = |x|$ for all x?

10. Suppose that $f(x)$ is defined and infinitely differentiable on an interval $(-r, r)$. Suppose that f satisfies the estimate $|f^{(n)}(x)| \leq M$ for all n and all $x \in (-r, r)$. Prove that the Taylor series of f converges to $f(x)$ for all $x \in (-r, r)$.

11. Find the radius of convergence of the series $\sum_{j=0}^{\infty} (j+1)(j+2)x^j$. Find the function to which the series converges.

12. Give examples which show that a Maclaurin series can either converge or diverge at the points $-R$ and R (independently) where R is the radius of convergence.

13. Suppose that we didn't know how to solve the differential equation $y'(t) = y(t)$, with initial condition $y(0) = y_o$. Let's try to write a power series $y(t) = \sum c_j t^j$ for the unknown solution. By differentiating the series,

show that we can make $y'(t) = y(t)$ if the coefficients satisfy

$$
\begin{aligned}
c_1 &= c_0, \\
2c_2 &= c_1, \\
3c_3 &= c_2,
\end{aligned}
$$

and so forth. Solve these relations to determine all the coefficients in terms of c_0. Determine c_0 from the initial condition and observe that we have found the solution. This example suggests that power series can be used to solve differential equations. This is discussed further in project 3.

6.5   Complex Numbers

We define the **complex numbers**, $\mathbb{C}$, to be the set of ordered pairs of real numbers (x, y) endowed with the following notions of addition and multiplication:

$$
\begin{aligned}
(x_1, y_1) + (x_2, y_2) &= (x_1 + x_2, y_1 + y_2) & (30) \\
(x_1, y_1)(x_2, y_2) &= (x_1 x_2 - y_1 y_2, x_1 y_2 + x_2 y_1). & (31)
\end{aligned}
$$

We regard two pairs as equal if and only if both components are equal. It is straightforward to check that addition and multiplication in $\mathbb{C}$ are commutative and associative. That is, if $z_1 = (x_1, y_1)$, $z_2 = (x_2, y_2)$, and $z_3 = (x_3, y_3)$, then

$$
\begin{aligned}
z_1 + z_2 &= z_2 + z_1 \\
(z_1 + z_2) + z_3 &= z_1 + (z_2 + z_3) \\
z_1 z_2 &= z_2 z_1 \\
(z_1 z_2) z_3 &= z_1 (z_2 z_3).
\end{aligned}
$$

Furthermore, the distributive law

$$
z_3(z_1 + z_2) = z_3 z_1 + z_3 z_2
$$

holds. The element $(0, 0)$ is called the zero of $\mathbb{C}$ and $(1, 0)$ is called the identity of $\mathbb{C}$ since

$$
\begin{aligned}
z + (0, 0) &= z \\
z(1, 0) &= z
\end{aligned}
$$

for all $z \in \mathbb{C}$. Given z_1 and z_2 in $\mathbb{C}$, the equation $z + z_1 = z_2$ has a unique solution z and the equation $zz_1 = z_2$ has a unique solution as long as $z_1 \neq (0,0)$. To check this last statement, let $z_1 = (x_1, y_1)$, $z_2 = (x_2, y_2)$, and $z = (x, y)$. Then $zz_1 = z_2$ if and only if

$$\begin{aligned} x_1 x - y_1 y &= x_2 \\ y_1 x + x_1 y &= y_2. \end{aligned}$$

Since the determinant of the coefficients on the left side is $x_1^2 + y_1^2$ the equations have a unique solution as long as x_1 and y_1 are not both zero. Thus, the complex numbers $\mathbb{C}$ satisfy the axioms for a field given in Section 1.1. The set of complex numbers is the same as the set of points in the Euclidean plane $\mathbb{R}^2$, and the definition of addition (30) corresponds to vector addition in the plane. But when we want to consider the plane as endowed with the special multiplication law (31), we denote it by $\mathbb{C}$ and call it the complex numbers.

Let R denote the special subset of $\mathbb{C}$ of elements of the form $(x, 0)$. It is easy to see that the operations of addition (30) and multiplication (31) take elements of R into itself and that R satisfies all the properties of a field. Thus R is a **subfield** of $\mathbb{C}$. Let ϕ be the function from the real numbers $\mathbb{R}$ to R defined by $\phi(x) = (x, 0)$. Then ϕ is a one-to-one function which maps $\mathbb{R}$ onto R. The notions of addition and multiplication on the two sets correspond to one another under ϕ; that is,

$$\phi(x_1)\phi(x_2) = (x_1, 0)(x_2, 0) = (x_1 x_2, 0) = \phi(x_1 x_2)$$

and similarly for addition. The function ϕ is said to be an **isomorphism** between the two fields $\mathbb{R}$ and R. By using ϕ to identify points of $\mathbb{R}$ with points of R, the real numbers can be regarded as a subset of $\mathbb{C}$. From now on, we do so, dropping the notation R, by saying that a complex number is real if it has the special form $(x, 0)$ for some real number x. Using the definition of multiplication (31), it is easy to see that

$$(x, y) = (x, 0) + (0, 1)(y, 0).$$

Thus, if we give the special complex number $(0, 1)$ the name i, we see that every complex number z can be written in the form

$$x + iy$$

where the real number x is called the **real part** of z, $x = Re(z)$, and the real number y is called the **imaginary part** of z, $y = Im(z)$.

It is possible to define the complex numbers by saying that they are the set of abstract objects of the form $x + iy$ where x and y are real numbers and i is a special object which satisfies $i^2 = -1$. We then define the operations of addition and multiplication on this set of objects to be what we get by adding and multiplying out, using the usual rules of arithmetic and the special rule $i^2 = -1$. This set of abstract objects is isomorphic to the complex numbers $\mathbb{C}$, as we have defined them.

If $z = x + iy$, we define the **absolute value** of z to be

$$|z| = \sqrt{x^2 + y^2},$$

so $|z|$ is just the Euclidean distance from the point (x, y) to the origin. Therefore, if $z_1 = x_1 + iy_1$ and $z_2 = x_2 + iy_2$, then $|z_1 - z_2|$ is the Euclidean distance between (x_1, y_1) and (x_2, y_2). In particular, the set of z which satisfy $|z - z_1| = c$ is a circle of radius c about (x_1, y_1). The absolute value satisfies several simple properties:

$$
\begin{aligned}
|z_1 + z_2| &\leq |z_1| + |z_2| \quad \text{(the triangle inequality)} \\
||z_1| - |z_2|| &\leq |z_1 - z_2| \\
|z_1 z_2| &= |z_1||z_2|.
\end{aligned}
$$

The second inequality follows from the triangle inequality, and the third statement is easy to verify directly (problem 2). To prove the triangle inequality, we square the left-hand side:

$$
\begin{aligned}
|z_1 + z_2|^2 &= (x_1 + x_2)^2 + (y_1 + y_2)^2 & (32) \\
&= (x_1^2 + y_1^2) + (x_2^2 + y_2^2) + 2(x_1 x_2 + y_1 y_2) & (33) \\
&\leq |z_1|^2 + |z_2|^2 + 2(x_1^2 + y_1^2)^{\frac{1}{2}} (x_2^2 + y_2^2)^{\frac{1}{2}} & (34) \\
&= (|z_1| + |z_2|)^2. & (35)
\end{aligned}
$$

Taking the square root of both sides gives the triangle inequality. In going from (33) to (34) we used the Cauchy-Schwarz inequality (problem 10(a) in Section 2.2). For any complex number $z = x + iy$, we define the **complex conjugate** $\bar{z}$ by $\bar{z} = x - iy$ and note that $|z|^2 = z\bar{z}$.

Definition. A sequence of complex numbers $\{z_n\}_{n=1}^{\infty}$ is said to **converge** to a limit z if, given $\varepsilon \geq 0$, there is an N such that

$$|z_n - z| \leq \varepsilon \quad \text{for all } n \geq N.$$

In this case we write $\lim_{n \to \infty} z_n = z$ or $z_n \to z$ as $n \to \infty$.

Thus, $z_n \to z$ if and only if, given any circle about z, the sequence gets inside the circle and stays inside after finitely many terms. If $z_n = x_n + iy_n$ and $z = x + iy$, then

$$|z_n - z|^2 = (x_n - x)^2 + (y_n - y)^2, \tag{36}$$

from which it follows that $z_n \to z$ if and only if $x_n \to x$ and $y_n \to y$. If for every given M there is an N so that $n \geq N$ implies $|z_n| \geq M$, we say that $\{z_n\}$ **converges to** ∞. Analogously to the real case, we say that the sequence $\{z_n\}$ is a **Cauchy sequence** if, given $\varepsilon \geq 0$, there is a N such that

$$|z_n - z_m| \leq \varepsilon \quad \text{if } n \geq N \text{ and } m \geq N.$$

❏ **Theorem 6.5.1** A sequence of complex numbers converges if and only if it is a Cauchy sequence.

Proof. Suppose that $\{z_n\}$ is a Cauchy sequence. Since

$$|z_n - z_m|^2 = (x_n - x_m)^2 + (y_n - y_m)^2, \tag{37}$$

we know that $|x_n - x_m| \leq |z_n - z_m|$ and $|y_n - y_m| \leq |z_n - z_m|$. It follows that $\{x_n\}$ and $\{y_n\}$ are Cauchy sequences in $\mathbb{R}$. By Theorem 2.4.2, $\{x_n\}$ and $\{y_n\}$ converge to finite limits x and y, respectively. If we define $z = (x, y)$, then (36) shows that $z_n \to z$. The converse argument is similar. ❏

Definition. Let $\{a_j\}$ be a sequence of complex numbers. The infinite series $\sum a_j$ is said to **converge** if the sequence of partial sums

$$S_n = \sum_{j=1}^{n} a_j$$

converges to a complex number S, in which case we call S the sum of the series and write $S = \sum_{j=1}^{\infty} a_j$. If $\sum_{j=1}^{\infty} |a_j| < \infty$, the series is said to **converge absolutely**.

Many of the theorems of Section 6.2 are true for infinite series of complex numbers with no change in proof. The geometric series (Example 1) converges if α is a complex number satisfying $|\alpha| < 1$ and the sum of the geometric series is $\frac{1}{1-\alpha}$. This follows from (31), which implies that $|\alpha^n| = |\alpha|^n \to 0$. Theorem 6.2.1 holds unchanged. The comparison test

(Theorem 6.2.2) makes no sense as it stands since there is no order relation among complex numbers; however, we can reformulate it as follows. The proof is virtually identical to the proof of Theorem 6.2.2.

❏ **Theorem 6.5.2**   Let $\{a_j\}$, $\{b_j\}$, and $\{c_j\}$ be sequences of complex numbers such that $|a_j| \leq |b_j| \leq |c_j|$ for all j.

(a) If $\sum |c_j|$ converges, then $\sum b_j$ converges absolutely.

(b) If $\sum |a_j|$ diverges, then $\sum b_j$ does not converge absolutely.

The hypotheses, conclusions, and proofs of the ratio and root tests (Theorems 6.2.3 and 6.2.4) refer only to the absolute values $|a_j|$, so they go over without change to the complex case.

Suppose now that $\{a_j\}$ is a sequence of complex numbers and z_o is a given complex number. We want to ask for which $z \in \mathbb{C}$ the power series

$$f(z) \;=\; \sum_{j=0}^{\infty} a_j (z - z_o)^j \tag{38}$$

converges. Note that where the right-hand side converges, it defines a function f that takes $\mathbb{C}$ into $\mathbb{C}$. To state the analogue of Theorem 6.4.1, we need several definitions. The radius of convergence is defined, as in the real case, in terms of the sequence $\{|a_j|\}$. The set $\{z \mid |z - z_o| < r\}$ is called the **open disk** of radius r about z_o, and $\{z \mid |z - z_o| \leq r\}$ is called the **closed disk** of radius r about z_o. The series is said to **converge uniformly** to f on a set $S \subseteq \mathbb{C}$ if, given $\varepsilon > 0$, there is an N so that

$$\left| f(z) - \sum_{j=0}^{n} a_j(z - z_o)^j \right| \;\leq\; \varepsilon \qquad \text{for all } n \geq N \text{ and all } z \in S.$$

❏ **Theorem 6.5.3**   Let R be the radius of convergence of (38). Then the series converges for all z in the open disk $\{z \mid |z - z_o| < R\}$ and diverges for all z outside the closed disk $\{z \mid |z - z_o| \leq R\}$. The series converges uniformly on each closed subdisk $\{z \mid |z - z_o| \leq r\}$ with $r < R$.

The proof of Theorem 6.5.3 is virtually identical to that of Theorem 6.4.1 and is omitted. Note that nothing is said about convergence on the circle $\{z \mid |z - z_o| = R\}$. In fact, series can converge at some points on the circle and diverge at others (see for example, problem 12 in Section 6.4).

Theorem 6.5.3 shows that the natural sets of convergence for complex power series are disks. The intersection of a disk with the real subset $\mathbb{R}$ of the complex numbers $\mathbb{C}$ is an interval or a point. This is why the natural domains of convergence of power series of a real variable x are intervals or points. In Chapter 8 we return to the question of which functions from $\mathbb{C}$ to $\mathbb{C}$ can be expressed by convergent power series. Using complex series, we can define the exponental function.

Example 1 (the exponential function)   The power series

$$e^z \ = \ \sum_{j=0}^{\infty} \frac{z^j}{j!} \tag{39}$$

converges for all complex numbers z since the radius of convergence is $R = \infty$. Thus we can use (39) as a definition of e^z. The exponential function satisfies

$$e^{z_1+z_2} \ = \ e^{z_1}e^{z_2}. \tag{40}$$

This can be proven by multiplying out the series for e^{z_1} and e^{z_2} and collecting terms (problem 12) or by using the fact that e^z is an entire analytic function, and (40) holds for real x and y (see problem 9 of Section 8.3). If θ is a real number then

$$e^{i\theta} \ = \ \sum_{j=0}^{\infty} \frac{(i\theta)^j}{j!} \tag{41}$$

$$= \ \sum_{j=0}^{\infty} (-1)^j \frac{\theta^{2j}}{(2j)!} \ + \ i \sum_{j=0}^{\infty} (-1)^j \frac{\theta^{2j+1}}{(2j+1)!} \tag{42}$$

$$= \ \cos\theta \ + \ i\sin\theta. \tag{43}$$

By using (43) and the complex conjugate $e^{-i\theta} = \cos\theta - i\sin\theta$, we can rearrange algebraically to find the usual formulas for $\sin\theta$ and $\cos\theta$ in terms of complex exponentials:

$$\cos\theta \ = \ \frac{e^{i\theta}+e^{-i\theta}}{2}, \qquad \sin\theta \ = \ \frac{e^{i\theta}-e^{-i\theta}}{2i}.$$

These representations of $\sin\theta$ and $\cos\theta$ will be very useful when we consider Fourier series in Chapter 9.

Example 2 (polar form)   Recall that the addition of complex numbers corresponds to vector addition in the plane. Using (40) we can figure out what is happening geometrically when we multiply complex numbers. For any complex number $z = x + iy$, we let θ denote the angle between the vector from the origin to the point (x, y) and the positive x axis. See Figure 6.5.1. This angle is traditionally called the **argument** of z. We write $\theta = Arg(z)$, noting that θ is only determined up to an integral multiple of 2π. The length of the vector is $|z| = \sqrt{x^2 + y^2}$. Thus we can write

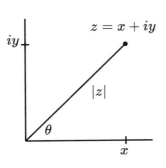

Figure 6.5.1

$$z = x + iy = |z|\cos\theta + i|z|\sin\theta = |z|e^{i\theta}.$$

which is called the polar form of z. Given two complex numbers, z_1 and z_2, we can compute their product by using (40) and the polar form:

$$z_1 z_2 = (|z_1|e^{i\theta_1})(|z_2|e^{i\theta_2}) = |z_1||z_2|e^{i(\theta_1+\theta_2)}.$$

Therefore, when we multiply complex numbers we multiply their absolute values and add their arguments.

Example 3 (roots of unity)   An important reason for the creation of the complex numbers was the desire to find a field, larger than the real number field, in which every number has a square root. Let $z = |z|e^{i\theta}$ be given. If $\omega = |\omega|e^{i\theta_1}$ is a square root of z, we must have

$$|z|e^{i\theta} = \omega^2 = |\omega|^2 e^{i2\theta_1}$$

Two complex numbers are equal if their absolute values are equal and their arguments are equal or differ by an integral multiple of 2π. Therefore $|\omega| = |z|^{\frac{1}{2}}$ and

$$\theta + m2\pi = 2\theta_1$$

for some integer m. Thus, $\theta_1 = \theta/2 + m\pi$. As m varies through the integers, θ_1 takes on only two distinct values modulo 2π, θ and $\theta + \pi$. Therefore, z has two square roots: $|z|^{\frac{1}{2}}e^{i\frac{\theta}{2}}$ and $|z|^{\frac{1}{2}}e^{i(\frac{\theta}{2}+\pi)}$. For example, $-1 = e^{i\pi}$, so the two square roots of -1 are $i = e^{i\frac{\pi}{2}}$ and $-i = e^{i\frac{3\pi}{2}}$.

If we want ω to be an nth root of z, the same reasoning leads to the requirement

$$\theta + m2\pi = n\theta_1.$$

Solving for θ_1 and letting m run through the integers, we find exactly n distinct choices, modulo 2π, for θ_1:

$$\theta_1 = \frac{\theta}{n} + \frac{m2\pi}{n}, \qquad m = 0, 1, 2, \ldots, n - 1.$$

Thus, every complex number has exactly n n^{th} roots. For example, the number 1 has n^{th} roots, $1, e^{i\frac{2\pi}{n}}, \ldots, e^{i\frac{(n-1)\pi}{n}}$. These numbers all have absolute value 1 and are equally spaced around the circle of radius 1 with center at the origin.

Problems

1. Verify from definitions (30) and (31) that complex addition is associative and complex multiplication is commutative.

2. For all z_1 and z_2 in $\mathbb{C}$, prove that $|z_1 z_2| = |z_1||z_2|$.

3. Describe the following regions in the complex plane.

    (a) $\{z \mid |z - i| \leq 1\}$.
    (b) $\{z \mid -1 \leq Im(z) < 1\}$.
    (c) $\{z \mid |z| > 2 \text{ and } Re(z) > 0\}$.
    (d) $\{z \mid Im(z^2) > 0\}$.
    (e) $\{z \mid z + \bar{z} = 2\}$.

4. Show how to express $\frac{2+3i}{1-6i}$ in the form $x + iy$ by multiplying numerator and denominator by the complex conjugate of $1 - 6i$.

5. Express the following complex numbers in polar form: (a) $1 + i$
    (b) $1 - i$ (c) -10 (d) $\sqrt{3} + i$.

6. Find the four fourth roots of i.

7. Suppose that $z_n \to z$. Prove that

    (a) $|z_n| \to |z|$.
    (b) $\bar{z}_n \to \bar{z}$.
    (c) $\{z_n\}$ is a bounded sequence.

8. Suppose that $z_n \to z$, $w_n \to w$, and $\beta \in \mathbb{C}$. Prove that

   (a) $z_n + w_n \to z + w$.

   (b) $z_n \cdot w_n \to z \cdot w$.

   (c) $\beta z_n \to \beta z$.

   (d) for each positive integer m, $z_n^m \to z^m$.

9. Let $\{a_j\}_{j=0}^m$ be complex numbers and for each z define $p(z) \equiv a_0 + a_1 z + a_2 z^2 + \ldots + a_m z^m$. Suppose that $z_n \to z$. Prove that

$$\lim_{n \to \infty} p(z_n) = p(z).$$

10. For all $z \in \mathbb{C}$, define

$$\sin z \equiv \frac{e^{iz} - e^{-iz}}{2i}, \qquad \cos z \equiv \frac{e^{iz} + e^{-iz}}{2}.$$

Find power series representations for $\sin z$ and $\cos z$ and verify that each series has radius of convergence $R = \infty$. Are $\sin z$ and $\cos z$ bounded functions on $\mathbb{C}$?

11. Prove that for $|z| > 1$,

$$\frac{1}{1 + z} = \sum_{j=1}^{\infty} \frac{(-1)^{j+1}}{z^j}.$$

12. Prove formula (40) by multiplying out the series on the right (using Theorem 6.2.6) and show by regrouping (using Theorem 6.2.5) that you get the series on the left.

6.6   Infinite Products and Prime Numbers

For any finite set of complex numbers, $a_1, \ldots, a_N$, we introduce a special symbol, $\prod$, to denote the product of the numbers in the set:

$$\prod_{j=1}^{N} a_j \equiv a_1 \cdot a_2 \cdot \ldots \cdot a_N.$$

Definition.   Let $\{a_n\}$ be an infinite sequence of nonzero complex numbers. The infinite product, $\prod a_n$, is said to **converge** if the sequence of partial products

$$P_n = \prod_{j=1}^{n} a_j$$

converges to a finite nonzero limit P as $n \to \infty$. In this case we define

$$P \equiv \prod_{j=1}^{\infty} a_j.$$

If $\{P_n\}$ converges to zero (or diverges to ∞) we say that the infinite product diverges to zero (or ∞). We require the a_n to be nonzero because if one a_n were zero, then all partial products beyond n would be zero automatically, no matter how wildly the sequence $\{a_n\}$ behaved. Similarly, if P were allowed to be zero, then "convergence" would not put strong conditions on the behavior of a_j for large j. For example, the sequence in which $a_j = \sqrt{j}$ for j odd and $a_j = \frac{1}{j}$ for j even diverges to zero. We need a criterion on the partial products $\{P_n\}$ which guarantees convergence to a nonzero limit.

❏ **Theorem 6.6.1**   The sequence of partial products $\{P_n\}$ converges to a nonzero limit P if and only if, given $\varepsilon > 0$, there exists an N so that

$$\left| \prod_{j=m}^{n} a_j - 1 \right| \leq \varepsilon \qquad \text{for all } n \geq m \geq N. \tag{44}$$

Proof.   Suppose $P_n \to P \neq 0$, and let $\varepsilon > 0$ be given. Since the absolute value is a continuous function (problem 7(a) of Section 6.5), we can choose N_1 so that $|P_n| \geq \frac{|P|}{2}$ for $n \geq N_1$. Thus,

$$|P_n - P_{m-1}| = \left| \prod_{j=1}^{n} a_j - \prod_{j=1}^{m-1} a_j \right|$$

$$= \left| \prod_{j=1}^{m-1} a_j \right| \left| \prod_{j=m}^{n} a_j - 1 \right|$$

$$\geq \frac{|P|}{2} \left| \prod_{j=m}^{n} a_j - 1 \right|.$$

Therefore,

$$\left| \prod_{j=m}^{n} a_j - 1 \right| \leq \frac{2}{P} |P_n - P_{m-1}| \qquad \text{for all } n \geq m > N_1.$$

Since $\{P_n\}$ is a Cauchy sequence, the right-hand side can be made $\leq \varepsilon$ by choosing $N \geq N_1$ large enough and requiring $n \geq m \geq N$. This proves (44).

Conversely, suppose that condition (44) holds. Then we can choose N so that $|\prod_{j=m}^{n} a_j - 1| \leq \frac{1}{2}$ for $n \geq m \geq N$. This implies, in particular, that

$$\frac{1}{2} \leq \left| \prod_{j=m}^{n} a_j \right| \leq \frac{3}{2}.$$

Thus,

$$|P_n - P_m| = \left| \prod_{j=1}^{N} a_j \prod_{j=N+1}^{m} a_j \left(\prod_{j=m+1}^{n} a_j - 1 \right) \right|$$

$$\leq \prod_{j=1}^{N} |a_j| \left(\frac{3}{2} \right) \left| \prod_{j=m+1}^{n} a_j - 1 \right|$$

for $n \geq m \geq N$; thus, it is clear that $|P_n - P_m|$ can be made small by choosing n and m sufficiently large. Therefore, $\{P_n\}$ converges. Furthermore,

$$\left| \prod_{j=1}^{n} a_j \right| = \prod_{j=1}^{N} |a_j| \left| \prod_{j=N+1}^{n} a_j \right| \geq \prod_{j=1}^{N} |a_j| \left(\frac{1}{2} \right),$$

so the product cannot converge to zero.   ❏

An immediate consequence of Theorem 6.6.1 is that $a_j \to 1$ is a necessary condition for convergence. This is analogous to the necessary condition $a_n \to 0$ for the convergence of the series $\sum a_j$. Since we must have $a_n \to 1$, it is convenient to write infinite products in the form

$$\prod_{j=1}^{\infty} (1 + b_j) \tag{45}$$

where $b_j \to 0$. If the b_j are real and nonnegative, then it is easy to say when the infinite product (45) converges.

❏ **Theorem 6.6.2**   Suppose that $b_j \geq 0$ for all j. Then $\prod(1 + b_j)$ converges if and only if $\sum b_j$ converges.

Proof.   First, suppose that $\sum b_j$ converges. It is easy to check that the function $f(x) = e^x - (1 + x)$ has its minimum at zero where it has the value zero. Thus, $(1 + x) \le e^x$ for all real numbers x. It follows that

$$\prod_{j=1}^{n}(1 + b_j) \;\le\; e^{\sum_{1}^{n} b_j} \tag{46}$$

for all n. Since $\sum b_j$ converges, the right-hand side of (46) is a bounded sequence. Thus the left-hand side is a bounded sequence, and since it is increasing, it converges by Theorem 2.4.3.

Conversely, suppose that $\prod(1 + b_j)$ converges. By multiplying out $\prod_{j=1}^{n}(1 + b_j)$ and using the positivity of the b_j, it is easy to see that

$$1 + \sum_{j=1}^{n} b_j \;\le\; \prod_{j=1}^{n}(1 + b_j)$$

for all n. Since $\prod(1 + b_j)$ converges the sequence on the right side is bounded. Thus, the sequence of partial sums on the left is bounded and therefore converges because it is increasing. ❏

Example 1   Since the harmonic series $\sum \frac{1}{j}$ diverges and the series $\sum \frac{1}{j^2}$ converges (Example 2 of Section 6.2), Theorem 6.6.1 guarantees that $\prod_{j=2}^{n}(1 - \frac{1}{j})$ diverges and $\prod_{j=2}^{n}(1 - \frac{1}{j^2})$ converges. We can verify this explicitly by computing the partial products:

$$\prod_{j=2}^{n}(1 - \frac{1}{j}) \;=\; \frac{2}{3} \cdot \frac{3}{4} \cdot \frac{4}{5} \cdot \ldots \cdot \frac{n-1}{n} \;=\; \frac{1}{n}$$

$$\prod_{j=2}^{n}(1 - \frac{1}{j^2}) \;=\; \frac{(1)(3)}{2^2} \cdot \frac{(2)(4)}{3^2} \cdot \frac{(3)(5)}{4^2} \cdot \ldots \cdot \frac{(n-1)(n+1)}{n^2} \;=\; \frac{n+1}{2n}.$$

Therefore, the first partial product approaches zero, and the second partial product converges to $\frac{1}{2}$ as $n \to \infty$.

We say that the infinite product $\prod(1 + b_j)$ **converges absolutely** if $\prod(1+|b_j|)$ converges. Theorem 6.6.2 gives a criterion for the convergence of the latter product, and the following theorem shows that absolute convergence implies convergence.

❏ **Theorem 6.6.3** If $\prod(1 + b_j)$ converges absolutely, then $\prod(1 + b_j)$ converges.

Proof. Let $P_n = \prod_{j=1}^{n}(1 + b_j)$ and $n > m$. By expanding the product $\prod_{j=m}^{n}(1 + b_j)$, subtracting 1, and using the triangle inequality, we find

$$\left| \prod_{j=m}^{n}(1 + b_j) - 1 \right| \le \prod_{j=m}^{n}(1 + |b_j|) - 1. \tag{47}$$

Since $\prod(1 + b_j)$ converges absolutely, Theorem 6.6.1 guarantees that the right hand side of (47) can be made smaller than any given $\varepsilon > 0$ by choosing n and m large enough. Thus, the same is true of the left-hand side, which proves, by Theorem 6.6.1, that the product converges. ❏

It might seem that the definition of absolute convergence for the product is unnatural. Why don't we say that $\prod(1 + b_j)$ converges absolutely if $\prod|1 + b_j|$ converges? Let θ be any nonzero real number and let $b_j = e^{ij\theta} - 1$. Then $\prod|1 + b_j|$ certainly converges since each term in the product equals 1, but $\prod(1 + b_j) = \prod e^{ij\theta}$ does not converge since the terms $e^{ij\theta}$ go around the unit circle in the complex plane as j increases and do not approach 1. The real definition above requires not that $|1 + b_j|$ be close to 1 but that the absolute value of the *difference* between $(1 + b_j)$ and 1 be small, which is a much stronger statement.

Example 2 The product $\prod_{j=1}^{\infty}(1 + \frac{(-1)^j}{j^2})$ converges absolutely since $\sum \frac{1}{j^2} < \infty$. However, the product $\prod_{j=1}^{\infty}(1 + \frac{(-1)^j}{j})$ does not converge absolutely since $\sum \frac{1}{j} = \infty$.

We now show how infinite products can be used in the study of prime numbers. The prime numbers are the set, $\mathbb{P}$, of positive integers that have no divisors besides themselves and the number 1. More than two thousand years ago, Euclid proved that $\mathbb{P}$ is an infinite set by the following argument. Suppose that $\mathbb{P}$ is a finite set and let $p_1, p_2, ..., p_K$ be a listing of the primes in $\mathbb{P}$. Then the number $q = p_1 p_2 ... p_K + 1$ must be a prime because it is not divisible by any of the p_i (since the remainder of the division is 1). But this leads to a contradiction since q is clearly larger than each of the p_i and therefore is not on our finite list of primes. Therefore $\mathbb{P}$ must be an infinite set. Throughout our discussion of prime numbers we shall use the fundamental theorem of arithmetic (Theorem 1.3.3).

Note that the fundamental theorem is used implicitly in Euclid's proof that $\mathbb{P}$ is infinite.

If we look at a list of prime numbers or compute how many there are between 1 and 100, between 101 and 200, and so forth, we find that the primes become sparse as the integers get larger. Let $\pi(n)$ denote the number of primes $\leq n$. The fact that $\mathbb{P}$ is infinite says simply that $\pi(n) \to \infty$ as $n \to \infty$. In order to characterize how sparse the prime numbers are as $n \to \infty$, we want to investigate how fast $\pi(n)$ goes to ∞. It is clear that $\pi(n) \leq n$ by definition. Suppose that n is even. There are $\frac{n}{2}$ multiples of 2 that are $\leq n$, only one of which (2 itself) is a prime. Thus,

$$\pi(n) \leq \frac{n}{2} + 1,$$

and a similar estimate holds for n odd. That is, less than half (approximately) of the integers $\leq n$ can be prime. Of course the multiples of 3 can not be prime either, and they constitute approximately $\frac{1}{3}$ of the integers $\leq n$. This suggests that less than $\frac{1}{6}$ of the integers $\leq n$ can be prime since $1 - \frac{1}{2} - \frac{1}{3} = \frac{1}{6}$. However, there is a problem with this reasoning since some multiples of 2 are also multiples of 3 and vice versa, and so we've subtracted too much. Nevertheless, this suggests that as n gets larger, $\pi(n)$ becomes smaller relative to n.

Lemma 1   The infinite product $\prod_{p \,\epsilon\, \mathbb{P}} \left(1 - \frac{1}{p}\right)$ diverges to zero.

Proof.   In the infinite product we will take the primes to be ordered in the natural way, $\mathbb{P} = \{2, 3, 5, 7, ...\}$, though it turns out that the ordering doesn't matter (problem 6). The statement of the lemma is equivalent to the statement that the product

$$\prod_{p \,\epsilon\, \mathbb{P}} \left(1 - \frac{1}{p}\right)^{-1} \tag{48}$$

diverges to ∞ since the corresponding partial products are reciprocals of each other. If $p_1, p_2, ..., p_K$ are the first K primes, then Q_K, the Kth partial product of (48), can be written

$$Q_K = \prod_{j=1}^{K} \left(1 - \frac{1}{p_j}\right)^{-1} \tag{49}$$

$$= \prod_{j=1}^{K} \left(1 + \frac{1}{p_j} + \left(\frac{1}{p_j}\right)^2 + ...\right) \tag{50}$$

since each term is a sum of a geometric series. Let $M > 0$ be given and choose N so that

$$\sum_{n=1}^{N} \frac{1}{n} \geq M$$

which we may do since the harmonic series diverges to ∞. By the fundamental theorem of arithmetic, each $n \leq N$ can be written uniquely as a finite product of powers of primes. Choose K large enough so that the finite product (49) is taken over all the primes that occur in the product representations of all $n \leq N$. Then, each term $\frac{1}{n}$ is contained in the sum obtained by multiplying out the series in the finite product (50). Note that we can multiply out these series because they are absolutely convergent (Theorem 6.2.5). Because the rest of the terms obtained from the product are positive, we see that

$$M \leq \sum_{n=1}^{N} \frac{1}{n} \leq \prod_{j=1}^{K} \left(1 + \frac{1}{p_j} + \left(\frac{1}{p_j} \right)^2 + \cdots \right).$$

Since $M > 0$ was arbitrary, we conclude that $Q_K \to \infty$ as $K \to \infty$. ❑

We introduce a function w defined on pairs of positive integers (m, r) to help us sort out the double-counting difficulty mentioned above. Define $w(m, r)$ to be the number of positive integers $\leq m$ that are *not* multiples of the first r prime numbers. Thus, for example, $w(12, 1) = 6$ since all the odd numbers are not multiples of 2, and $w(12, 4) = 2$ since only 1 and 11 are not multiples of 2,3,5, or 7.

Lemma 2 For any real x, let $[x]$ denote the greatest integer $\leq x$. Then, for all m and r,

$$w(m, r) = m - \sum_{i_1 \leq r} \left[\frac{m}{p_{i_1}} \right] + \sum_{i_1 < i_2 \leq r} \left[\frac{m}{p_{i_1} p_{i_2}} \right] - $$

$$\cdots + (-1)^r \left[\frac{m}{p_1 p_2 \cdots p_r} \right]. \qquad (51)$$

Proof. Notice that for each choice of $i_1, i_2, ..., i_k$

$$\left[\frac{m}{p_{i_1} p_{i_2} \cdots p_{i_k}} \right]$$

is just the number of multiples of $p_{i_1} p_{i_2} \ldots p_{i_k}$ which are $\leq m$. Thus, a fixed integer $n \leq m$ is counted in the term

$$\sum_{i_1 < i_2 < \ldots < i_k} \left[\frac{m}{p_{i_1} p_{i_2} \ldots p_{i_k}} \right]$$

a number of times equal to the number of different products $p_{i_1} p_{i_2} \ldots p_{i_k}$ that divide n. So, if n is not a multiple of any of the p_i, it is counted once in the first term (m) and not counted in any of the other terms. Suppose n is a multiple of exactly one of the first r primes. Then, n is counted once in the first term (m) and counted once (with a minus sign) in the second term, so the net count is zero. In general, suppose that n is a multiple of exactly j of the first r primes. Then, n is counted once in the first term, n is counted j times in the second term, n is counted $\binom{j}{2}$ times in the second term because there are $\binom{j}{2}$ ways in which n is a multiple of two primes, and so forth. So, the net number of times that n is counted (including minus signs) is

$$1 - j + \binom{j}{2} - \binom{j}{3} + \ldots + (-1)^j \binom{j}{j} = (1-1)^j = 0.$$

Thus, each n that is not a multiple of any p_i is counted once by the expression on the right of (51). Each n which is a multiple of at least one p_i is counted zero times (net) by the expression on the right of (51). Thus the sum on the right of (51) is equal to the number of integers $\leq m$ that are not multiples of the first r primes. That is, the sum equals $w(m, r)$.
❏

❑ **Theorem 6.6.4** Let $\pi(m)$ denote the number of primes $\leq m$. Then,

$$\lim_{m \to \infty} \frac{\pi(m)}{m} = 0. \tag{52}$$

Proof. The set of primes $\leq m$ is contained in the union of the first r primes and the set of numbers $\leq m$ that are not multiples of the first r primes. Thus, for every m and r

$$\pi(m) \quad \leq \quad r + w(m, r) \tag{53}$$

Notice that replacing $[x]$ by x always makes an error ≤ 1. Therefore, since there are

$$1 + \binom{r}{} + \binom{r}{2} + \ldots + \binom{r}{r} = (1+1)^r = 2^r$$

terms in expression (51) for $w(m, r)$, we have

$$\pi(m) \quad \leq \quad r + 2^r + m - \sum_{i \leq r} \frac{m}{p_{i_1}} + \sum_{i_1 < i_2 \leq r} \frac{m}{p_{i_1} p_{i_2}} - \ldots + (-1)^r \frac{m}{p_1 p_2 \ldots p_r}$$

$$= \quad r + 2^r + m \prod_{i=1}^{r} \left(1 - \frac{1}{p_i}\right).$$

This estimate is true for all positive integers m and r. Now, let $\varepsilon > 0$ be given. By Lemma 1 we can choose an r (henceforth fixed) so that $\prod_{i=1}^{r} \left(1 - \frac{1}{p_i}\right) \leq \frac{\varepsilon}{2}$. Thus,

$$\pi(m) \leq r + 2^r + m\frac{\varepsilon}{2}$$

or

$$\frac{\pi(m)}{m} \leq \frac{r}{m} + \frac{2^r}{m} + \frac{\varepsilon}{2}$$

for all m. For m large enough,

$$\frac{r}{m} + \frac{2^r}{m} \leq \frac{\varepsilon}{2}$$

and so

$$\frac{\pi(m)}{m} \leq \varepsilon,$$

which proves the theorem.   ❏

The above theorem is a simple example of the use of analysis in number theory. Much more is known about the detailed behavior of $\pi(m)$. The Prime Number Theorem, which states that

$$\lim_{m \to \infty} \frac{\pi(m)}{m/\ln m} = 1,$$

was originally conjectured by C. F. Gauss (1777 – 1855). It was proven in 1896 independently by J. Hadamard and C.-J. de la Vallee Poussin, using advanced techniques from the theory of analytic functions. Some of the basic ideas of analytic function theory are developed in Chapter 8.

Problems

1. Prove that

$$\prod_{j=2}^{\infty} \left(1 - \frac{2}{j(j+1)}\right) = \frac{1}{3}.$$

2. Prove that for every complex number z satisfying $|z| < 1$,

$$\prod_{j=0}^{\infty}(1 + z^{2^j}) = \frac{1}{1-z}.$$

3. Let z be a complex number.

   (a) Show that there is an N such that $n \geq N$ implies

   $$|e^{-zn} - (1 - \frac{z}{n})| \leq \frac{C}{n^2}$$

   for some constant C.

   (b) Use the estimate in (a) to prove the convergence of the infinite product

   $$\prod_{j=1}^{\infty}(1 + \frac{z}{n})e^{-zn}.$$

4. Let $\{a_j\}$ be a sequence of real numbers such that $|a_j| < 1$ for each j. Prove that $\prod_{j=1}^{\infty}(1 + a_j)$ converges if and only if $\sum \ln(1 + a_j)$ converges. Hint: relate the partial sum and product.

5. Use the result of problem 4 to show that the infinite product

$$\prod_{j=2}^{\infty}\left(1 + \frac{(-1)^j}{j}\right)$$

   converges but does not converge absolutely.

6. Let S be an infinite subset of the positive integers and let b_s be a function defined on S so that $0 \leq b_s < 1$ for all $s \in S$. Show that if the infinite product

$$\prod_{s \in S}(1 - b_s)$$

   diverges to zero for a particular ordering of S then it diverges to zero for all orderings of S.

7. The Riemann zeta function was defined in problem 12 of Section 6.3. Prove that for all $x > 1$

$$\zeta(x) = \prod_{p \in \mathbb{P}}\left(1 - \frac{1}{p^x}\right)^{-1}.$$

8. (a) Show that for all nonintegral x, the following infinite product converges:

$$\pi x \prod_{j=1}^{\infty} \left(1 - \frac{x^2}{j^2}\right).$$

(b) Show that the limiting function is continuous and that by defining it to be zero at the integers it can be extended to be a continuous function on $\mathbb{R}$. Hint: show that the partial products converge uniformly on appropriate sets.

(c) Generate the graphs of several partial products and use the graphs to guess what the limiting function is.

9. (a) Write a computer program which computes $\pi(n)$ for any given n.

(b) Use the program to provide numerical evidence for Theorem 6.6.4 and the Prime Number Theorem.

Projects

1. The purpose of this project is to develop the theory of **alternating series**. A series is called an alternating series if the signs of the terms alternate. For example,

$$1 - \frac{1}{2} + \frac{1}{3} - \frac{1}{4} + \frac{1}{5} - \frac{1}{6} + \frac{1}{7} - \dots$$

is the alternating harmonic series. If the first term is positive, we can write the alternating series as $\sum_{j=0}^{\infty}(-1)^j c_j$, where $c_j \geq 0$ for all j. We will suppose that the sequence $\{c_j\}$ is non-increasing; that is, $c_j \geq c_{j+1}$ for all j.

(a) Let $S_n = \sum_{j=0}^{n}(-1)^j c_j$. Prove that for each $n \geq 2$, the number S_n lies between S_{n-1} and S_{n-2}.

(b) Use the Bolzano-Weierstrass theorem to show that a subsequence S_{n_k} converges.

(c) Suppose, in addition, that $c_j \to 0$ as $j \to \infty$. Prove that the whole sequence $\{S_n\}$ converges. This is known as the **Alternating Series Theorem**.

(d) Does the alternating harmonic series converge? Does it converge absolutely?

(e) Let $S = \lim_{n\to\infty} S_n$. Prove that

$$|S - S_n| \leq S_{n+1}.$$

Note that this gives us a very easy way to estimate how close a partial sum is to the sum of an alternating series. How many terms of the alternating geometric series do we have to take to be within 10^{-4} of the limit?

(f) Prove that for each x we can choose a J so that the Maclaurin series for $\sin x$ and $\cos x$ satisfy the hypotheses for the Alternating Series Theorem for $j \geq J$. Use the alternating series remainder to estimate how close $1 - \frac{x^2}{2!} + \frac{x^4}{4!}$ is to $\cos x$ on the interval $[-1, 1]$. Compare your estimate to the one you get by using Taylor's theorem.

2. Polynomials are easy to integrate and power series are "almost" polynomials. This gives us a new way to approximate certain integrals.

(a) What is the Maclaurin series for the function e^{-x^2}?

(b) Use the first three nonzero terms of the Maclaurin series to approximate the integral $\int_0^1 e^{-x^2}\, dx$.

(c) Using the alternating series error estimate from Project 1, estimate the error in your approximation.

(d) How many terms of the series would you have to take to be sure that your estimate of the integral is within 10^{-4} of the correct answer? Compare the computational effort involved with the numerical methods discussed in Section 3.4.

(e) Suppose that we want to estimate the integral $\int_0^\infty e^{-x^2}\, dx$ to within 10^{-4}. First, choose N so that $\int_N^\infty e^{-x^2}\, dx \leq 10^{-4}/2$. Hint: $e^{-x^2} \leq e^{-x}$ for $x \geq 1$. Then estimate $\int_0^N e^{-x^2}\, dx$. Note: The exact value of $\int_0^\infty e^{-x^2}\, dx$ can be computed analytically. See problem 9 of Section 10.3.

3. The purpose of this project is to show how power series can be used to find or approximate the solutions of certain differential equations.

(a) Use the idea in problem 13 of Section 6.4 to find a power series solution of the initial value problem

$$y'(t) = y(t) + t, \qquad y(0) = 2. \tag{54}$$

After determining the coefficients, you should be able to express the series in terms of functions that you know. Check to be sure that your function solves (54).

(b) Use the same idea to find a power series solution of

$$y''(t) + \frac{1}{t}y'(t) + y(t) = 0$$

that satisfies the condition $y(0) = 1$. Verify that the series that you found converges for all t and satisfies the differential equation. The solution, $J_0(t)$, is called a **Bessel function**.

(c) Power series can also be used for nonlinear equations. Use it to find the first four terms of the solution of

$$y'(t) = y(t)^2, \qquad y(0) = 2.$$

Note that when you square the series for $y(t)$, the lower-order terms are easy to calculate. Check to make sure that the terms you found coincide with the series expansion of the solution found in Example 2 of Section 7.1.

4. The purpose of this project is to develop the idea of decimal expansions. We shall discuss decimal expansions for numbers in the interval $[0, 1]$. The extension to all real numbers is straightforward.

   (a) Let $\{a_n\}_{n=1}^{\infty}$ be a sequence, each of whose terms is an integer $0, 1, 2, \ldots, 8$, or 9. Prove that $\sum_{n=1}^{\infty} a_n 10^{-n}$ converges to a real number $x \in [0, 1]$. We define $.a_1 a_2 a_3 \ldots \equiv \sum_{n=1}^{\infty} a_n 10^{-n}$ and call $.a_1 a_2 a_3 \ldots$ a **decimal expansion** for x.

   (b) Suppose that $x \in [0, 1]$ has the form $x = k 10^{-n}$ for some positive integers k and n such that $k < 10^n$. Prove that x has a decimal expansion which has only 0's after the nth term. Prove that x also has a decimal expansion which has only 9's after the nth term.

   (c) Suppose that $x \in [0, 1]$ does not have the form $x = k 10^{-n}$ for any positive integers k and n. Then x is in one of the open intervals $(0, \frac{1}{10}), (\frac{1}{10}, \frac{2}{10}), \ldots, (\frac{9}{10}, 1)$. If x is in the mth interval, define $a_1 = m - 1$ and notice that $0 \le (x - .a_1 000\ldots) < \frac{1}{10}$. Now divide the mth interval up into 10 parts each of length $\frac{1}{100}$, and use this to choose a_2. Continue in this manner and prove that $.a_1 a_2 a_3 \ldots$ is a decimal expansion for x.

   (d) Prove that if two decimal expansions, $.a_1 a_2 a_3 \ldots$ and $.b_1 b_2 b_3 \ldots$, are distinct and neither ends in a string of 9's, then the numbers which they represent are different. Hint: let n be the first integer such that $a_n \ne b_n$. Then, either $a_n > b_n$ or $a_n < b_n$. If the first is true, prove that $.a_1 a_2 a_3 \ldots > .b_1 b_2 b_3 \ldots$.

   (e) Prove that every $x \in (0, 1)$ that is of the form $k 10^{-n}$ has exactly two decimal expansions, one ending in 0's and the other in 9's.

   (f) Prove that every $x \in (0, 1)$ that is not of the form $k 10^{-n}$ has a unique decimal expansion that does not end in a string of 0's or 9's.

CHAPTER 7

Differential Equations

In many applications of mathematics, the fundamental assumption states that the rate of change (i.e., the derivative) of an important function is equal to an expression involving other quantities and (sometimes) the function itself. The problem is to "solve" the resulting differential equation and find the function. Because of the importance of this problem, it is not surprising that the study of differential equations has played a central role in mathematics for the past three centuries. Indeed, the branch of mathematics which we call analysis began when Newton invented calculus so that he could write down the differential equations for the motions of the planets.

In the first section we shall see that, under mild assumptions, differential equations always have unique solutions if we specify the value of the unknown function at an initial point. This solution may only exist for a short time, however. Therefore, in Section 7.2 we give conditions which guarantee that solutions are global, that is, that they exist for all times. If the differential equation is very simple, one may be able to find an explicit expression for the solution. More often, one needs machine computation to solve the differential equation approximately. In Section 7.3, we prove an error estimate for Euler's method, the simplest approximation algorithm. Throughout, the idea is to show in a simple way the usefulness of the analytic concepts that we have developed. For further study of differential equations, see [2], [4], [21], and [7].

7.1 Local Existence

To see what kind of theorem we could try to prove, it is useful to start with a simple, familiar example.

Example 1 Consider the differential equation

$$y'(t) = 2y(t).$$

For every choice of the constant c, the function $y(t) = ce^{2t}$ satisfies the differential equation. Thus, the differential equation has a whole family of solutions. Often one refers to both the function $y(t)$ and to its graph (which is a curve in the $t - y$ plane) as a "solution." If the value of $y(t)$ at a particular time $t = t_o$ is given, then c is determined. To see this, suppose $y(t_o) = y_o$. Then, $y_o = y(t_o) = ce^{2t_o}$, so $c = y_o e^{-2t_o}$. Thus, given any point, (t_o, y_o), in the plane, there is a function, $y(t)$, which solves the differential equation and whose graph passes through the point. The condition $y(t_o) = y_o$ is called an **initial condition** because the value of y is being specified at the "initial" time t_o. Note, however, that the solution $y(t)$ is determined for times before t_o, as well as for times after t_o.

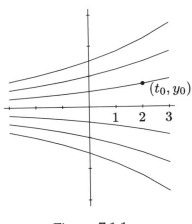

Conversely, suppose that $y_1(t) = c_1 e^{2t}$ and $y_2(t) = c_2 e^{2t}$ are two solutions which are equal at some time t_1. Then, $c_1 e^{2t_1} = c_2 e^{2t_1}$, from which it follows that c_1 must equal c_2. Therefore, the solutions are equal for all times t. Thus, distinct solutions can never have the same value at the same time. In terms of the geometry of solution curves in the plane, we can restate what we have proven as follows: every point in the plane has a solution curve going through it, and distinct solution curves never cross. Several of these solution curves are shown in Figure 7.1.1.

Figure 7.1.1

We would like to show that the initial-value problem for a general first-order differential equation

$$y'(t) = f(t, y(t)), \qquad y(t_o) = y_o \tag{1}$$

has the same nice properties as the solutions in Example 1. Here f is a function of two independent variables, t_o is a real number, and the value of y at t_o is specified. The proof is quite difficult for two reasons. First, for a general f we have no way of exhibiting explicit functions which solve the differential equation, as we did in Example 1. This is even true for quite simple functions f, such as $f(t, y) = \sin(2y)$ or $f(t, y) = t^2 + y\sin(2y)$. Thus, we will have to *prove* that solutions exist and have the right properties rather than just checking the properties of a given family of functions, as we did in Example 1. The second difficulty is that solutions may exist only for short times as shown by the following example.

Example 2 Solving the differential equation

$$y'(t) = y(t)^2, \qquad y(0) = y_o$$

by the method of separation of variables (project 1 of Chapter 4), we find that

$$y(t) = \frac{1}{\frac{1}{y_o} - t}$$

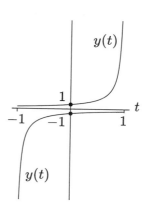

if $y_o \neq 0$. If $y_o = 0$, the solution is $y(t) \equiv 0$. For $y_o > 0$, $y(t) \to +\infty$ as $t \nearrow \frac{1}{y_o}$. So, not only does the solution exist for only a finite time interval, but also the amount of time it exists depends on the initial condition y_o. Notice, however, that the solution exists for all negative times. For $y_o < 0$, the situation is just the opposite. The solution exists for all positive times but diverges to $-\infty$ as $t \searrow \frac{1}{y_o}$. The cases $y_o = \pm 1$ are shown in Figure 7.1.2.

Figure 7.1.2

❏ **Theorem 7.1.1** Let f be continuously differentiable in a square

$$S \equiv [t_o - \delta, t_o + \delta] \times [y_o - \delta, y_o + \delta]$$

centered at (t_o, y_o). Then there is a $T \leq \delta$ and a unique, continuously differentiable function $y(t)$ defined on $[t_o - T, t_o + T]$ so that (1) holds.

Proof. If $y(t)$ is continuously differentiable, then by the Fundamental Theorem of Calculus,

$$y(t) - y_o = \int_0^t y'(s)\, ds$$

and so if $y(t)$ satisfies (1),

$$y(t) \; = \; y_o + \int_0^t f(s, y(s)) \, ds. \tag{2}$$

We will solve (2) by an iteration method and then show that the solution satisfies (1). Notice that (2) allows us to reformulate both the differential equation and the initial condition into a single condition on the function $y(t)$. Let $T > 0$ be a number such that $T \leq \delta$, and let $y_1(t) \equiv y_o$. We define functions $y_n(t)$ inductively for $n \geq 1$ by

$$y_{n+1}(t) \; = \; y_o + \int_{t_o}^t f(s, y_n(s)) \, ds. \tag{3}$$

We shall show that if T is small enough, this definition makes sense and that the resulting sequence of functions, $\{y_n(t)\}$, is a Cauchy sequence in $C[t_o - T, t_o + T]$. For simplicity, we denote the interval $[t_o - T, t_o + T]$ by I_T. Throughout, we shall denote the sup norm on $C[I_T]$ by $\| \cdot \|_{\infty, T}$ to emphasize the dependence on T.

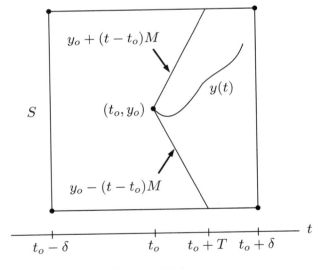

Figure 7.1.3

Suppose that $0 < T \leq \delta/M$ where $M \equiv \max_S |f(t, y)|$. Since f is continuous on S, $M < \infty$ (Theorem 4.6.1). We shall first prove that each function $y_n(t)$ is continuous on I_T and its values lie in $[y_o - \delta, y_o + \delta]$. Since $y_1(t) \equiv y_o$, this is certainly true for $n = 1$. Suppose that it is true for y_n.

Then, since the values of y_n lie in $[y_o - \delta, y_o + \delta]$, we know that $f(s, y_n(s))$ is well defined on I_T. Furthermore, $f(s, y_n(s))$ is continuous on I_T since the composition of continuous functions is continuous (problem 4 of Section 4.6). Thus, the integral of $f(s, y_n(s))$ from t_o to t is a continuous function of t (problem 13 of Section 3.3). It follows that y_{n+1} is continuous, and for $t \in I_T$,

$$
\begin{aligned}
|y_{n+1}(t) - y_o| &\leq \left| \int_{t_o}^t |f(s, y_n(s))| \, ds \right| \\
&\leq M|t - t_o| \\
&\leq MT \\
&\leq \delta,
\end{aligned}
$$

so the values of y_{n+1} lie in $[y_o - \delta, y_o + \delta]$. By induction, each function y_n is continuous on I_T and its values on I_T lie in $[y_o - \delta, y_o + \delta]$.

By hypothesis, f is continuously differentiable on S, so

$$
K \equiv \max_S \left| \frac{\partial f}{\partial y} \right| < \infty.
$$

For all β_1 and β_2 in $[y_o - \delta, y_o + \delta]$, the Mean Value Theorem implies that there is a ξ between β_1 and β_2 so that

$$
|f(t, \beta_1) - f(t, \beta_2)| \leq \left| \frac{\partial f}{\partial y}(t, \xi)(\beta_1 - \beta_2) \right| \tag{4}
$$

$$
\leq K|\beta_1 - \beta_2|. \tag{5}
$$

Using (5), we estimate

$$
\begin{aligned}
|y_{n+1}(t) - y_n(t)| &\leq \left| \int_{t_o}^t |f(s, y_n(s)) - f(s, y_{n-1}(s))| \, ds \right| \\
&\leq \left| \int_{t_o}^t K|y_n(s) - y_{n-1}(s)| \, ds \right| \\
&\leq K|t - t_o| \, \|y_n - y_{n-1}\|_{\infty, T} \\
&\leq KT\|y_n - y_{n-1}\|_{\infty, T}.
\end{aligned}
$$

Thus, taking the supremum of the left-hand side over all $t \in I_T$, we obtain

$$
\|y_{n+1} - y_n\|_{\infty, T} \leq \alpha \|y_n - y_{n-1}\|_{\infty, T} \tag{6}
$$

where $\alpha = KT$. If we choose $T < \min\{\frac{\delta}{M}, \frac{1}{K}, \delta\}$, then $\alpha < 1$. It follows in exactly the same way as in the proof of Theorem 5.4.1 that $\{y_n\}$ is a Cauchy sequence in the *sup* norm in $C[I_T]$. Thus, by Theorem 5.3.3, there is a continuous function $y(t)$ on I_T such that $y_n \to y$ uniformly. One can also use the contraction mapping principle to prove the existence of y (see problem 5). Since estimate (5) holds,

$$|f(s, y_n(s)) - f(s, y(s))| \leq K|y_n(s) - y(s)|$$

for all $s \varepsilon I_T$. It follows that $f(t, y_n(s))$ converges uniformly to $f(t, y(s))$ on I_T. Therefore, using Theorem 5.2.2 and (3), we find

$$
\begin{aligned}
y(t) &= \lim_{n\to\infty} y_{n+1}(t) \\
&= \lim_{n\to\infty} \left(y_o + \int_{t_o}^{t} f(s, y_n(s))\, ds \right) \\
&= y_o + \int_{t_o}^{t} f(s, y(s))\, ds.
\end{aligned}
$$

The function y therefore satisfies (2) and $y(t_o) = y_o$. Since $f(s, y(s))$ is continuous, the Fundamental Theorem of Calculus implies that $y(t)$ is continuously differentiable and (1) holds.

To see that $y(t)$ is unique, suppose that $z(t)$ is another continuously differentiable function on I_T that satisfies (1). Then $z(t)$ satisfies the integral equation (2) also. Subtracting the integral equation for $y(t)$ from the integral equation for $z(t)$, we find

$$z(t) - y(t) = \int_{t_o}^{t} (f(s, z(s)) - f(s, y(s)))\, ds \tag{7}$$

The same estimates as above show that $\|z - y\|_{\infty,T} \leq \alpha\|z - y\|_{\infty,T}$. But, since $\alpha < 1$, this can only be true if $\|z - y\|_{\infty,T} = 0$, which implies that $z(t) = y(t)$ for all t in the interval. Thus, $y(t)$ is unique. ❏

Theorem 7.1.1 is called a *local* existence theorem because we have proven the existence of a solution of (1) only on a small time interval $[t_o - T, t_o + T]$ containing t_o. That is, there is a solution curve only near the point (t_o, y_o) in the plane. We saw in Example 2 that one cannot expect existence for all times without additional hypotheses on f. This is the subject of Section 7.2. The hypothesis that f is continuously differentiable can be weakened considerably. The same proof shows that a unique $y(t)$ exists locally if f is uniformly Lipschitz continuous (problem 8). We assumed that f is continuously differentiable for t larger than t_o and t smaller than t_o, so we obtained a solution curve on both sides of t_o. If one knows only that f is continuously differentiable for $t \geq t_o$ then the same proof gives local existence of a unique solution $y(t)$ on the interval $[t_o, t_o + T]$.

Note the interesting line of argument in the proof. We showed the existence of a continuous function $y(t)$ that satisfied the integral equation (2). It followed automatically (from the Fundamental Theorem of Calculus) that $y(t)$ was continuously differentiable. This same line of reasoning can be used to prove that $y(t)$ is, in fact, infinitely often continuously differentiable if f is infinitely often continuously differentiable (problem 9). Ordinary differential equations are a special case of a class of partial differential equations, called elliptic equations, for which this kind of differentiability holds. Although Theorem 7.1.1 gives local existence and uniqueness for a single first-order differential equation, a very similar proof works for first order systems of equations; see project 1. Furthermore, higher-order equations can be reduced to systems of first-order equations (problem 11), so the ideas of Theorem 7.1.1 are very general.

If f is only continuous, it can be shown that a local solution exists by other methods. However, in that case, uniqueness may be lost, as the following example shows.

Example 3   Consider the initial-value problem

$$y'(t) = \sqrt{y(t)}, \qquad y(0) = 0.$$

The method of separation of variables yields the solution $y(t) = t^2/4$ on $[0, \infty)$, but the function $y(t) \equiv 0$ is also a solution. The function $f(y) = \sqrt{y}$ is continuous of $[0, \infty]$. Its derivative, $\frac{\partial f}{\partial y} = \frac{1}{2}\frac{1}{\sqrt{y}}$, is continuous for $y > 0$ but not at $y = 0$. Thus, if the initial condition were $y(0) = y_o$ with $y_o > 0$, then Theorem 7.1.1 would guarantee a unique local solution. But if $y_o = 0$, the theorem cannot be used because the hypothesis on f is not satisfied.

We saw in Example 1 that distinct solution curves cannot cross. It is a very important consequence of Theorem 7.1.1 that this is true in the general case too. This fact is often used in proofs of global existence; see Example 3 in Section 7.2.

❑ **Theorem 7.1.2**   Let $y(t)$ and $z(t)$ be continuously differentiable solutions of (1) on an interval $[a, b]$. If there is one point $t_o \in [a, b]$ such that $y(t_o) = z(t_o)$, then $y(t) = z(t)$ for all $t \in [a, b]$.

Proof.   Suppose $y(t_o) = z(t_o)$ at a point $t_o \in [a, b]$. Then, both $y(t)$ and $z(t)$ satisfy (1) and the same initial condition at t_o. Thus, the uniqueness statement in Theorem 7.1.1 implies that $y(t) = z(t)$ for all t in some interval I_T containing t_o.   See Figure 7.1.4.   Let $t_1 = \sup\{t \,|\, y(t) = z(t)\}$ and suppose that $t_1 < b$.   We know that $z(t_1) = y(t_1)$ since z and y are continuous functions which are equal in the interval $[t_o, t_1)$. However, the local existence theorem would then guarantee that $y(t) = z(t)$ for all t in some interval containing t_1, which would contradict the definition of t_1. Thus, we must have $t_1 \geq b$, so $y(t) = z(t)$ on the interval $[t_o, b]$. The proof that $y(t) = z(t)$ on the interval $[a, t_o]$ is similar.   ❑

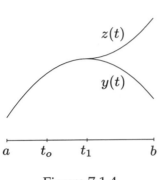

Figure 7.1.4

Finally, we show that the solution which we have constructed in Theorem 7.1.1 depends continuously on the initial value y_o.

❑ **Theorem 7.1.3**   Let f, S, M, and K be as in Theorem 7.1.1, and let $y(t)$ be the solution of (1) on the interval I_T where $T < \min\{\frac{\delta}{M}, \frac{1}{K}\}$. Suppose that $\overline{y}_o$ satisfies $|\overline{y}_o - y_o| \leq \delta/2$, and define $\overline{T} = T/2$. Then the solution $\overline{y}(t)$ of (1) satisfying $\overline{y}(0) = \overline{y}_o$ exists on the interval $I_{\overline{T}}$ and satisfies

$$|\overline{y}(t) - y(t)| \;\leq\; e^{Kt}\, |\overline{y}_o - y_o|. \qquad (8)$$

Proof.   By Theorem 7.1.1, the solution $\overline{y}(t)$ exists in some small time interval about t_o. Since $\overline{y}(t)$ is continuous, it's graph cannot escape from S immediately and as long as it remains in S, $\overline{y}(t)$ satisfies the estimate

$$|\overline{y}(t) - \overline{y}_o| \;\leq\; M(t - t_o).$$

Thus,

$$
\begin{aligned}
|\overline{y}(t) - y_o| \;&\leq\; |\overline{y}(t) - \overline{y}_o| + |\overline{y}_o - y_o| \\
&\leq\; M(t - t_o) + \delta/2 \\
&\leq\; \delta/2 + \delta/2
\end{aligned}
$$

if $|t - t_o| \leq \delta/2M$. Therefore, the solution $\bar{y}(t)$ remains in the rectangle for $t \in I_{\bar{T}}$. Now define

$$\sigma(t) = (y(t) - \bar{y}(t))^2.$$

Since both solutions are continuously differentiable, $\sigma(t)$ is continuously differentiable and

$$
\begin{aligned}
\sigma'(t) &= 2(y(t) - \bar{y}(t))(y'(t) - \bar{y}'(t)) & (9) \\
&= 2(y(t) - \bar{y}(t))(f(t, y(t)) - f(t, \bar{y}(t))) & (10) \\
&= 2f_y(t, \beta)(y(t) - \bar{y}(t))^2 & (11) \\
&\leq 2K\sigma(t), & (12)
\end{aligned}
$$

where we used the Mean Value Theorem in step (11). Since $\sigma(t)$ satisfies this differential inequality, Proposition 7.2.2 (proven in the next section) guarantees that

$$\sigma(t) \leq e^{2Kt}\sigma(0).$$

Taking the square root of both sides yields (8). $\square$

Problems

1. Use the method of separation of variables to find explicit solutions of the following initial-value problems. In both cases, sketch the graphs of the solutions curves for several different values of y_o.

    (a) $y'(t) = ty(t)$, $y(0) = y_o$.
    (b) $y'(t) = -2ty(t)^3$, $y(0) = y_o$.

2. Consider the initial-value problem

$$y'(t) = \sin(y(t)), \qquad y(0) = 1.$$

    Let $\delta = 1$. Determine values for the constants M and K used in the proof of Theorem 7.1.1. Prove that a solution exists on the interval $[-\frac{1}{2}, \frac{1}{2}]$.

3. Consider the initial-value problem

$$y'(t) = t^2 + y(t)\sin(2y(t)), \qquad y(1) = 3.$$

    Let $\delta = 1$. Determine values for the constants M and K used in the proof of Theorem 7.1.1. On what interval I_T does Theorem 7.1.1 guarantee that a solution exists?

4. Consider the initial-value problem

$$y'(t) \;=\; y(t)^2, \qquad y(0) = 2.$$

Let $\delta = 1$. Determine values for the constants M and K used in the proof of Theorem 7.1.1. On what interval I_T does Theorem 7.1.1 guarantee that a solution exists?

5. Show how to use the contraction mapping principle to avoid mimicking the convergence argument of Theorem 5.4.1 in the proof of Theorem 7.1.1.

6. Let $y(t)$ be the solution of the initial-value problem in problem 2. Let $\bar{y}(t)$ be the solution of the same differential equation with the initial condition $\bar{y}(0) = 1.05$. Estimate $|y(t) - \bar{y}(t)|$ on the interval $[-\frac{1}{4}, \frac{1}{4}]$.

7. Suppose that the two solutions, $y(t)$ and $\bar{y}(t)$, described in problem 6 exist for all times $t \in \mathbb{R}$. Use the proof of Theorem 7.1.3 to estimate $|y(t) - \bar{y}(t)|$.

8. Suppose that $f(t, y)$ is uniformly Lipschitz continuous in y in the square S. That is, assume that the M in the definition of Lipschitz continuous can be chosen uniformly for all $t \in I_T$. Explain carefully why the proof of Theorem 7.1.1 still works.

9. Suppose that f is infinitely often continuously differentiable in S. Prove that the solution $y(t)$, whose existence was shown in Theorem 7.1.1, is infinitely often continuously differentiable on I_T.

10. For each of the following functions, $f(t, y)$, find the set of initial values $y(0) = y_o$ for which Theorem 7.1.1 or problem 8 guarantees a unique local solution $y(t)$:

   (a) $f(t, y) = t^2 \sin y$.

   (b) $f(t, y) = \sqrt{y - 2}$.

   (c) $f(t, y) = (y)^{\frac{1}{3}}$.

   (d) $f(t, y) = (\cos t)\sqrt{1 - y^2}$.

   (e) $f(t, y) = y^{-1} \sin y$.

11. Let g be a function of $n + 1$ variables. Show that the nth order differential equation

$$y^{(n)}(t) \;=\; g(t, y(t), y'(t), y''(t), \ldots, y^{(n-1)}(t))$$

can be converted to a system of first-order equations by the change of variables $z_0(t) = y(t), z_1(t) = y', \ldots, z_{n-1}(t) = y^{(n-1)}(t)$. What hypotheses on g and what type of initial conditions do you think are necessary to prove a local existence result for the system? What form do these initial conditions take in terms of $y(t)$?

12. Suppose that f is a continuously differentiable function on $\mathbb{R}^2$. In Theorem 7.1.1 a solution of (1) was constructed on the interval $[t_o - T, t_o + T]$. Take $(t_o + T, y(t_o + T))$ as the initial point and explain carefully why the same proof shows that there is a $T_1 > 0$ so that the solution can be extended to the interval $[t_o - T, t_o + T + T_1]$. If we repeat this process, does this prove that the solution can be extended to the whole interval $[t_o - T, \infty)$?

7.2   Global Existence

Theorem 7.1.1 guarantees that solutions of differential equations exist locally, that is, for short times. We would like to derive conditions on f which guarantee that solutions exist for all times. Example 2 in Section 7.1 shows that the solution of an ordinary differential equation may go to $+\infty$ at a finite time t_1 if f grows quickly in y. The solution ceases to exist at t_1 in the sense that there is no continuously differentiable function on a interval containing t_1 that equals the given solution to the left of t_1. Are there other ways in which a solution could stop being a solution? Perhaps solutions could oscillate faster and faster or remain bounded and suddenly become non-differentiable. The following theorem shows that if f is continuously differentiable everywhere, then the only way solutions stop being solutions is by going to $+\infty$ or $-\infty$ in finite time.

❏ **Theorem 7.2.1**   Let f be continuously differentiable on $\mathbb{R}^2$, and let $y(t)$ be the local solution of

$$y'(t) = f(t, y(t)), \qquad y(t_o) = y_o$$

given by Theorem 7.1.1. Define

$$t_1 \equiv \sup\{s \mid y(t) \text{ is a solution on } [t_o, s)\}$$

and suppose $t_1 < \infty$. Then, as $t \nearrow t_1$, either $y(t) \to \infty$ or $y(t) \to -\infty$.

Proof.   We will show that if $y(t)$ does not converge either to $+\infty$ or $-\infty$, then $y(t)$ can be extended past t_1, contradicting the definition of t_1. If $y(t) \to \infty$ as $t \nearrow t_1$, then for every $N > 0$ there is μ such that $y(t) \geq N$ for all $t \geq t_1 - \mu$. Similarly, if $y(t) \to -\infty$ as $t \nearrow t_1$ then for every $N > 0$ there is μ so that $y(t) \leq -N$ for all $t \geq t_1 - \mu$. Therefore, if $y(t)$ doesn't converge either to $+\infty$ or to $-\infty$, there is an $N > 0$ so that every interval $[t_1 - \mu, t_1)$ contains at least one point $\bar{t}_\mu$ such that $-N \leq y(\bar{t}_\mu) \leq N$.

Consider the initial-value problem

$$z'(t) \;=\; f(t, z(t)), \qquad z(\bar{t}_\mu) = y(\bar{t}_\mu). \tag{13}$$

We will choose μ and the corresponding $\bar{t}_\mu$ below. Let S^* be the square $\{(t,y) \mid |t| \le |t_1| + 1, |y| \le N + 1\}$ and define

$$M^* \;\equiv\; \max_{S^*} |f(t,y)|, \qquad K^* \;\equiv\; \max_{S^*} \left| \frac{\partial f}{\partial y}(t,y) \right|.$$

For each μ, the δ in Theorem 7.1.1 can be taken to be 1 because, by hypothesis, f is continuously differentiable on the whole plane. If we do so, then, for *each* μ the corresponding square S_μ of Theorem 7.1.1 is contained in S^* and thus M_μ and K_μ are less than M^* and K^*, respectively. Now, the length of time of existence, T, of the local solution constructed in Theorem 7.1.1 depended only on the the constants M and K.

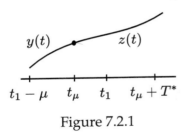

Figure 7.2.1

Therefore, if we choose T^* so that $T^* < 1/M^*$ and $T^* < 1/K^*$; then the solution of each of the initial-value problems (13) exists for a time T^* independent of μ (since M^* and K^* are independent of μ). Choose μ small enough so that $t_1 - \mu + T^* > t_1$. Then, since $t_\mu \ge t_1 - \mu$, we have

$$t_\mu + T^* \;\ge\; t_1 - \mu + T^* \;>\; t_1.$$

Thus, the solution $z(t)$ of (13) exists on the interval $[t_\mu, t_\mu + T^*)$, which contains t_1. See Figure 7.2.1. By the uniqueness proven in Theorem 7.1.1, the solution $z(t)$ coincides with $y(t)$ on the interval $[t_\mu, t_1)$ where both solutions exist. Therefore $z(t)$ extends the solution $y(t)$ past t_1 which contradicts the definition of t_1. Thus, either $y(t) \to \infty$ as $t \nearrow t_1$ or $y(t) \to -\infty$ as $t \nearrow t_1$. $\qquad \Box$

This theorem is so useful because if we can show that a solution can't go to $+\infty$ or $-\infty$ in finite time, then the solution must exist for all times t. In this case we say that the differential equation has a **global solution**.

Example 1   Consider the initial-value problem

$$y'(t) \;=\; \sin\left(g(t, y(t))\right), \qquad y(t_o) = y_o$$

where g is some continuously differentiable function of two variables. Even if g is very simple (for example $g(t, y) = y$), we can't "solve" this equation in the sense of writing down a solution in terms of elementary functions. But we can say that the solution exists for all times t. The reason is simple. Whatever the function $y(t)$ is, we know from the equation that

$$|y'(t)| = |\sin(g(t, y(t)))| \leq 1$$

for all times t. Thus in a time interval of length T, $y(t)$ cannot increase or decrease by more than T, so $y(t)$ cannot approach $+\infty$ or $-\infty$ in finite time. Theorem 7.2.1 implies, therefore, that the solution is global.

Even when $y'(t)$ grows, $y(t)$ can't go to infinity in finite time if $y'(t)$ doesn't grow too fast. For example, the solution of $y'(t) = y(t)$ with y(0) = 1, is the function $y(t) = e^t$, so $y'(t)$ does indeed grow exponentially. Nevertheless, the solution exists for all times. Suppose that we consider the initial-value problem

$$y'(t) = \frac{y(t)}{1 + y(t)^2}, \qquad y(0) = y_o. \tag{14}$$

As long as $y_o > 0$, $y(t)$ will increase and remain positive. From (14) it follows that $y'(t) \leq y(t)$ for all t, so it seems reasonable to guess that the solution of (14) is less than e^t and therefore can't go to infinity in finite time. This is true and is so useful that we state it separately as a proposition.

Proposition 7.2.2 Let $y(t)$ be a continuously differentiable function on a finite interval $[a, b]$. Suppose that $y(t)$ satisfies $y'(t) \leq My(t)$ for all $t \in [a, b]$. Then $y(t) \leq y(a)e^{M(t-a)}$ on $[a, b]$.

Proof. Define $x(t) \equiv y(t)e^{-Mt}$. Then, for all $t \in [a, b]$,

$$x'(t) = e^{-Mt}(y'(t) - My(t)) \leq 0,$$

so, by the Fundamental Theorem of Calculus, $x(t) \leq x(a)$. Substituting for x in terms of y gives the desired inequality. ❏

The proposition enables us to prove an extremely useful theorem that allows us to compare the solutions of two different equations if we can compare the functions on the right-hand sides.

❑ **Theorem 7.2.3.**   Suppose that f and g are continuously differentiable functions of two variables that satisfy

$$f(t, y) \leq g(t, y) \tag{15}$$

for all points (t, y) in the strip $a \leq t \leq b$, $-\infty \leq y \leq \infty$. Suppose that $y(t)$ and $z(t)$ satisfy the differential equations

$$y'(t) = f(t, y(t)), \qquad z'(t) = g(t, z(t))$$

on the interval $[a, b]$. Then,

(a) If $y(a) \leq z(a)$, then $y(t) \leq z(t)$ for all $t \in [a, b]$.

(b) If $y(b) \leq z(b)$, then $y(t) \geq z(t)$ for all $t \in [a, b]$.

Proof.   We shall prove (a); the proof of (b) is similar. Suppose that $y(a) \leq z(a)$ and that there is a t_2 in the interval $[a, b]$ such that $y(t_2) > z(t_2)$. We will show that this leads to a contradiction. Let t_1 be the supremum of the set of $t \in [a, t_2]$ such that $y(t) \leq z(t)$. The hypothesis $y(a) \leq z(a)$ shows that the set is nonempty. Since y and z are continuous functions, we know that $y(t_1) = z(t_1)$. In particular, $t_1 < t_2$. On the interval $[t_1, t_2]$, we define $h(t) \equiv y(t) - z(t)$. Since y and z are continuous on $[t_1, t_2]$, they are bounded. Thus, there is a constant B so that $|y(t)| \leq B$ and $|z(t)| \leq B$ for all $t \in [t_1, t_2]$. Let K be the supremum of $|\frac{\partial f}{\partial y}|$ on the rectangle $[t_1, t_2] \times [-B, B]$. Then for $t \in [t_1, t_2]$,

$$
\begin{align}
h'(t) &= y'(t) - z'(t) \tag{16} \\
&= f(t, y(t)) - g(t, z(t)) \tag{17} \\
&\leq f(t, y(t)) - f(t, z(t)) \tag{18} \\
&\leq K|y(t) - z(t)| \tag{19} \\
&= K(y(t) - z(t)) \tag{20} \\
&= Kh(t). \tag{21}
\end{align}
$$

We used the hypothesis (15) in step (18) and the Mean Value Theorem in step (19). Since $h'(t) \leq Mh(t)$ on $[t_1, t_2]$, Proposition 7.2.2 assures us that $h(t) \leq h(t_1)e^{Mt}$. But $h(t)$ is nonnegative and $h(t_1) = 0$, so $h(t) = 0$ for all $t \in [t_1, t_2]$. Thus, $y(t_2) \leq z(t_2)$. Since this violates our assumption about t_2, the proof is complete.   ❑

Theorems 7.2.1 and 7.2.3 can be combined to give useful conditions for existence on long time intervals.

❑ **Theorem 7.2.4**   Let f, g, and h be continuously differentiable functions
of two variables that satisfy $g(t,y) \leq f(t,y) \leq h(t,y)$ for all $a \leq t \leq b$
and $-\infty \leq y \leq \infty$. Suppose $t_o \in [a,b]$. Suppose that solutions $x(t)$ and
$z(t)$ of the initial-value problems

$$x'(t) = g(t, x(t)), \qquad x(t_o) = y_o \qquad (22)$$

and

$$z'(t) = h(t, z(t)), \qquad z(t_o) = y_o \qquad (23)$$

exist on the interval $[a,b]$. Then, the solution of

$$y'(t) = f(t, y(t)), \qquad y(t_o) = y_o \qquad (24)$$

exists on the interval $[a,b]$. Furthermore, $x(t) \leq y(t) \leq z(t)$ for all $t \in [t_o, b]$
and $z(t) \leq y(t) \leq x(t)$ for all $t \in [a, t_o]$.

Proof.   By Theorem 7.1.1, the equation (24) has a local solution $y(t)$
which exists near $t = t_o$. By Theorem 7.2.1, either $y(t)$ exists on $[a,b]$ or
it goes to $+\infty$ or to $-\infty$ as t approaches some $t_1 \in [a,b]$. Suppose $t_1 > t_o$.
On the interval $[t_o, t_1)$, part (a) of Theorem 7.2.3 guarantees that $y(t)$ is
bounded above by $z(t)$ and bounded below by $x(t)$, so $y(t)$ cannot go to
$+\infty$ or to $-\infty$ as $t \nearrow t_1$. Therefore, the solution exists for all $t \in [t_o, b]$ and
the estimate $x(t) \leq y(t) \leq z(t)$ holds. A similar proof, using part (b) of
Theorem 7.2.3, shows that the solution exists for all $t \in [a, t_o]$ and that the
estimate $z(t) \leq y(t) \leq x(t)$ holds there.   ❑

Example 2   Consider the initial-value problem

$$y'(t) = 2t^2(\cos t)(\sin y(t)) + 5y(t), \qquad y(0) = y_o. \qquad (25)$$

Choose $M > 0$. For $t \in [-M, M]$,

$$-2M^2 + 5y \leq 2t^2(\cos t)(\sin y(t)) + 5y \leq 2M^2 + 5y.$$

By separation of variables, one can solve easily each of the equations
$y'(t) = -2M^2 + 5y(t)$ and $y'(t) = 2M^2 + 5y(t)$ with initial condition $y(0) =$
y_o and observe that their solutions exist on $[-M, M]$. By Theorem 7.2.4,
a solution of (25) exists on the interval $[-M, M]$. Since M is arbitrary, the
solution exists globally.

Example 3   The fact that orbits cannot cross (Theorem 7.1.2) can some-times be used to show that solutions exist on infinite time intervals. Consider the initial-value problem

$$y'(t) \ = \ -y(t)^2, \qquad y(0) = y_o > 0.$$

Since $y'(t) \leq 0$, the solution is decreasing and therefore cannot approach $+\infty$. On the other hand, the t-axis is the orbit of the solution that is identically zero for all t. Since $y(t)$ starts positive and cannot cross the t axis, the solution remains positive. Thus it cannot approach $-\infty$. Since the solution can not approach either $+\infty$ or $+\infty$ on the interval $[0, \infty)$, by Theorem 7.2.1 it exists on the entire interval $[0, \infty)$.

It is worthwhile to emphasize two points. First, even when a general criterion like Theorem 7.2.4 does not hold, a differential equation may have global solutions. One must use other, more detailed properties of f to prove it. Second, a differential equation may have global solutions for some initial conditions but not for others. Both these points are illustrated by problem 3.

Problems

1. Solve the following initial-value problem explicitly and determine for which $p > 0$ the solution goes to ∞ in finite time.

$$y'(t) \ = \ y(t)^p, \qquad y(0) = 1.$$

   Note: for $x > 0$ and $p > 0$ we define $x^p = e^{p \ln x}$.

2. Solve the initial-value problem in Example 3 explicitly and verify that the solution exists on $[0, \infty)$ and satisfies $0 < y(t) \leq y_o$.

3. Consider the initial-value problem

$$y'(t) \ = \ (y(t) - 1)(y(t) - 2), \qquad y(0) = y_o.$$

   (a) What are the solutions if $y_o = 1$ or $y_o = 2$?

   (b) Prove that the solution is global if $1 < y_o < 2$.

   (c) Prove, without solving explicitly, that the solution goes to $+\infty$ in finite time if $y_o > 2$.

4. Prove global existence for the initial-value problem

$$y'(t) \ = \ \sin y(t), \qquad y(0) = y_o.$$

5. Prove global existence for the solution of the initial-value problem

$$y'(t) \quad = \quad 2y(t) \; + \; \frac{1}{4} \sin y(t), \qquad y(0) = y_o.$$

6. Without solving explicitly, prove global existence for the initial-value problem

$$y'(t) \quad = \quad t \; + \; y(t), \qquad y(0) = y_o.$$

Hint: for $t \in [-N, N]$, the inequality $-N + y \le t + y \le N + y$ holds.

7. Let g be a bounded continuously differentiable function on $\mathbb{R}^2$. Show that the solution, $y_\varepsilon(t)$, of

$$y'(t) \quad = \quad y(t) \; + \; \varepsilon g(t, y(t)), \qquad y(0) = 1$$

exists globally and converges uniformly to e^t on each finite interval as $\varepsilon \to 0$. Hint: estimate $y_\varepsilon(t)$ from above and below.

8. Suppose that $0 \le \varepsilon \le y_o$. Prove that the solution of the initial-value problem

$$y'(t) \quad = \quad (y(t))^2 \; + \; \varepsilon \sin y(t), \qquad y(0) = y_o$$

converges to $+\infty$ in finite time.

9. Suppose that all the hypotheses of Theorem 7.2.3 hold except that the hypothesis that f and g are continuously differentiable is weakened to the statement that both f and g are continuous and one of them is uniformly Lipschitz continuous. Prove that the conclusion still holds.

7.3   The Error Estimate for Euler's Method

Most differential equations cannot be solved analytically in terms of familiar elementary functions like polynomials, trigonometric functions, exponentials, and so forth. Analytical expressions are extremely valuable when they exist because they allow one to see explicitly the dependence of the solution on the parameters of the problem, for example, the initial condition or coefficients. When closed-form expressions do not exist, sometimes power series or transform methods allow representations of the solution. But, in general, most differential equations have to be solved approximately by machine computation, and an important part of the design of algorithms is the proof of error estimates. In this section we show how to estimate the error in Euler's method, the simplest numerical method for approximating the solutions of differential equations.

Suppose that we wish to approximate the solution of the differential equation

$$y'(t) \ = \ f(t, y(t)), \qquad y(a) = y_o \qquad (26)$$

on the interval $[a, b]$. We divide the interval into N equal parts of length $h \equiv (b - a)/N$ by setting $t_0 = a, t_1 = a + h, t_2 = a + 2h, \ldots, t_N = b$. Set $y_0 \equiv y_o$. We know the value of y at t_0, and the differential equation tells us the value of y' at t_0, namely, $f(t_0, y(t_0))$. Thus it is natural to approximate $y(t)$ on the interval $[t_0, t_1]$ by $y_0 + (t - t_0)f(t_0, y(t_0))$ since this straight line has the same value and slope as $y(t)$ at t_0. This gives us the approximation

$$y_1 \equiv y_0 \ + \ hf(t_0, y(t_0))$$

for the value of $y(t_1)$ at t_1. Using y_1, we can approximate $y'(t_1) = f(t_1, y(t_1)) \approx f(t_1, y_1)$, and this enables us to define the straight line approximation $y_1 + (t - t_1)f(t_1, y_1)$ on the interval $[t_1, t_2]$. This second straight line gives us the approximation

$$y_2 \equiv y_1 \ + \ hf(t_1, y(t_1))$$

for the value of $y(t_2)$. Continuing in this manner, we define recursively

$$y_{n+1} \ \equiv \ y_n \ + \ h \, f(t_n, y(t_n)) \qquad (27)$$

for $n = 0, \ldots, N - 1$. Connecting the points (t_n, y_n) by straight lines gives the polygonal approximation to $y(t)$ first used by Euler and known as **Euler's method.**

Example 1   We will use Euler's method to approximate the solution of

$$y'(t) \ = \ -2y(t) + 5t, \qquad y(0) = 5 \qquad (28)$$

on the interval $[0, 2]$. We will use 8 subintervals, so $h = .25$, and $t_n = .25n$. The recursion relation (28) for the approximate values at the points t_n is $y_{n+1} = y_n + (.25)(-2y_n + 5t_n)$. Carrying out the recursion starting with $t_0 = 0$ and $y_0 = 5$, we get the table of values and the polygonal graph shown in Figure 7.3.1. This equation has a simple closed-form solution, so for comparison we have listed its values at the points t_n in the right-hand column and drawn its graph.

n	t_n	y_n	$y(t_n)$
0	0	5	5
1	.25	2.5	3.17
2	.5	1.56	2.30
3	.75	1.41	2.02
4	1.0	1.64	2.10
5	1.25	2.07	2.39
6	1.5	2.60	2.81
7	1.75	3.17	3.31
8	2.0	3.77	3.86

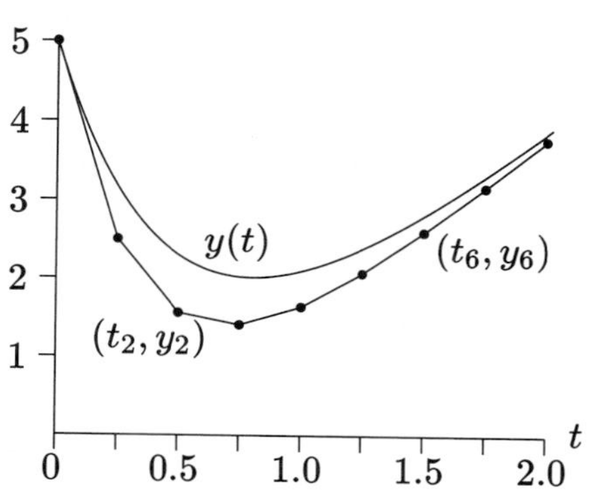

Figure 7.3.1

If the solution of (26) is a straight line, Euler's method gives the solution exactly. By the Fundamental Theorem of Calculus (or by Taylor's
theorem), the deviation of $y(t)$ from a straight line can be estimated if we
can bound the second derivative of y. Thus, we expect that error estimates for Euler's method should involve bounds on the second derivative of $y(t)$, that is, on the first derivatives of f.

❏ **Theorem 7.3.1** Let f be a continuously differentiable function of two
variables on a rectangle R of the form $R \equiv [a, b] \times [y_o - c, y_o + c]$. Suppose
that the solution $y(t)$ of (26) exists on the interval $[a, b]$ and that the graph
of $y(t)$ lies in the rectangle. Let N, h, $\{t_n\}$, and $\{y_n\}$ be as defined above
and suppose that the points $\{(t_n, y_n)\}$ all lie within R. Then, for each n,

$$|y(t_n) - y_n| \quad \leq \quad \frac{Lh}{2K}(e^{(b-a)K} - 1), \qquad (29)$$

where $K \equiv \sup_R |\frac{\partial f}{\partial y}|$ and $L \equiv \sup_R |\frac{\partial f}{\partial t} + f\frac{\partial f}{\partial y}|$.

Proof.   The numbers $\{y_n\}$ are defined by the recursion relation (27).
We can also write a recursion relation for the numbers $\{y(t_n)\}$ as follows. Since $y(t)$ and f are continuously differentiable, the composition
$f(t, y(t))$ is also continuously differentiable. By (26), $y'(t)$ is continuously
differentiable, so $y(t)$ is twice continuously differentiable. Therefore, by
Taylor's theorem,

$$y(t_{n+1}) = y(t_n) + hy'(t_n) + \frac{h^2}{2!}y''(\xi_n)$$

for some point ξ_n between t_n and t_{n+1}. Since $y(t)$ satisfies (26), we can rewrite this as

$$y(t_{n+1}) \;=\; y(t_n) \;+\; hf(t_n, y(t_n)) \;+\; \frac{h^2}{2!}\left[\frac{\partial f}{\partial t} \;+\; f\frac{\partial f}{\partial y}\right](\xi_n, y(\xi_n)). \quad (30)$$

In the third term on the right, all three functions are evaluated at the point $(\xi_n, y(\xi_n))$. Subtracting (27) from (30) and taking absolute values, we find

$$|y(t_{n+1}) - y_{n+1}| \;\leq\; |y(t_n) - y_n| + h|f(t_n, y(t_n)) - f(t_n, y_n)| + \frac{h^2}{2!}L \quad (31)$$

$$\leq\; |y(t_n) - y_n| + hK|y(t_n) - y_n| + \frac{h^2}{2!}L, \quad (32)$$

where we used the Mean Value Theorem in the second step. Throughout, we used the hypotheses that the points $(t_n, y(t_n))$ and (t_n, y_n) are in the rectangle. For simplicity, we write the error at the nth step as $E_n \equiv |y(t_n) - y_n|$ and set $A \equiv (1 + hK)$ and $B \equiv \frac{h^2}{2!}L$. Then (32) can be written

$$E_{n+1} \;\leq\; AE_n + B.$$

Iterating this inequality, and using the partial sum of the geometric series gives

$$E_{n+1} \;\leq\; B(1 + A + A^2 + \dots + A^n)$$

$$\leq\; B\frac{A^{n+1} - 1}{A - 1}$$

$$\leq\; \frac{Lh}{2K}((1 + Kh)^{n+1} - 1).$$

To get an estimate that is independent of n, we use the estimate $1 + Kh \leq e^{Kh}$, which follows easily from the power series representation for e^{Kh} or by the Fundamental Theorem of Calculus (problem 3). Since

$$(1 + Kh)^{n+1} \;\leq\; (e^{Kh})^{n+1} \;\leq\; (e^{Kh})^N \;=\; e^{KhN} \;=\; e^{K(b-a)},$$

we obtain

$$|y(t_n) - y_n| \;\leq\; \frac{Lh}{2K}(e^{K(b-a)} - 1)$$

which is what we set out to prove.   ❏

Euler's method is called a first-order method because the error bound decreases proportionally to the first power of the step size h. We note that the right-hand side of (29) is an upper bound for the error; the actual error may be much less. In addition, we used the rather crude estimate $1 + x \ \leq \ e^x$ in order to get a bound independent of n. Nevertheless, it is true that Euler's method is not a very efficient method in that one must make h very small (thus the number of intervals, N, very large) in order to approximate the solution well. As in the case of the numerical estimation of integrals discussed in Section 3.4, there are serious drawbacks to choosing h too small because of the tiny but real round-off error that may occur with each computational step. This is explored further in project 2. Thus, the design of higher-order methods and the proof of error bounds have played (and play) an important role in the study of ordinary and partial differential equations. The proofs of the error estimates for higher-order algorithms are more complicated than the one above but use the same analytical ideas.

A natural question has probably come to mind. What is this mysterious rectangle R which occurs in the hypotheses of Theorem 7.3.1? Since we are using numerical techniques precisely because we *can't* solve the equation explicitly, how can we know what R is? The answer is that we must derive estimates on the solution by using the ideas of Section 7.2.

Example 2   Suppose that we want to use Euler's method to approximate the solution of

$$y'(t) \ = \ \sin y(t), \qquad y(0) = 3 \qquad\qquad (33)$$

on the interval $[0, 2]$. Then $f(t, y) = \sin y$, so $\frac{\partial f}{\partial t} = 0$ and $\frac{\partial f}{\partial y} = \cos y$. Thus, whatever the rectangle R is, the constants K and L in Theorem 7.3.1 can be taken to equal 1. Therefore, estimate (29) is

$$|y(t_n) - y_n| \ \leq \ \frac{h}{2}(e^2 - 1).$$

If we want the Euler's method points (t_n, y_n) to be within 10^{-3} of the true solution, we can guarantee that if we choose h small enough so that

$$\frac{h}{2}(e^2 - 1) \ \leq \ 10^{-3}.$$

In Example 2 we didn't need to determine a suitable R because the function f and its derivatives were uniformly bounded everywhere. That is not the case in the following example.

Example 3 Let's use Euler's method to approximate the solution of

$$y'(t) \ = \ ty(t)\sin y(t), \qquad y(0) = 3$$

on the interval $[0,2]$. Since $f(t,y) = ty\sin y$, we have $\frac{\partial f}{\partial t} = y\sin y$ and $\frac{\partial f}{\partial y} = t\sin y + ty\cos y$. Notice that the curve $\{(t,0)\,|\,0 \le t \le \infty\}$ is an orbit, so, by Theorem 7.1.2, the solution, $y(t)$, remains positive. Thus, for $0 \le t \le 2$,

$$-2y \ \le \ ty\sin y \ \le \ 2y,$$

so by Theorem 7.2.4, $y(t)$ exists on the interval $[0,2]$ and

$$3e^{-2t} \ \le \ y(t) \ \le 3e^{2t}$$

Notice that we do not know what the solution $y(t)$ is, but we have estimates on it from above above and below. Thus, the solution curve $y(t)$ remains in the rectangle

$$R \ \equiv \ [0,2] \times [0, 3e^4].$$

To see that the same is true for the Euler's method iterates, we estimate

$$y_{n+1} \ = \ y_n + h t_n y_n \sin y_n \ \le \ (1 + 2h)y_n,$$

so

$$y_n \ \le \ (1 + 2h)^n 3 \ \le \ 3e^{2hN} \ \le \ 3e^4.$$

Similarly,

$$y_{n+1} \ = \ y_n + h t_n y_n \sin y_n \ \ge \ (1 - 2h)y_n$$

so if $h \le \frac{1}{2}$, we have $y_n \ge 0$ for all n. Thus, the points (t_n, y_n) also remain in the rectangle R. In the rectangle, we know that $|f(t,y)| \le 6e^4$, so we can estimate

$$K \ \equiv \ \sup_R \left|\frac{\partial f}{\partial y}\right| \ \le \ 2 + 6e^4$$

and

$$L \ \equiv \ \sup_R \left|\frac{\partial f}{\partial t} + f\frac{\partial f}{\partial y}\right| \ \le \ 3e^4 + 6e^4(2 + 6e^4).$$

Using these estimates for K and L, we can determine, by Theorem 7.3.1 and (29), how small we must choose h to guarantee that the Euler's method approximations are within any given distance from the true values, $y(t_n)$.

Problems

1. Use Euler's method to approximate the solution to the following initial-value problem on the interval $[0, 2]$:

$$y'(t) = ty(t) - 2, \quad y(0) = 3.$$

2. Use Euler's method to approximate the solution to the following initial-value problem on the interval $[0, 2]$:

$$y'(t) = t^2 + e^{-y(t)^2}, \quad y(0) = 0.$$

3. Use the Fundamental Theorem of Calculus to prove that $1 + x \leq e^x$ for all $x \geq 0$.

4. Use Theorem 7.3.1 and the methods of Example 3 to determine how small h must be chosen so that the Euler points y_n are within 10^{-4} of the true values $y(t_n)$ for the differential equation in problem 1 if $y(0) = 4$.

5. Use Theorem 7.3.1 and the methods of Example 3 to determine how small h must be chosen so that the Euler points y_n are within 10^{-4} of the true values $y(t_n)$ for the differential equation in problem 2.

6. Use Euler's method to approximate the solution to the following initial-value problem on the interval $[0, 2]$:

$$y'(t) = y(t)^2, \quad y(0) = 1.$$

Did you see the blowup of the solution at $t = 1$? Why not?

7. Use Euler's method to approximate the solution to the following initial-value problem on the interval $[0, 6]$:

$$y'(t) = \sin(e^t), \quad y(0) = 1.$$

Do you think that Euler's method gives a good approximation? Why or why not? What is the size of the error bound given by (29) for your choice of h?

8. Design an Euler's method to solve a pair of coupled ordinary differential equations:

$$
\begin{aligned}
y'(t) &= f(t, y(t), z(t)), & y(t_o) &= y_o \\
z'(t) &= g(t, y(t), z(t)), & z(t_o) &= z_o.
\end{aligned}
$$

9. In a certain model of competing species, $N_1(t)$ and $N_2(t)$ represent the amount of species 1 and species 2, respectively, in some normalized units. Suppose that $N_1(t)$ and $N_2(t)$ satisfy the following differential equations:

$$
\begin{aligned}
N_1'(t) &= N_1(t)(1 - N_1(t) - \frac{1}{2}N_2(t)), & N_1(0) &= n_1 \\
N_2'(t) &= N_2(t)(1 - N_2(t) - \frac{1}{2}N_1(t)), & N_2(0) &= n_2.
\end{aligned}
$$

Use the Euler's method scheme from problem 8 to investigate the behavior of solution curves $(N_1(t), N_2(t))$ in the $N_1 - N_2$ plane for different choices of (n_1, n_2).

10. In a certain model of predator-prey interactions, the numbers of the two species, $N_1(t)$ and $N_2(t)$, satisfy the following differential equations:

$$\begin{aligned}
N_1'(t) &= N_1(t) - N_1(t)N_2(t), & N_1(0) &= n_1 \\
N_2'(t) &= -N_2(t) + N_1(t)N_2(t), & N_2(0) &= n_2.
\end{aligned}$$

Use the Euler's method scheme from problem 8 to investigate the behavior of solution curves $(N_1(t), N_2(t))$ in the $N_1 - N_2$ plane for different choices of (n_1, n_2). In fact, all solutions are periodic in t. Is that what you found? Why not?

Projects

1. We have seen that some differential equations can be solved explicitly, and we have analyzed a simple numerical method for finding approximate solutions to others. We can also obtain information about solutions by using analytical tools without solving explicitly or using numerical methods. We will illustrate some of the simplest ideas in the case of the logistic equation

$$y'(t) = ay(t)(b - y(t)), \qquad y(0) = y_o.$$

Here a, b, and y_o are positive and $y(t)$ represents the population size at time t in some units.

(a) First we suppose that $y_o < b$. Explain why $y(t)$ is always increasing. Explain why there can be no time t at which $y(t) = b$. (Hint: recall Theorem 7.1.2.) Explain why the solution $y(t)$ exists for all positive times and is unique.

(b) Again suppose that $y_o < b$. Explain why $c \equiv \lim_{t \to \infty} y(t)$ exists. Prove that if $c < b$, then $y(t)$ will be eventually higher than c, giving a contradiction. Conclude that $y(t) \to b$ as $t \to \infty$.

(c) Again suppose that $y_o < b$. Compute $y''(t)$ in terms of $y(t)$ and use the result to help you draw an accurate sketch of the solution.

(d) Suppose that $y_o > b$. Use the ideas in (a), (b), and (c) to show that the solution exists for all positive times, and draw an accurate graph of it.

2. The purpose of this project is to analyze the trade-off between small step size and round-off error in Euler's method. Every time the computer calculates an Euler iterate, it makes an error whose size is bounded above

by a number ε, which depends on the machine being used. Thus, instead of computing the iterates given by (27) in Section 7.3, the machine is really computing the iterates $\bar{y}_n$ that are given by the recursion relation

$$\bar{y}_{n+1} \;=\; \bar{y}_n + f(t_n, \bar{y}(t_n)) + \varepsilon_n, \tag{34}$$

where we know that the numbers ε_n satisfy $|\varepsilon_n| \leq \varepsilon$.

(a) Using the terminology and hypotheses of Theorem 7.3.1, except that we replace (27) in Section 7.3 by (34) above and y_n by $\bar{y}_n$, prove that

$$E_{n+1} \leq (1 + hK)E_n + \frac{h^2}{2!}L + \varepsilon.$$

(b) Iterate this inequality to prove that for all $n \leq N$,

$$E_n \;\leq\; \left(\frac{Lh}{2K} + \frac{\varepsilon}{hK} \right)(e^{(b-a)K} - 1).$$

(c) Explain why the error bound cannot be made arbitrarily small no matter how we choose h. How should one choose h to make the error bound as small as possible?

(d) Suppose that $\varepsilon = 10^{-8}$. What is the maximum accuracy you can get by using Euler's method for the differential equations in problems 1 and 2 of Section 7.3?

3. The purpose of this project is to show how a local existence and uniqueness result analogous to Theorem 7.1.1 can be proved for systems. We will consider the initial-value problem for the system

$$
\begin{aligned}
y'(t) &= f(t, y(t), z(t)), & y(t_o) &= y_o & (35)\\
z'(t) &= g(t, y(t), z(t)), & z(t_o) &= z_o & (36)
\end{aligned}
$$

where f and g are continuously differentiable functions of three independent variables. Let $I_T = [t_o - T, t_o + T]$, as in Section 7.1. We denote the set of pairs of continuous functions on I_T with values in $\mathbb{R}$ by $C(I_T : \mathbb{R}^2)$. If $(f, g) \in C(I_T : \mathbb{R}^2)$, we define

$$\|(f, g)\|_\infty \;\equiv\; \|f\|_\infty + \|g\|_\infty.$$

According to problem 7 in Section 5.3, $C(I_T : \mathbb{R}^2)$ is a complete normed linear space.

(a) Show that if $y(t)$ and $z(t)$ satisfy (35) and (36), then they also satisfy a pair of integral equations.

(b) Follow the proof of Theorem 7.1.1 to show that if T is small enough, the integral equations can be solved by iteration.

(c) Show that the solutions of the integral equations are continuously differentiable and satisfy the differential equations.

(d) Show that the solutions are unique.

4. Suppose that $y(t)$ and $z(t)$ satisfy (35) and (36) on a time interval $a \leq t \leq b$. The curve $\{(y(t), z(t)) \mid t \in [a, b]\}$ in the $y - z$ plane is called an **orbit**. Prove the analogue of Theorem 7.1.2 by showing that if two orbits over the time interval $[a, b]$ cross, then they are identical.

5. Suppose that $y(t)$ and $z(t)$ satisfy (35) and (36) on a time interval $a < t < b$. The solution pair $(y(t), z(t))$ is said to go to ∞ at finite time b if, for every M, there is a $t_M < b$ so that

$$\sqrt{y(t)^2 + z(t)^2} \geq M \qquad \text{for } t_M \leq t < b.$$

We define similarly what it means for $(y(t), z(t))$ to go to ∞ at finite time a.

(a) Explain geometrically what these definitions mean.

(b) Prove the analogue of Theorem 7.2.1. That is, show that if the local solution of (35) and (36) does not go to ∞ in finite time, then the solution exists for all times.

(c) Let $n_1 > 0$ and $n_2 > 0$, and let $N_1(t)$ and $N_2(t)$ be the local solutions of the differential equations in problem 9 of Section 7.3, which are guaranteed to exist by project 3. Prove that the solution must stay in the positive orthant. Hint: use the result of project 4.

(d) Prove that the solutions $N_1(t)$ and $N_2(t)$ exist for all positive times. Hint: what are the signs of $N_1'(t)$ and $N_2'(t)$ if either $N_1(t)$ or $N_2(t)$ is large?

CHAPTER 8

Complex Analysis

In this chapter we develop the rudiments of the theory of analytic functions of one complex variable. We have three main purposes. The first is to show that the ideas of classical real analysis (the Riemann integral, series, interchanging limits and integration, etc.) play a fundamental role in complex analysis. Second, we want to answer the question about which functions of a real variable are equal to their Taylor series. Finally, we want to introduce the class of analytic functions because it plays a crucial role in both classical and modern analysis and in the applications of analysis. Throughout, we use the properties of real-valued functions of two variables, discussed in Section 4.6. For the further development of the subject and many applications, see [6] and [23].

8.1 Analytic Functions

The analytic functions are a special class of functions which are defined on (subsets of) the complex numbers and take values in the complex numbers. The natural domains of definition of analytic functions are open sets in the complex plane. We say that a set D is **open** if for every $z \epsilon D$ there is a circle about z so that every complex number in the interior of the circle is also in D. Thus, for example, the set $\{z \epsilon \mathbb{C} \mid Re(z) > 0\}$ is open while the set $\{z \epsilon \mathbb{C} \mid Re(z) \geq 0\}$ is not. Other examples are given in problem 1.

Definition. Let D be an open set in $\mathbb{C}$. A function f defined on D, taking values in $\mathbb{C}$, is called **analytic** if for every $z \epsilon D$ the limit

$$\lim_{h \to 0} \frac{f(z + h) - f(z)}{h} \equiv f'(z) \tag{1}$$

exists.

In the definition of analytic, each h is a complex number. Saying that $h \to 0$ means that $|h| \to 0$. Thus h can approach zero in many different ways. For example, if $h_n = \frac{1}{n}$, then the sequence $\{h_n\}$ approaches zero along the positive real axis. If $h_n = -i\frac{1}{n}$, then the sequence $\{h_n\}$ approaches zero along the negative imaginary axis. Let θ be a fixed small real number and define $h_n = \frac{1}{n}e^{in\theta}$. Then, the complex numbers $\{h_n\}$ spiral in toward zero. Thus, the statement that the limit in (1) exists means that the limit is independent of the way in which $h \to 0$. We shall see that this is a very strong restriction and that analytic functions have very special properties.

Recall that every complex number $z = x + iy$ is specified by the pair of real numbers (x, y). Since the value of f at z is a complex number, it is specified by giving its real and imaginary parts, u and v. Since the value of f depends on z and therefore on x and y, u and v will depend on x and y. Thus, specifying the function f is the same as specifying two real-valued functions, u and v, defined on (a subset of) $\mathbb{R}^2$ so that

$$f(z) = u(x, y) + iv(x, y).$$

The functions $u(x, y)$ and $v(x, y)$ are called the **real** and **imaginary** parts of the function f.

Example 1 Suppose f is the function on $\mathbb{C}$ which squares each complex number; that is, $f(z) = z^2$. Then, since $f(z) = (x+iy)^2 = (x^2-y^2)+2xyi$, we have $u(x, y) = x^2 - y^2$ and $v(x, y) = 2xy$.

As we shall see, analyticity puts certain restrictions on u and v. In what follows, we discuss only functions f whose real and imaginary parts are continuously differentiable. First, suppose that $h \to 0$ through real numbers; that is, $h = \lambda$ where λ is real. Then,

$$\frac{f(z+h) - f(z)}{h} = \frac{(u(x+\lambda, y) + iv(x+\lambda, y)) - (u(x,y) + iv(x,y))}{\lambda}$$

$$= \frac{u(x+\lambda, y) - u(x, y)}{\lambda} + i\frac{v(x+\lambda, y) - v(x, y)}{\lambda}.$$

Thus,

$$\lim_{h \to 0} \frac{f(z+h) - f(z)}{h} = u_x(x, y) + iv_x(x, y), \qquad (2)$$

where u_x and v_x are the partial derivatives of u and v with respect to x. On the other hand, suppose that $h \to 0$ through purely imaginary numbers. That is, $h = i\mu$ where μ is real and $\mu \to 0$. Then,

$$\frac{f(z+h) - f(z)}{h} = \frac{(u(x,y+\mu) + iv(x,y+\mu)) - (u(x,y) + iv(x,y))}{i\mu}$$

$$= \frac{v(x,y+\mu) - v(x,y)}{\mu} - i\frac{u(x,y+\mu) - u(x,y)}{\mu},$$

so

$$\lim_{h \to 0} \frac{f(z+h) - f(z)}{h} = v_y(x,y) - iu_y(x,y). \tag{3}$$

According to the definition, if f is analytic, (2) and (3) must be the same, so

$$u_x(x,y) + iv_x(x,y) = v_y(x,y) - iu_y(x,y).$$

Since two complex numbers are equal if and only if their real and imaginary parts are equal,

$$u_x(x,y) = v_y(x,y) \tag{4}$$
$$u_y(x,y) = -v_x(x,y). \tag{5}$$

Equations (4) and (5) are known as the **Cauchy-Riemann equations**. They are not only necessary but, under suitable hypotheses, also sufficient for f to be analytic.

❏ **Theorem 8.1.1** Suppose that u and v are continuously differentiable real-valued functions defined on a domain $D \subseteq \mathbb{C}$. Then $f(z) = u(x,y) + iv(x,y)$ is analytic on D if and only if the Cauchy-Riemann equations hold in D.

Proof. We have already shown that if f is analytic, then the Cauchy-Riemann equations hold. We shall prove the converse. Suppose that the Cauchy-Riemann equations hold in D. Let z be in D and let $h_n = x_n + iy_n$ be any sequence such that $z + h_n \in D$ for all n and $h_n \to 0$. It follows that $|h_n| \to 0$, and this implies that $x_n \to 0$ and $y_n \to 0$. For simplicity of notation, we suppose that $z = 0$; the proof for a general z is the same but the expressions for the difference quotients are larger.

$$\frac{f(h_n) - f(0)}{h_n} = \frac{u(x_n, y_n) - u(0,0)}{x_n + iy_n} + i\frac{v(x_n, y_n) - v(0,0)}{x_n + iy_n} \tag{6}$$

$$= \frac{u(x_n, y_n) - u(0, y_n) + u(0, y_n) - u(0, 0)}{x_n + i y_n} \tag{7}$$

$$+ i \frac{v(x_n, y_n) - v(x_n, 0) + v(x_n, 0) - v(0, 0)}{x_n + i y_n} \tag{8}$$

$$= \frac{u_x(\tau_1, y_n) x_n + u_y(0, \tau_2) y_n}{x_n + i y_n} + i \frac{v_y(x_n, \tau_3) y_n + v_x(\tau_4, 0) x_n}{x_n + i y_n} \tag{9}$$

$$= \frac{u_x(\tau_1, y_n) x_n + i u_x(x_n, \tau_3) y_n}{x_n + i y_n} + i \frac{v_x(\tau_4, 0) x_n + i v_x(0, \tau_2) y_n}{x_n + i y_n} \tag{10}$$

$$\longrightarrow \quad u_x(0, 0) + i v_x(0, 0). \tag{11}$$

In going from (7) and (8) to (9), we used the Mean Value Theorem four times. In going from (9) to (10), we used the Cauchy-Riemann equations and rearranged the terms. From the Mean Value Theorem, we know that τ_1, τ_2, τ_3, and τ_4 all converge to zero as $x_n \to 0$ and $y_n \to 0$. Using the continuity of u_x and v_x, the convergence in the last step follows from the result in problem 2.   ❏

Example 2 (polynomials)   Suppose that $f(z) = c = c_1 + i c_2$ is a constant function. Since all the partial derivatives of $u = c_1$ and $v = c_2$ are zero, the Cauchy-Riemann equations hold and the function is analytic. Now, suppose $f(z) = z = x + iy$. Then $u(x, y) = x$ and $v(x, y) = y$, and it is easy to check that the Cauchy-Riemann equations hold; thus, $x + iy$ is analytic in the whole complex plane, $\mathbb{C}$. Furthermore, sums, products, quotients, and the composition of analytic functions are again analytic (except where the denominators vanish). These results can be proved in the same way in which we proved that sums, products, quotients, and compositions of differentiable functions are again differentiable (Theorems 4.1.2 and 4.1.3). Thus, polynomials in z and quotients of polynomials in z are analytic except where the denominators vanish. The fact that such functions are analytic can also be proved directly; see problem 3.

Note that if $p(x, y)$ and $q(x, y)$ are polynomials in x and y, then

$$f(z) = p(x, y) + i q(x, y)$$

will be a well-defined continuously differentiable (in fact, infinitely often differentiable) function from $\mathbb{C}$ to $\mathbb{C}$. But, in general, f will *not* be analytic because the Cauchy-Riemann equations won't hold. For example, the function $f(z) = x - iy$ is not analytic. In this case, $u(x, y) = x$ and $v(x, y) = -y$, so $u_x = 1 \neq -1 = v_y$. If, however, f is a polynomial in $(x + iy)$, then, by the argument above, f is analytic and the Cauchy-Riemann equations hold. For example, suppose $f(z) = z^2 = x^2 - y^2 + i(2xy)$. In this case, $u = x^2 - y^2$ and $v = 2xy$, so

$$u_x = 2x = v_y \quad \text{and} \quad u_y = -2y = -v_y.$$

Thus $f(z) = z^2$ is analytic.

Example 3 (series) Since polynomials in z are analytic it is natural to ask about power series in z. According to Theorem 6.5.3, the power series

$$f(z) \;=\; \sum_{j=0}^{\infty} a_j (z - z_o)^j \tag{12}$$

converges uniformly in every closed disk about z_0 that is smaller than the radius of convergence R of the series. And according to Theorem 6.4.2, the function to which a power series converges is infinitely often differentiable inside the radius of convergence, and the derivatives can be computed by differentiating the series term by term. Although Theorem 6.4.2 was proved for real power series, the same proof shows that the result holds for series in powers of z. Thus, in order to check whether the limit of the series in (12) is analytic, we can just differentiate the series term by term. Let u_j and v_j be the real and imaginary parts of $a_j(z - z_o)^j$. Then $f(z) = \sum u_j + i \sum v_j$ and

$$\left(\sum u_j \right)_x = \sum (u_j)_x = \sum (v_j)_y = \left(\sum v_j \right)_y,$$

since $(u_j)_x = (v_j)_y$ for each j because $a_j(z - z_o)^j$ is analytic. Similarly, $(\sum u_j)_y = -(\sum v_j)_x$. Thus, the real and imaginary parts of f satisfy the Cauchy-Riemann equations so the function f, defined by (12), is analytic inside its radius of convergence.

For example, since the power series for e^z has radius of convergence $R = \infty$, e^z is analytic on $\mathbb{C}$. Similarly,

$$\sin z \equiv \frac{e^{iz} - e^{-iz}}{2i}, \quad \text{and} \quad \cos z \equiv \frac{e^{iz} + e^{-iz}}{2}$$

are analytic functions on $\mathbb{C}$ whose restrictions to the real axis are $\sin x$ and $\cos x$.

Problems

1. Which of the following subsets of $\mathbb{C}$ are open?

    (a) $\{z \in \mathbb{C} \,|\, |z| \leq 1\}$.

    (b) $\{z \in \mathbb{C} \,|\, |z| > 1\}$.

    (c) $\{z \in \mathbb{C} \,|\, Re(z) > 0 \text{ and } Im(z) > 0\}$.

    (d) $\{z \in \mathbb{C} \,|\, Re(z) > 0 \text{ and } Im(z) \geq 0\}$.

    (e) $\{z \in \mathbb{C} \,|\, p(z) \neq 0\}$, where p is a polynomial in z.

2. Suppose that $\{x_n\}$ and $\{y_n\}$ are sequences of real numbers such that $x_n + iy_n \neq 0$ for any n and $x_n + iy_n \to 0$. Suppose that $\{\alpha_n\}$ and $\{\beta_n\}$ are sequences of real numbers such that $\alpha_n \to \gamma$ and $\beta_n \to \gamma$. Prove that

$$\lim_{n \to \infty} \frac{\alpha_n x_n + i\beta_n y_n}{x_n + iy_n} = \gamma.$$

    Hint: write $\alpha_n = (\alpha_n - \alpha) + \alpha$ and $\beta_n = (\beta_n - \beta) + \beta$.

3. For each of the following functions, find the real and imaginary parts in terms of x and y and verify that the Cauchy-Riemann equations hold on the indicated domain:

    (a) $f(z) = z^3$ on $\mathbb{C}$.

    (b) $f(z) = \frac{1}{z}$ on $\{z \in \mathbb{C} \,|\, z \neq 0\}$.

    (c) $f(z) = \sum_{j=0}^{\infty} z^j$ on $\{z \in \mathbb{C} \,|\, |z| < 1\}$.

4. Use the Cauchy-Riemann equations to determine which of the following functions are analytic:

    (a) $x^2 - y^2 + i(2xy)$ (b) $e^x \cos y + ie^x \sin y$ (c) $x^2 + iy^2$

    (d) $x - iy$ (e) $\ln\sqrt{x^2 + y^2} + i\arcsin\frac{y}{x}$ (f) $\frac{1}{x - iy - 5}$

5. We say that a function, f, which takes $\mathbb{C}$ to $\mathbb{C}$ is **continuous** if $z_n \to z_o$ implies $\lim_{n \to \infty} f(z_n) = f(z_o)$.

    (a) Prove that f is continuous on $\mathbb{C}$ if and only if its real and imaginary parts are continuous functions on $\mathbb{R}^2$.

    (b) Prove that if f is analytic, then f is continuous.

6. Suppose that f and g are analytic functions on $\mathbb{C}$. Prove that $f + g$ and fg are analytic and that $(f + g)'(z) = f'(z) + g'(z)$ and $(f(z)g(z))' = f'(z)g(z) + f(z)g'(z)$.

7. Suppose that f and g are analytic functions on $\mathbb{C}$. Prove that $f \circ g$ is analytic and $(f(g(z))' = f'(g(z))g'(z)$. Hint: write out the real and imaginary parts of $f \circ g$ in terms of the real and imaginary parts of f and g.

8. Say where each of the following functions is analytic and compute its derivative:

    (a) $f(z) = (\cos z)^2 + e^{2z}$.

    (b) $f(z) = e^{1/z}$.

    (c) $f(z) = e^{1/\sin z}$.

9. Let $f(z) = \frac{1}{1-z}$. Where is f analytic?

    (a) Find a power series representation for f in the region $\{z \,|\, |z| < 1\}$.

    (b) Find representation for f as an infinite series of powers of $\frac{1}{z}$ valid in the region $\{z \,|\, |z| > 1\}$.

    (c) Find a power series representation for f around the point $z = i$. what is its radius of convergence? Hint: write

$$f(z) = (1 - i)^{-1} \left(1 - \frac{z - i}{1 - i}\right)^{-1}.$$

    (d) Find a power series representation for $g(z) = \left(\frac{1}{1-z}\right)^2$ around the point $z = i$.

10. What is the radius of convergence of the series

$$(z - 1) - \frac{1}{2}(z - 1)^2 + \frac{1}{3}(z - 1)^3 - \frac{1}{4}(z - 1)^4 + \ldots?$$

Define $\ln z$ to be this series in the region of convergence. Is $\ln z$ analytic there? Compute $(\ln z)'$.

8.2 Integration on Paths

We can extend the notions of integration and differentiation easily to complex-valued functions of a real variable. If $g(t)$ is such a function, then we say that g is **continuous** if $\lim_{n \to \infty} g(t_n) = g(t_o)$ whenever $t_n \to t_o$. We say that g is **continuously differentiable** if the limit of the difference quotient $h^{-1}(g(t + h) - g(t))$ exists and is continuous for each t. We can write $g(t) = h(t) + ik(t)$, where $h(t)$ and $k(t)$ are the real and imaginary parts of $g(t)$. It is easy to check (problem 1) that g is continuous if and only if h and k are continuous and g is differentiable if and only if h and k are differentiable, in which case $g'(t) = h'(t) + ik'(t)$. If g is continuous, we define its integral on finite intervals by

$$\int_a^b g(t)\, dt = \int_a^b h(t)\, dt + i \int_a^b k(t)\, dt.$$

Recall that any complex number can be written in polar form $x + iy = z = e^{i\theta}|z|$. Note that $|x| \leq |z|$ and $Re\{e^{-i\theta}z\} = |z|$. If we let $\theta = Arg\left(\int_a^b g(t)\,dt\right)$, then

$$\left|\int_a^b g(t)\,dt\right| = Re\{e^{-i\theta}\int_a^b g(t)\,dt\} \tag{13}$$

$$= \int_a^b Re\{e^{-i\theta}g(t)\}\,dt \tag{14}$$

$$\leq \int_a^b |g(t)|\,dt, \tag{15}$$

where we used Theorem 3.3.4 in the last step.

Example 1 (the Fourier transform)   Let f be a real-valued continuous function on $\mathbb{R}$ such that the improper Riemann integral $\int_{-\infty}^{\infty}|f(t)|\,dt$ exists. According to problem 12 of Section 3.6, it follows that the improper Riemann integrals, $\int_{-\infty}^{\infty}(\sin t)f(t)\,dt$ and $\int_{-\infty}^{\infty}(\cos t)f(t)\,dt$ exist also. For every $x \in \mathbb{R}$, we define

$$\hat{f}(x) \equiv \frac{1}{\sqrt{2\pi}}\int_{-\infty}^{\infty} e^{-ixt}f(t)\,dt$$

$$= \frac{1}{\sqrt{2\pi}}\int_{-\infty}^{\infty}(\cos xt)f(t)\,dt - i\int_{-\infty}^{\infty}(\sin xt)f(t)\,dt.$$

The function $\hat{f}$ is called the **Fourier transform** of f. The function f can sometimes be recovered from its Fourier transform by the formula

$$f(t) \equiv \frac{1}{\sqrt{2\pi}}\int_{-\infty}^{\infty} e^{ixt}\hat{f}(x)\,dx.$$

Detailed analysis of the Fourier transform requires more advanced techniques (like the Lebesgue integral) than we have at our disposal. However, we do want to highlight one important property of the Fourier transform. Suppose that f is zero outside the interval $[a,b]$. For every complex number $z = x + iy$, we define

$$\hat{f}(z) \equiv \frac{1}{\sqrt{2\pi}}\int_{\infty}^{\infty} e^{-izt}f(t)\,dt$$

$$= \frac{1}{\sqrt{2\pi}}\int_a^b e^{-ixt}e^{yt}f(t)\,dt.$$

Notice that the integral makes sense since the interval is finite and the integrand is a continuous function of t; thus, $\hat{f}(z)$ is a well-defined function on $\mathbb{C}$. The real and imaginary parts of $\hat{f}(z)$ are, respectively,

$$u(x,y) \;=\; \frac{1}{\sqrt{2\pi}} \int_a^b e^{yt}(\cos xt) f(t)\,dt$$

$$v(x,y) \;=\; -\frac{1}{\sqrt{2\pi}} \int_a^b e^{yt}(\sin xt) f(t)\,dt.$$

Using Theorem 5.2.4, one can compute the partial derivatives of u and v by differentiating under the integral sign. When one does so (problem 2), one finds that u and v satisfy the Cauchy-Riemann equations. Thus, $\hat{f}$ is the restriction to the real axis of a function that is analytic in the entire complex plane. This property of the Fourier transform plays an important role in the theory of signal transmission in electrical engineering and in scattering theory in physics.

We shall now define the integral of a complex-valued continuous function f on a smooth curve C of finite length in the complex plane. C is called a **smooth curve** if there is an interval $[a,b]$ in $\mathbb{R}$ and continuously differentiable real-valued functions $x(t)$ and $y(t)$ on $[a,b]$ so that $z(t) = x(t) + iy(t)$ sweeps out the curve from it's starting point, $\omega = z(a)$, to its ending point, $\tau = z(b)$, and $|z'(t)| \neq 0$. We call $z(t)$ a **parameterization** of the curve. The **length** of the curve, L, is defined to be

$$L \;\equiv\; \int_a^b |z'(t)|\,dt \;=\; \int_a^b \sqrt{(x'(t))^2 + (y'(t))^2}\,dt,$$

a formula which should be familiar from Calculus. By analogy with the Riemann integral of real-valued continuous functions on $\mathbb{R}$, we want the integral of the complex-valued function f over the smooth curve C to be given approximately by the Riemann sum

$$\int_C f(z)\,dz \;\approx\; \sum_{j=1}^N f(z_j)(z_j - z_{j-1}),$$

where the points $z_0 = \omega$, $z_N = \tau$, and $z_1, z_2, \ldots, z_{N-1}$ lie along the curve in order between z_0 and z_N. See Figure 8.2.1.

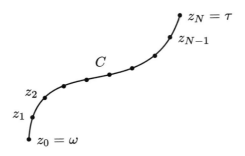

Figure 8.2.1

Let $\{t_0, t_1, ..., t_N\}$ be a partition of the interval $[a, b]$ and let $z_j = z(t_j)$. Then,

$$\sum_{j=1}^{N} f(z_j)(z_j - z_{j-1}) = \sum_{j=1}^{N} f(z(t_j))(z(t_j) - z(t_{j-1})) \qquad (16)$$

$$= \sum_{j=1}^{N} f(z(t_j)) \frac{z(t_j) - z(t_{j-1})}{t_j - t_{j-1}} (t_j - t_{j-1}) \quad (17)$$

$$\approx \sum_{j=1}^{N} f(z(t_j)) z'(t_j)(t_j - t_{j-1}). \qquad (18)$$

Notice that (18) is a Riemann sum for $\int_a^b f(z(t))z'(t)\, dt$, which is an integral of a continuous function on a finite interval on the real line, since f is continuous and $z(t)$ is continuously differentiable. This gives us the idea for the following definition.

Definition.   Let C be a smooth curve of finite length in the complex plane and let f be continuous on an open set containing C. Let $z(t), a \leq t \leq b$, be a parameterization of C. Then, we define

$$\int_C f(z)\, dz \equiv \int_a^b f(z(t))z'(t)\, dt. \qquad (19)$$

It can be shown that this definition is independent of the parameterization of the curve C chosen. Note, however, that a parameterization sweeps out the curve in a given direction (from ω to τ). Definition (19) shows that if two parameterizations differ in direction, then the corresponding integrals will differ by a minus sign. It is straightforward to prove (see problems 7 and 8) that the integral has the other properties that we expect. For example, the integral of a sum of functions is the sum of the integrals and

$$\left| \int_C f(z)\,dz \right| \;\leq\; (\max_{z\,\epsilon\,C} |f(z)|) L. \tag{20}$$

We can also write $\int_C f(z)\,dz$ in terms of real line integrals in $\mathbb{R}^2$. If $u(x,y)$ and $v(x,y)$ are the real and imaginary parts of f, then

$$\int_C f(z)\,dz \;=\; \int_a^b f(z(t))z'(t)\,dt$$

$$= \int_a^b (u(x(t),y(t)) + iv(x(t),y(t)))(x'(t) + iy'(t))\,dt$$

$$= \int_a^b (u(x(t),y(t))x'(t) - v(x(t),y(t))y'(t))\,dt$$

$$+\, i \int_a^b (u(x(t),y(t))y'(t) + v(x(t),y(t))x'(t))\,dt$$

$$= \int_C u\,dx - v\,dy \;+\; i \int_C u\,dy + v\,dx.$$

Example 2 Let C be the straight line from $(0,0)$ to $(1,2)$ and let $f(z) = z^2$. Then, $z(t) = t + 2it$, $0 \leq t \leq 1$, is a parameterization of C. Since $z'(t) = (1 + 2i)$, we have

$$\int_C f(z)\,dz \;=\; \int_0^1 (t + 2it)^2(1 + 2i)\,dt \;=\; (1 + 2i)^3 \int_0^1 t^2\,dt \;=\; \frac{1}{3}(1 + 2i)^3.$$

Example 3 Let C be the arc of the unit circle ($|z| = 1$) from 1 to i and let $f(z) = z$. We can parameterize C by $z(t) = \cos t + i \sin t = e^{it}$ on the interval $[0, \frac{\pi}{2}]$. Thus, $z'(t) = ie^{it}$, so

$$\int_C z\,dz \;=\; \int_0^{\frac{\pi}{2}} e^{it} i e^{it}\,dt$$

$$= \int_0^{\frac{\pi}{2}} i e^{2it}\,dt$$

$$= \frac{1}{2}(e^{i\pi} - 1) \;=\; -1$$

Example 4   Let z_o be any complex number and let C be the circle of radius r centered at z_o, taken in the counterclockwise direction. We can parameterize C by choosing $z(\theta) = z_o + re^{i\theta}$ for $0 \le \theta \le 2\pi$. Here θ plays the role of the parameter t. Since $z'(\theta) = ire^{i\theta}$,

$$\int_C (z - z_o)^n \, dz = \int_0^{2\pi} (re^{i\theta})^n ire^{i\theta} \, d\theta$$

$$= \int_0^{2\pi} i(e^{i\theta})^{n+1} r^{n+1} \, d\theta$$

$$= \begin{cases} 0 & \text{if } n \ne -1 \\ 2\pi i & \text{if } n = -1 \end{cases}$$

Thus, if $p(z) = \sum_{j=0}^N a_j (z - z_o)^j$ is a polynomial in powers of $z - z_o$,

$$\int_C p(z) \, dz = \int_C \sum_{j=0}^N a_j (z - z_o)^j \, dz$$

$$= \sum_{j=0}^N a_j \int_C (z - z_o)^j \, dz$$

$$= 0.$$

Since any polynomial in z can be written as a polynomial in $z - z_o$ by making the substitution $z = (z - z_o) + z_o$, we have shown that the integral of any polynomial on any circle in $\mathbb{C}$ is zero. Furthermore, this line of reasoning can also be used for functions that can be represented by power series in $z - z_o$. If $f(z) = \sum_{j=0}^\infty a_j (z - z_o)^j$ and the series has a radius of convergence larger than r, then the series converges uniformly on C. Thus, by the analogue of Theorem 5.2.2 (see problem 11),

$$\int_C f(z) \, dz = \int_C \lim_{N \to \infty} \sum_{j=0}^N a_j (z - z_o)^j \, dz$$

$$= \lim_{N \to \infty} \int_C \sum_{j=0}^N a_j (z - z_o)^j \, dz$$

$$= 0.$$

Since we know that a series in powers of $z - z_o$ is analytic inside its radius of convergence (Example 3 of Section 8.1), this result is a special case of Cauchy's theorem, which is proved in the next section.

Problems

1. Let $g(t) = h(t) + ik(t)$ be a complex-valued function of a real variable t. Define $\bar{g}(t) = h(t) - ik(t)$ and note that $h(t) = \frac{1}{2}(g(t) + \bar{g}(t))$.

   (a) Show that g is continuous if and only if h and k are continuous.

   (b) Show that g is continuously differentiable if and only if h and k are continuously differentiable, in which case $g'(t) = h'(t) + ik'(t)$.

2. Explain carefully why one can differentiate under the integral sign in computing the partial derivatives of the real and imaginary parts of the Fourier transform of a continuous function that vanishes outside of a finite interval (see Example 1). Verify that the Cauchy-Riemann equations hold.

3. Let f be the function which equals 1 on the interval $[a, b]$ and equals 0 elsewhere. Compute the Fourier transform of f and verify explicitly that it is analytic in the entire complex plane.

4. Let f be an analytic function. Show that the real and imaginary parts, u and v, satisfy **Laplace's equation**:

$$u_{xx} + u_{yy} = 0 = v_{xx} + v_{yy}.$$

5. Let C be the path from $(0, 0)$ to $(2, 4)$ in $\mathbb{R}^2$ that follows the graph of the function $y = x^2$. Compute the following line integrals:

   (a) $\int_C x\,dx + y\,dy$ \qquad (b) $\int_C x\,dx - y^2\,dy$ \qquad (c) $\int_C y^2\,dx - xe^{xy}\,dy$

6. Let C be the straight line from $(1, 1)$ to $(3, 0)$. Compute

   (a) $\int_C z\,dz$ \qquad (b) $\int_C z^2\,dz$ \qquad (c) $\int_C |z|\,dz$

7. Let f and g be continuous functions on $\mathbb{C}$ and let C be a smooth curve of finite length. Prove that

$$\int_C (f(z) + g(z))\,dz \;=\; \int_C f(z)\,dz \;+\; \int_C g(z)\,dz.$$

8. Let C be a smooth curve of finite length in $\mathbb{C}$ and f be a continuous function on C. Prove that

$$\left| \int_C f(z)\,dz \right| \;\le\; \left(\max_{z \,\in\, C} |f(z)| \right) L$$

   where L is the arc length of C.

9. Let C be the circle of radius 1 with center at the origin. Evaluate the integral

$$\int_C \frac{1}{z(z-2)}\, dz.$$

Hint: write $\frac{1}{z(z-2)} = \frac{1}{2}\left\{ \frac{1}{z-2} - \frac{1}{z} \right\}$ and use the ideas in Example 4.

10. Let C be the circle of radius R with center at the origin. Find an upper bound for

$$\left| \int_C \frac{e^z}{z}\, dz \right|.$$

11. Let C be a smooth curve of finite length in $\mathbb{C}$ and f be a continuous function on C. Suppose that $\{f_n\}$ is a sequence of continuous functions that converges uniformly to f on C. Then

$$\int_C f(z)\, dz = \lim_{n\to\infty} \int_C f_n(z)\, dz.$$

Hint: use problem 8.

8.3   Cauchy's Theorem

In this section we prove two theorems, Cauchy's Theorem and Cauchy's Integral Formula which are the basis for the usefulness of analytic function theory. We define a **contour** to be a finite collection of smooth curves attached end to end (for example, the boundary of a rectangle). A contour is called **simple** if it does not intersect itself except possibly at the first and last points in which case it is called a simple **closed** contour. An example of a simple, closed contour is shown in Figure 8.3.1.

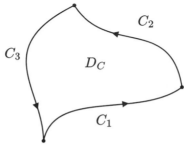

Figure 8.3.1

We give a proof of Cauchy's theorem which depends on Green's theorem from multivariable calculus. Direct proofs are available; see [6] or [23].

❑ **Theorem 8.3.1 (Cauchy's Theorem)** Let f be an analytic function in an open domain $D \subseteq \mathbb{C}$. Then

$$\int_C f(z)\, dz = 0 \tag{21}$$

for every simple closed contour C in D whose interior contains only points of D.

Proof. Let D_C denote the region of $\mathbb{C}$ consisting of C and the points inside C; see Figure 8.3.1. Since f is analytic, u and v are continuously differentiable on D_C. Thus, Green's theorem implies that

$$\int_C f(z)\, dz = \int_C u\, dx - v\, dy + i \int_C u\, dy + v\, dx$$

$$= \int_{D_C} (-v_x - u_y)\, dxdy + i \int_{D_C} (u_x - v_y)\, dxdy.$$

Since f is analytic, u and v satisfy the Cauchy-Riemann equations, $u_x = v_y$ and $u_y = -v_x$, on D_C. Thus, each of the last two integrals equals zero.

❑

Example 1 If f is a polynomial in z, then f is analytic, so (21) is true for all simple closed contours C in $\mathbb{C}$. If $f(z) = \sum_{j=0}^{\infty} a_j (z - z_o)^j$, then (21) holds for any simple closed contour inside the radius of convergence since f is analytic there. We computed these results explicitly in the case where C is a circle in Example 4 of Section 8.2.

Example 2 The function $f(z) = \frac{1}{z - z_o}$ is analytic everywhere except $z = z_o$. Let C be any circle with center at z_o traversed in the counterclockwise direction. We calculated in Example 4 of Section 8.2 that

$$\int_C \frac{1}{z - z_o}\, dz = 2\pi i. \tag{22}$$

This shows that if the interior of a simple closed contour contains a single point at which f is not analytic, then the conclusion of Cauchy's theorem may not hold.

Example 3 The function

$$f(z) = \frac{1}{z(z - 2)} = \frac{1}{2} \left\{ \frac{1}{z - 2} - \frac{1}{z} \right\}$$

is analytic except at the points $z = 0$ and $z = 2$. The calculation in Example 4 of Section 8.2 and Cauchy's theorem allow us to evaluate the following integrals immediately. All curves are traversed in the counterclockwise direction. If C_1 is the circle of radius 1 about the point $2i$, then by Cauchy's theorem,

$$\int_{C_1} f(z)\, dz = 0.$$

If C_2 is the circle of radius 1 about the point 0, then by Example 4 and Cauchy's theorem,

$$\int_{C_2} f(z)\, dz = -\frac{1}{2} \int_{C_2} \frac{dz}{z} + \frac{1}{2} \int_{C_2} \frac{dz}{z - 2} = -\pi i + 0.$$

If C_3 is the circle of radius $\frac{1}{2}$ about the point 1, then by Cauchy's theorem,

$$\int_{C_3} f(z)\, dz = 0.$$

We remark that we are assuming the Jordan Curve Theorem: every simple closed continuous curve has an interior and an exterior. This looks obvious for any simple example but is not easy to prove in general. For a discussion of the Jordon Curve Theorem, see [23].

We now prove the most striking consequence of Cauchy's Theorem. If f is analytic, its value at any point inside of C is an appropriate weighted average of its values on C. In particular, the values of f inside of C are all determined by the values on C. And this is true for *any* contour C just as long as f is analytic on and inside of C.

❏ **Theorem 8.3.2 (Cauchy's Integral Formula)** Let f be an analytic function in an open domain $D \subseteq \mathbb{C}$. Let C be a simple closed contour C in D traversed in the counterclockwise direction whose interior contains only points of D. Then for every z_o inside C,

$$f(z_o) = \frac{1}{2\pi i} \int_C \frac{f(z)}{z - z_o}\, dz. \tag{23}$$

Proof. Suppose that $\varepsilon > 0$ is small enough so that $C(z_o, \varepsilon)$, the circle of radius ε centered at z_o, lies entirely inside of C. We will first show that the integral of $f(z)/(z - z_o)$ on C is equal to the integral of $f(z)/(z - z_o)$ on $C(z_o, \varepsilon)$. Choose a point w on $C(z_o, \varepsilon)$ and let L denote the radial line segment from w to the first intersection point, τ, of the line with C. These curves are indicated in Figure 8.3.2. Let C_1 be the contour consisting of C (from τ to τ), followed by $-L$, followed by $-C(z_o, \varepsilon)$ (from w to w), followed by L. We indicate that a curve is traversed in the opposite direction from its defined direction by a minus sign in front of the symbol for the curve.

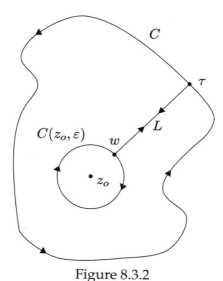

Figure 8.3.2

Since $f(z)/(z - z_o)$ is analytic on and everywhere inside of C_1, Cauchy's theorem implies

$$
\begin{aligned}
0 &= \int_{C_1} \frac{f(z)}{z - z_o} \, dz \\
&= \left\{ \int_C + \int_{-L} + \int_{-C(z_o, \varepsilon)} + \int_L \right\} \frac{f(z)}{z - z_o} \, dz \\
&= \int_C \frac{f(z)}{z - z_o} \, dz - \int_{C(z_o, \varepsilon)} \frac{f(z)}{z - z_o} \, dz,
\end{aligned}
$$

which shows that the integral of $f(z)/(z - z_o)$ on C and the integral of $f(z)/(z - z_o)$ on $C(z_o, \varepsilon)$ are equal. Notice that this equality is true for all small ε.

We now estimate the difference between the right-hand and left-hand sides of (23).

$$\left| f(z_o) - \frac{1}{2\pi i} \int_C \frac{f(z)}{z - z_o}\, dz \right| \;=\; \left| f(z_o) - \frac{1}{2\pi i} \int_{C(z_o,\varepsilon)} \frac{f(z)}{z - z_o}\, dz \right|$$

$$= \left| f(z_o) - \frac{f(z_o)}{2\pi i} \int_{C(z_o,\varepsilon)} \frac{dz}{z - z_o} - \frac{1}{2\pi i} \int_{C(z_o,\varepsilon)} \frac{f(z) - f(z_o)}{z - z_o}\, dz \right|$$

$$= \frac{1}{2\pi} \left| \int_{C(z_o,\varepsilon)} \frac{f(z) - f(z_o)}{z - z_o}\, dz \right|$$

$$\leq \left(\max_{z\, \epsilon\, C(z_o,\varepsilon)} \frac{|f(z) - f(z_o)|}{\varepsilon} \right) (2\pi\varepsilon)$$

$$= 2\pi \left(\max_{z\, \epsilon\, C(z_o,\varepsilon)} |f(z) - f(z_o)| \right).$$

In the second step we used the result of Example 2 to cancel the first two terms on the right. In the last step we used the estimate proved in problem 8 of Section 8.2. Since f is continuous, the right-hand side converges to zero as $\varepsilon \to 0$. However, the left-hand side does not depend on ε, so we conclude that (23) holds.   ❏

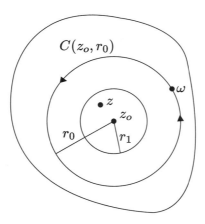

Figure 8.3.3

In Example 3 of Section 8.1, we showed that inside its radius of convergence a power series in powers of $z - z_o$ converges to an analytic function. Now we can show the converse.

❏ **Theorem 8.3.3** Let f be an analytic function in an open domain $D \subseteq \mathbb{C}$. Let z_o be a point of D, and let $C(z_o, r_0)$ be a circle about z_o such that the circle and its interior contain only points of D. Then f may be represented by a power series in $z - z_o$ that converges uniformly on any closed disc $\{z \mid |z - z_o| \leq r\}$ such that $r < r_0$.

Proof. Choose r_1 so that it satisfies $r < r_1 < r_0$. Let z be a point in the disk $\{\tau \mid |\tau - z_o| \leq r_1\}$ and let w be a point on $C(z_o, r_0)$. Since $r_1 < r_0$, there is an α independent of z and w so that

$$\frac{|z - z_o|}{|w - z_o|} \leq \alpha < 1. \tag{24}$$

Thus,

$$\frac{1}{w - z} = \frac{1}{w - z_o} \frac{1}{1 - \frac{z - z_o}{w - z_o}}$$

$$= \frac{1}{w - z_o} \sum_{j=0}^{\infty} \left(\frac{z - z_o}{w - z_o}\right)^j,$$

where the series converges uniformly for w on $C(z_o, r_0)$. Therefore, by Cauchy's Integral Formula and the result of problem 11 of Section 8.2,

$$f(z) = \frac{1}{2\pi i} \int_{C(z_o, r_0)} \frac{f(w)}{w - z} dw \tag{25}$$

$$= \frac{1}{2\pi i} \int_{C(z_o, r_0)} \frac{f(w)}{w - z_o} \lim_{N \to \infty} \sum_{j=0}^{N} \left(\frac{z - z_o}{w - z_o}\right)^j dw \tag{26}$$

$$= \lim_{N \to \infty} \sum_{j=0}^{N} \left(\frac{1}{2\pi i} \int_{C(z_o, r_0)} \frac{f(w)}{(w - z_o)^{j+1}} dw\right) (z - z_o)^j \tag{27}$$

Since the series in (27) converges for $z \in \{z \mid |z - z_o| \leq r_1\}$, we know, by Theorem 6.5.3, that the radius of convergence of the series is $\geq r_1$. Thus the series converges uniformly on $\{z \mid |z - z_o| \leq r\}$ if $r < r_1$. ❏

Corollary 8.3.4 On its domain of definition, an analytic function is infinitely often differentiable. Further, the series (27) is the Taylor series of f about z_o, and the derivatives of f are given by the formulas

$$f^{(n)}(z_o) = \frac{n!}{2\pi i} \int_C \frac{f(w)}{(w - z_o)^{n+1}} dw \tag{28}$$

where C is any simple closed contour about z_o in D whose interior contains only points of D.

Proof.   The same proof as in Theorem 6.4.2 shows that a complex power
series is infinitely differentiable inside its radius of convergence and the
derivatives can be computed by differentiating the series term by term.
Differentiating (27) term by term and setting $z = z_o$ gives formula (28)
in the case where $C = C(z_o, r_0)$. Thus, the coefficient of the jth power
in (27) is $f^{(j)}(z_o)/j!$, so (27) is just the Taylor series of f. Formula (28)
holds for general C satisfying the hypotheses by an argument similar to
the argument at the beginning of the proof of Theorem 8.3.2.   ❑

We have shown that analytic functions of a complex variable have
very special properties. What is not obvious is that analytic function theory is extremely useful in many problems which at first glance do not
seem to have anything to do with complex variables or analytic functions. Here are some examples:

(a) In Section 6.4 we asked which real-valued functions of a real vari-
able, $f(x)$, are equal to their Taylor series, $\sum \frac{1}{j!} f^{(j)}(x_o)(x - x_o)^j$ in an
interval $(x_o - r, x_o + r)$. We now know that this is true if and only if $f(x)$
is the restriction to the real axis of a function of the complex variable z
that is analytic in the disk $\{z \mid |z - x_0| < r\}$.

(b) Cauchy's theorem can sometimes be used to evaluate the integrals
of real-valued functions on the real line. See project 1.

(c) The Prime Number Theorem, which we stated at the end of Section 6.6, describes the asymptotic behavior of the function $\pi(m)$ whose
value is the number of primes $\leq m$. It doesn't seem likely that this could
have anything to do with complex numbers or analytic functions. However, the Riemann zeta function, which we introduced in problem 12 of
Section 6.3, is the restriction to the set $\{x \mid x > 1\}$ of an analytic function (see problem 10). The properties of $\zeta(z)$ played a central role in the
original proofs of the Prime Number Theorem.

(d) The Fourier transform, defined in Example 1 of Section 8.2, has
played an important role in mathematics and physics for almost 200
years. Although it was introduced as a technique for solving partial differential equations, it has many other uses. For example, it is used in
the theory of group representations and is the main tool for proving the
Central Limit Theorem in probability theory. In the example, we showed
that the Fourier transform of a function which is zero outside of an interval is the restriction to the real axis of an analytic function on $\mathbb{C}$. It turns

out that the size of the region where f is nonzero can be characterized in terms of the growth properties of $\hat{f}$ in the imaginary directions. As mentioned in the example, this plays an important role in the theory of signal transmission in electrical engineering and in scattering theory in high-energy physics.

(e) Let A be a linear transformation from $\mathbb{R}^n$ to $\mathbb{R}^n$. If we choose a basis for $\mathbb{R}^n$, then A is represented by an $n \times n$ matrix with real entries. Let I denote the identity matrix. Except for finitely many values of z, the eigenvalues of A, the matrix $(zI - A)$ is invertible. It turns out that the function $z \rightarrow (zI - A)^{-1}$ is an analytic matrix-valued function of z except at the eigenvalues. A generalization of this fact plays an important role in the analysis of linear transformations on Hilbert and Banach spaces. These transformations, in turn, play a central role in operations research and quantum mechanics.

Problems

1. Let C_1, C_2, and C_3 be circles with center the origin and radii 1, 3, and 5, respectively, traversed counterclockwise. Use Cauchy's Theorem and the Cauchy integral formula to evaluate the following integrals:

   (a) $\int_{C_1} \frac{1}{z-2} \, dz$ (b) $\int_{C_3} \frac{1}{z-2} \, dz$ (c) $\int_{C_1} \frac{1}{z^3-2} \, dz$

   (d) $\int_{C_3} \frac{1}{(z-6i)(z-1)} \, dz$ (e) $\int_{C_1} \frac{\cos z}{z} \, dz$ (f) $\int_{C_3} \frac{e^{z^2}}{z-i} \, dz$

2. Compute $\int_{C_1} \frac{\sin z}{z^5} \, dz$ where C_1 is as defined in problem 1. Hint: use the power series for $\sin z$.

3. Let f be the function

$$f(z) \equiv \frac{1}{(z-1)(z-4)}.$$

   Use an argument similar to that in Theorem 8.3.2 to show that

$$\int_{C_3} f(z) \, dz = \int_{C_4} f(z) \, dz + \int_{C_5} f(z) \, dz$$

   where C_3 is as defined in problem 1, and C_4 and C_5 are small counterclockwise circles about the points 1 and 4. Use Cauchy's Integral Formula to evaluate the two integrals on the right.

4. Suppose that f is analytic in the disk $\{z \in \mathbb{C} \,|\, |z| < R\}$ and that $f'(z) = 0$ for all z in the disk. Prove that f is constant in the disk. Hint: use power series.

5. Suppose that f is analytic in the entire complex plane and bounded; that is, $|f(z)| \leq M$ for some M. Use (28) and (20) to prove that $f^{(n)}(0) = 0$ for all $n \geq 1$. Conclude from this that f is a constant. This is known as **Liouville's Theorem.**

6. Let C_1 be the straight line path from i to 1 and let C_2 be the path that goes in a straight line from i to 0 and then in a straight line from 0 to 1.

   (a) Compute $\int_{C_1} z \, dz$ and $\int_{C_2} z \, dz$ and show that they are the same. How could you have predicted this by using Cauchy's theorem?

   (b) Compute $\int_{C_1} \bar{z} \, dz$ and $\int_{C_2} \bar{z} \, dz$ and show that they are not the same.

7. (a) Suppose that f is analytic in the open unit disk $\{z \,|\, |z| < 1\}$ and suppose that $f(z) = 0$ when z is real. Prove that $f(z) \equiv 0$ in the disk. Hint: what can you say about the Taylor coefficients of f at 0?

   (b) Suppose that f is analytic in the unit disk $\{z \,|\, |z| < 1\}$ and that $f(z_n) = 0$ at a sequence of points $\{z_n\}$ that have a limit point in the disc. Prove that $f(z) \equiv 0$ in the disk.

8. Suppose that f and g are both analytic functions on $\mathbb{C}$ and that $f(z) = g(z)$ if z is real. Prove that $f(z) = g(z)$ for all z.

9. Use the fact that $e^z e^w = e^{z+w}$ for z and w real to prove that the same equality must hold for all $z \in \mathbb{C}$ and all $w \in \mathbb{C}$.

10. If $b > 0$, define $b^z \equiv e^{z \ln b}$. Show that the Riemann ζ function,

$$\zeta(z) = \sum_{j=1}^{\infty} \frac{1}{j^z},$$

is a well-defined analytic function in the region $\{z \,|\, Re\{z\} > 1\}$.

Projects

1. The purpose of this project is to show how the theory of analytic functions can be used to evaluate improper Riemann integrals on $\mathbb{R}$. The integral

$$\int_{-\infty}^{\infty} \frac{1}{1 + x^2} \, dx$$

can be computed by elementary means. To see how to compute it by using the theory of analytic functions, notice that $f(x)$ is the restriction to $\mathbb{R}$ of the function $f(z) = (1 + z^2)^{-1}$, which is analytic everywhere in $\mathbb{C}$ except at $z = i$ and $z = -i$.

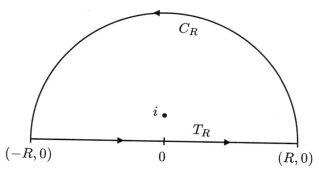

(a) Let Γ_R be the contour in the above figure, consisting of a first piece, T_R, traversing the real line from $x = -R$ to $x = R$, and a second piece, C_R, which traverses counterclockwise from $(R, 0)$ to $(-R, 0)$ along the circle of radius R with center at the origin. Suppose $R > 1$. By writing $f(z) = (z + i)^{-1}(z - i)^{-1}$ and using the Cauchy integral formula, explain why

$$\int_{T_R} f(z)\, dz + \int_{C_R} f(z)\, dz = \int_{\Gamma_R} f(z)\, dz = \pi.$$

(b) Use (20) to prove that $\left| \int_{C_R} f(z)\, dz \right| \to 0$ as $R \to \infty$.

(c) Explain why the improper Riemann integral $\int_{-\infty}^{\infty} \frac{1}{1+x^2}\, dx$ exists and equals π.

(d) Use the same ideas to evaluate

$$\int_{-\infty}^{\infty} \frac{1}{(1 + x^2)(4 + x^2)}\, dx.$$

2. The purpose of this project is to prove the **Fundamental Theorem of Algebra**, namely, that every polynomial

$$p(z) \equiv a_0 + a_1 z + a_2 z^2 + \ldots + a_n z^n$$

has at least one root if $n \geq 1$. Since we are assuming that p has order n, we know that $a_n \neq 0$. Note that the coefficients $\{a_j\}$ may be complex. We shall outline a proof by contradiction. Suppose that there is no z_o so that $p(z_o) = 0$.

(a) Explain why $\frac{1}{p(z)}$ is analytic in the entire complex plane.

(b) Prove that there exists an r_0 so that

$$|a_0| + |a_1|r + |a_2|r^2 + \ldots + |a_{n-1}|r^{n-1} \leq \frac{1}{2}|a_n|r^n$$

for all $r \geq r_0$.

(c) Prove that

$$\left| \frac{1}{p(z)} \right| \leq \frac{2}{|a_n||z|^n}$$

if $|z| \geq r_0$.

(d) Prove that $\frac{1}{p(z)}$ is bounded in the entire complex plane.

(e) Use Liouville's theorem to conclude that $\frac{1}{p(z)}$ is constant and explain why this gives a contradiction.

Remark: to see the power of analytic function theory, try to prove the Fundamental Theorem directly.

CHAPTER 9

Fourier Series

Fourier analysis has played an important role in mathematics since the early part of the 19th century. In Section 9.1 we show Fourier's technique for solving a partial differential equation describing heat flow. His calculation posed a question for mathematicians which proved to be both difficult and exceptionally fruitful. Fourier series are formally defined in Section 9.2 where several examples are given. The theorems on pointwise convergence and mean-square convergence are proved in Sections 9.3 and 9.4, respectively.

9.1   The Heat Equation

It this section we investigate the flow of heat in a thin metal bar of length L centimeters. We assume that the temperature, u, is constant in each cross section, so u depends only on the time t and the distance x along the bar. Heat energy is measured in calories. The specific heat of a material, c, is the number of calories needed to raised 1 gram 1 degree centigrade. Thus the heat per unit length at x is $c\rho A u(t, x)$, where A is the area of the cross section and ρ is the density of the metal. For simplicity, we assume that the bar is homogeneous and a perfect cylinder, so c, ρ, and A are constants. It follows that the total heat in the segment of the bar between $x = a$ and $x = b$ is

$$x = 0 \qquad a \qquad\qquad b \qquad x = L$$

Figure 9.1.1

$$H_{a,b}(t) = \int_a^b c\rho A u(t, x)\, dx,$$

and thus the rate of change of heat in the segment is

$$H'_{a,b}(t) = \int_a^b c\rho A u_t(t, x)\, dx. \tag{1}$$

We will now calculate this rate of change in another way. We assume that the bar is insulated on its sides so that heat can only flow along the bar (or out the ends). A fundamental physical principle is that the rate of heat flow at any point in a material is proportional to the gradient of the temperature. In our case, since heat flows only along the x-axis, the rate of heat flow in the positive x direction (in calories per unit time per unit length per unit cross section) at x is $-\alpha u_x(t, x)$, where the constant $\alpha > 0$ is determined by the properties of the material. We take the minus sign in front of α so that heat flows from hot to cold. Returning now to the segment of bar between a and b we see that the heat flowing *into* the segment at b is $\alpha A u_x(t, b)$ and the heat flowing *into* the segment at a is $-\alpha A u_x(t, a)$. Thus, the net rate of change of heat in the segment is $\alpha A u_x(t, b) - \alpha A u_x(t, a)$. By the Fundamental Theorem of Calculus, we can write this net rate of change as

$$\alpha A u_x(t, b) - \alpha A u_x(t, a) \;=\; \int_a^b \alpha A u_{xx}(t, x)\, dx. \tag{2}$$

Both the right-hand side of (1) and the right-hand side of (2) represent the net rate of change of heat in the segment, so setting them equal and rearranging, we find

$$\int_a^b u_t(t, x) - \frac{\alpha}{c\rho} u_{xx}(t, x)\, dx \;=\; 0. \tag{3}$$

The constant $\kappa = \frac{\alpha}{c\rho}$ is called the diffusivity. Since (3) holds for all choices of a and b, the integrand in (3) must be zero if it is continuous (problem 14 of Section 3.3). Thus,

$$u_t(t, x) - \kappa u_{xx}(t, x) \;=\; 0. \tag{4}$$

This partial differential equation is called the **heat equation**. The temperatures along the bar are given at time $t = 0$,

$$u(0, x) \;=\; f(x), \tag{5}$$

and we suppose that the ends of the bar are held at temperature zero at all times (for example, by placing them in an ice bath). Thus,

$$u(t, 0) \;=\; 0 \;=\; u(t, L). \tag{6}$$

The mathematical problem is to find the function, $u(t, x)$, defined on the half-strip, $0 \le x \le L$, $t \ge 0$, so that the partial differential equation (4), the initial condition (5), and the boundary conditions (6) all hold.

For the moment we forget about the initial conditions and look for solutions of (4) and (6) that have the special form $u(t, x) = X(x)T(t)$, which is a function of x times a function of t. If we substitute $X(x)T(t)$ into (4), carry out the differentiations and rearrange algebraically, we find

$$\frac{T'(t)}{\kappa T(t)} = \frac{X''(x)}{X(x)}. \tag{7}$$

Since t and x are independent variables, this can only be true for all t and all x if both sides are constant; we call the constant $-\lambda$. Thus, $T(t)$ satisfies the differential equation

$$T'(t) + \lambda \kappa T(t) = 0, \tag{8}$$

and $X(x)$ satisfies the differential equation

$$X''(x) + \lambda X(x) = 0. \tag{9}$$

The general solution of (9) is $X(x) = a_1 \cos \sqrt{\lambda}x + a_2 \sin \sqrt{\lambda}x$. However, since the boundary conditions (6) must be satisfied, it is easy to see that we must have $a_1 = 0$ and $\sin \sqrt{\lambda}L = 0$. Thus λ cannot be any constant but must satisfy $\sqrt{\lambda}L = \pm n\pi$ for some integer n. Since the function $\sin x$ is odd, the solutions for n negative are minus the solutions for n positive, and the solution for $n = 0$ is identically zero. Thus, if u is not identically zero, the possible λ's are $\lambda_n \equiv (\frac{n\pi}{L})^2$, $n = 1, 2, 3, \ldots$, and $X(x)$ is one of the functions

$$X_n(x) = b_n \sin \frac{n\pi x}{L}$$

where b_n is a constant. For each n, the solution of (8) is a constant times $e^{-\lambda_n \kappa t}$. Thus, each of the functions

$$u_n(t, x) = b_n e^{-\lambda_n \kappa t} \sin \frac{n\pi x}{L}$$

satisfies the partial differential equation and the boundary conditions. This can be checked directly by computing the partial derivatives of u_n. Note that the heat equation is linear; that is, only the first power of the unknown function u or its derivatives occurs. It follows, since differentiation is a linear operation, that linear combinations of the functions u_n are again solutions. This suggests that we try to write the general solution $u(t, x)$ of the heat equation with the boundary conditions (6) in the form

$$u(t, x) = \sum_{n=1}^{\infty} b_n e^{-\lambda_n \kappa t} \sin \frac{n\pi x}{L}. \tag{10}$$

Assuming that the series converges, such a $u(t, x)$ satisfies the heat equation and the boundary conditions. But what about the initial condition (5)? If we set $t = 0$ in (10), we see that

$$f(x) \quad = \quad \sum_{n=1}^{\infty} b_n \sin \frac{n\pi x}{L}. \tag{11}$$

Here is the question posed by Joseph Fourier (1768 – 1830): given f, can we choose the coefficients $\{b_n\}$ so that (11) holds? Although Fourier did not answer the question, he showed that if the answer is yes, then the coefficients b_n are determined by simple formulas. The family of functions $\{\sin \frac{n\pi x}{L}\}$ satisfies (problem 1)

$$\int_0^L \sin \frac{n\pi x}{L} \sin \frac{m\pi x}{L} \, dx \quad = \quad \begin{cases} 0 & \text{if } n \neq m \\ \frac{L}{2} & \text{if } n = m. \end{cases}$$

We shall use these special properties to compute formulas for the coefficients $\{b_n\}$. Suppose that we know that the series on the right of (11) converges uniformly to $f(x)$ on the interval $[0, L]$. Multiplying both sides of (11) by $\sin \frac{m\pi x}{L}$ and integrating, we find

$$\int_0^L f(x) \sin \frac{m\pi x}{L} \, dx \quad = \quad \int_0^L \left(\sum_{n=1}^{\infty} b_n \sin \frac{n\pi x}{L} \right) \sin \frac{m\pi x}{L} \, dx$$

$$= \quad \sum_{n=1}^{\infty} b_n \int_0^L \sin \frac{n\pi x}{L} \sin \frac{m\pi x}{L} \, dx$$

$$= \quad \frac{b_m L}{2},$$

where we used Theorem 6.3.2 to exchange the sum and the integral. We conclude that

$$b_n \quad = \quad \frac{2}{L} \int_0^L f(x) \sin \frac{n\pi x}{L} \, dx. \tag{12}$$

Thus, *if f* can be represented in the form (11), we have a formula for the coefficients $\{b_n\}$. This, in turn, allows us to write down an explicit formula (10) for the solution of the heat equation that satisfies the initial and boundary conditions.

Example 1 Suppose that $f(x) = x(L - x)$. To evaluate the integral (12), we integrate by parts:

$$
\begin{aligned}
b_n &= \frac{2}{L} \int_0^L x(L - x) \sin \frac{n\pi x}{L} \, dx \\[2mm]
&= \frac{2}{L} \left\{ \left(-\frac{L}{n\pi} \right) x(L - x) \cos \frac{n\pi x}{L} \right\}_0^L + \frac{2}{n\pi} \int_0^L (L - 2x) \cos \frac{n\pi x}{L} \, dx \\[2mm]
&= \frac{2L}{(n\pi)^2} \left\{ (L - 2x) \sin \frac{n\pi x}{L} \right\}_0^L + \frac{4L}{(n\pi)^2} \int_0^L \sin \frac{n\pi x}{L} \, dx \\[2mm]
&= \frac{4L^2}{n^3 \pi^3} (1 - (-1)^n).
\end{aligned}
$$

Therefore, $b_n = 0$ if n is even and $b_n = \frac{8L^2}{n^3\pi^3}$ if n is odd. Thus $u(t, x)$ is given by the series

$$
\begin{aligned}
u(t, x) = \; &\frac{8L^2}{\pi^3} e^{-(\pi/L)^2 \kappa t} \sin \frac{\pi x}{L} + \frac{8L^2}{3^3 \pi^3} e^{-(3\pi/L)^2 \kappa t} \sin \frac{3\pi x}{L} \\[2mm]
&+ \frac{8L^2}{5^3 \pi^3} e^{-(5\pi/L)^2 \kappa t} \sin \frac{5\pi x}{L} + \cdots.
\end{aligned}
$$

Notice that for $t > 0$, the series converges very fast because of the exponential factors. This is the key fact which is exploited in the theorem below. When $t = 0$, the series is

$$
\begin{aligned}
\frac{8L^2}{\pi^3} \sin \frac{\pi x}{L} &+ \frac{8L^2}{3^3 \pi^3} \sin \frac{3\pi x}{L} + \frac{8L^2}{5^3 \pi^3} \sin \frac{5\pi x}{L} + \cdots \\[2mm]
&= 8L^2 \sum_{n=0}^{\infty} \frac{1}{(2n + 1)^3 \pi^3} \sin \frac{(2n + 1)\pi x}{L}.
\end{aligned}
$$

If $u(t, x)$ satisfies the initial condition (5), then this series must converge to the simple polynomial $x(L - x)$ for each x in $[0, L]$. Because of the cubic power, the series certainly converges. But does it converge to $x(L - x)$? We shall see in Section 9.3 that the answer is yes.

As in this example, we can prove, in general, that the function $u(t, x)$ defined by (10) satisfies the partial differential equation and the boundary condition when $t > 0$. Not only that, the solution $u(t, x)$ is infinitely often continuously differentiable in both t and x for $t > 0$.

❑ **Theorem 9.1.1** Suppose that f is piecewise continuous on $[0, L]$, and let $u(t, x)$ be defined by (10) where the b_n are given by (12). Then for $t > 0$, $u(t, x)$ is infinitely often continuously differentiable in both t and x and satisfies the heat equation (4) and the boundary conditions (6).

Proof. According to Theorem 6.3.3, we can differentiate series (10) with respect to x if each term is continuously differentiable in x and the series which results from term-by-term differentiation: namely,

$$\sum_{n=1}^{\infty} b_n e^{-\lambda_n \kappa t} \left(\frac{n\pi}{L}\right) \cos \frac{n\pi x}{L}$$

converges uniformly. Since f is piecewise continuous on $[0, L]$, we know that $|f(x)| \leq M$ for some M. It follows by estimating the integral in (12) that $|b_n| \leq 2M$. Fix $t_o > 0$ and define $M_n = b_n \frac{n\pi}{L} e^{-\lambda_n \kappa t_o}$. Since

$$e^{-\lambda_n \kappa t_o} = \frac{1}{1 + (\lambda_n \kappa t_o) + \frac{1}{2!}(\lambda_n \kappa t_o)^2 + \dots} \leq \frac{2!}{(\lambda_n \kappa t_o)^2}$$

and $\lambda_n = \left(\frac{n\pi}{L}\right)^2$, we can estimate

$$M_n \leq 2M \left(\frac{n\pi}{L}\right) \frac{2!}{(\lambda_n \kappa t_o)^2} \leq \frac{C}{n^3}.$$

This estimate shows that $\sum M_n < \infty$. Thus, by the Weierstrass M-test, the differentiated series converges uniformly for all x. Notice that the estimates hold uniformly for all $t \geq t_o$ and that the terms in the differentiated series are continuous functions. Thus we have proved that $u(t, x)$ is continuously differentiable in x in the region $-\infty < x < \infty, t_o \leq t$. Since t_o was an arbitrary positive number, it follows that $u(t, x)$ is continuously differentiable in x for $t > 0$ and the derivative can be computed by differentiating the series for $u(t, x)$ term by term.

Similarly, the term-by-term derivative with respect to t,

$$\sum_{n=1}^{\infty} -\lambda_n \kappa b_n e^{-\lambda_n \kappa t} \sin \frac{n\pi x}{L},$$

also converges uniformly in x and $t > t_o > 0$. Thus u is continuously differentiable in both x and t for all x and $t > 0$, and the derivatives can be computed by differentiating the series term by term.

By repeating these arguments, one can show that u is infinitely often continuously differentiable in the region $-\infty < x < \infty, t > 0$ and the derivatives can be computed by term-by-term differentiation (problem 2). Differentiating the series for u term by term allows one to verify that u satisfies the heat equation (4). In addition, u satisfies the boundary condition (6) since each term in the series does. ❑

Theorem 9.1.1 expresses a very important "smoothing" property of the heat equation. Even if the initial temperature distribution, f, is only continuous (or even has jump discontinuities), the solution of the heat equation will be infinitely often differentiable for $t > 0$. This is a general property of a class of partial differential equations called parabolic equations. Note that we have not yet answered the question of whether the series converges to $f(x)$ when $t = 0$ so that the initial condition (5) is satisfied.

We derived (4) by analyzing heat flow in a bar, but the same partial differential equation arises in many other contexts. Albert Einstein (1879 – 1955) showed that if $u(t, x)$ is the density of a gas along the real line at time t, then under appropriate physical assumptions, $u(t, x)$ will satisfy (4). Thus, the equation (4), which also occurs in chemical and biological contexts, is often called the diffusion equation. A variant of the heat equation occurs in the Black-Scholes model for stock-option pricing.

Problems

1. Use the expression $\sin \theta = (2i)^{-1}(e^{i\theta} - e^{-i\theta})$ to prove that

$$\int_0^L \sin \frac{n\pi x}{L} \sin \frac{m\pi x}{L} \, dx = \begin{cases} 0 & \text{if } n \neq m \\ \frac{L}{2} & \text{if } n = m \end{cases}$$

2. Fill in the details in the proof of Theorem 9.1.1.

3. Let $f(x) = x(L - x)$, as in Example 1. Verify by explicit differentiation that the function $u(t, x)$ defined there satisfies the heat equation.

4. Let $f(x) = x(L - x)$, as in Example 1, and suppose, for simplicity, that $L = \pi$ and $\kappa = 1$. Let $S_n \equiv \sum_{j=1}^n b_n e^{-\lambda_n t} \sin nx$, where the b_n are the same as those computed in Example 1.

    (a) Generate the graphs of f, S_1, S_3, and S_5 on the interval $[0, \pi]$ when $t = 0$. Does it look as if the series for $u(0, x)$ is converging to f?

    (b) Find an upper bound on the error we make if we replace u by S_5.

    (c) Compare the graphs of S_5 for $t = 0, .5, 1, 2$, and 5. Is this the way you would expect $u(t, x)$ to behave?

5. Let $L = \pi$ and $\kappa = 1$ and define $f(x) = x$ on the interval $[0, \pi]$.

   (a) Compute the coefficients b_n. Hint: use formula (12) and integrate by parts.

   (b) Show by explicit differentiation that the function u defined by (10) satisfies the heat equation for $t > 0$ and explain why the boundary conditions (6) hold for all $t \geq 0$.

   (c) Explain why the series for u at $t = 0$ cannot converge to f for *all* $x \in [0, \pi]$.

6. Let S_n be the nth partial sum of the series for the solution u of problem 5.

   (a) Generate enough graphs of $S_1, S_2, S_3, \ldots$ to convince yourself that, for $t = 0$, the series converges to f for all $x \in [0, \pi]$ except $x = \pi$.

   (b) Compare the graphs of S_5 for $t = 0, .5, 1, 2$, and 5. Is this the way you would expect $u(t, x)$ to behave?

7. Consider heat flow in a bar where we do not prescribe the temperature at the ends but instead assume that the ends are insulated.

   (a) Explain why the right boundary conditions are

$$u_x(t, 0) \;\; = \;\; 0 \;\; = \;\; u_x(t, L). \tag{13}$$

   (b) Use the methods of the section to derive the formal solution

$$u(t, x) = \sum_{n=0}^{\infty} a_n e^{-\lambda_n \kappa t} \cos \frac{n\pi x}{L},$$

   where $a_n = \frac{2}{L} \int_0^L f(x) \cos \frac{n\pi x}{L} \, dx$ and $\lambda_n = \left(\frac{n\pi}{L}\right)^2$, if $n > 0$, and $a_0 = \frac{1}{L} \int_0^L f(x) \, dx,$.

   (c) Explain why the same arguments as in Theorem 9.1.1 show that u is infinitely differentiable in x and t for $t > 0$ and u satisfies the heat equation and the boundary conditions (13).

   (d) By using the partial differential equation, prove that $\int_0^L u(t, x) \, dx$ is independent of t. Why is that reasonable?

   (e) What happens to the solution as $t \to \infty$?

8. Suppose that the ends of the bar are kept at temperature zero and that $\beta cu(t, x)$ units of heat are added to the bar per gram per unit time by some internal chemical reaction where β is a constant. Show that u should satisfy the partial differential equation

$$u_t(t, x) - \kappa u_{xx}(t, x) - \beta u(t, x) \;\; = \;\; 0. \tag{14}$$

Using the methods of the section, write down a formal solution to (14) which satisfies the boundary condition (6) and the initial condition (5). How does the behavior of the solution as $t \to \infty$ depend on β?

9. Suppose that the ends of the bar are kept at temperature zero and that $cg(x)$ units of heat are added to the bar per gram per unit time at time t. Show that u should satisfy the partial differential equation

$$u_t(t, x) - \kappa u_{xx}(t, x) = g(x).$$

Suppose that the initial temperatures are given by $f(x)$ and assume that g can be written in the form $g(x) = \sum_{n=1}^{\infty} c_n \sin \frac{n\pi x}{L}$. Suppose that the solution $u(t, x)$ has the form

$$u(t, x) = \sum_{n=1}^{\infty} T_n(t) \sin \frac{n\pi x}{L}$$

for some unknown functions $\{T_n(t)\}$. For each n, find and solve the ordinary differential equation that must be satisfied by $T_n(t)$.

10. Consider a rectangular plate with coordinates $(0, 0)$, $(a, 0)$, (a, b), and $(0, b)$ for the vertices. Assume that the temperature $u(t, x, y)$ satisfies the two-dimensional heat equation

$$u_t(t, x) - \kappa u_{xx}(t, x) - \kappa u_{yy}(t, x) = 0.$$

Suppose that the boundaries of the rectangle are kept at temperature zero and that the initial temperatures are given by a function f that can be written as

$$f(x, y) = \sum_{n=1}^{\infty} \sum_{m=1}^{\infty} b_{m,n} \sin \frac{n\pi x}{a} \sin \frac{m\pi y}{b}.$$

Using the methods of the section, find functions $T_{n,m}(t)$ so that this initial-boundary value problem has a solution of the form

$$u(t, x, y) = \sum_{n=1}^{\infty} \sum_{m=1}^{\infty} T_{n,m}(t) \sin \frac{n\pi x}{a} \sin \frac{m\pi y}{b}.$$

9.2 Definitions and Examples

We want to investigate under what conditions a real or complex-valued function on $\mathbb{R}$, $f(x)$, can be written in the form

$$f(x) = a_0 + \sum_{m=1}^{\infty} a_m \cos mx + b_m \sin mx. \tag{15}$$

Note that if $\sum |a_m| < \infty$ and $\sum |b_m| < \infty$ then the series on the right certainly converges for each x. If we define

$$S_n(x) \equiv a_0 + \sum_{m=1}^{n} a_m \cos mx + b_m \sin mx,$$

then (15) means that the partial sums $S_n(x)$ converge to $f(x)$. Each partial sum is a periodic function of period 2π; that is, $S_n(x + 2\pi) = S_n(x)$. It is easy to prove from this (problem 1) that if the series converges, then f must be periodic of period 2π. From now on we assume that f has this property. For calculations, it is often convenient to rewrite the series and partial sums in terms of complex exponentials. Using the formulas for $\sin x$ and $\cos x$ in terms of complex exponentials (Example 1 of Section 6.5), we find

$$S_n(x) \;=\; a_0 + \sum_{m=1}^{n} a_m \frac{e^{imx} + e^{-imx}}{2} + b_m \frac{e^{imx} - e^{-imx}}{2i}$$

$$=\; a_0 + \sum_{m=1}^{n} \frac{1}{2}(a_m - ib_m)e^{imx} + \sum_{m=1}^{n} \frac{1}{2}(a_m + ib_m)e^{-imx}.$$

Thus, if we define

$$c_m \;\equiv\; \begin{cases} \frac{1}{2}(a_m - ib_m) & \text{if } m > 0 \\ a_0 & \text{if } m = 0 \\ \frac{1}{2}(a_{-m} + ib_{-m}) & \text{if } m < 0 \end{cases}$$

then

$$S_n(x) \;=\; \sum_{m=-n}^{n} c_m e^{imx}.$$

Therefore, (15) holds if and only if

$$f(x) \;=\; \sum_{m=-\infty}^{\infty} c_m e^{imx}, \tag{16}$$

where the infinite sum on the right means the limit of the partial sums $\sum_{m=-n}^{n} c_m e^{imx}$ as $n \to \infty$. Some of the Fourier series which we shall study are not absolutely convergent, so it is important to specify which sequence of partial sums we mean.

If (16) holds and if the series converges uniformly, then the coefficients $\{c_m\}$ are determined. To see this, multiply both sides of (16) by e^{-inx} and integrate term by term. Then,

$$\int_{-\pi}^{\pi} f(x)e^{-inx}\,dx \;=\; \sum_{m=-\infty}^{\infty} c_m \int_{-\pi}^{\pi} e^{i(m-n)x}\,dx$$

$$=\; 2\pi c_n$$

since the integral is zero unless $n = m$. This shows that the coefficient c_n is determined for each integer n. If we define

$$c_m \equiv \frac{1}{2\pi} \int_{-\pi}^{\pi} f(x) e^{-imx}\, dx, \tag{17}$$

then the series on the right side of (16) is called the **Fourier series** of f and the c_m are called the **Fourier coefficients**. Since (15) is just a rewriting of (16), it is also called the Fourier series of f. The coefficients $\{a_m\}_{m=0}^{\infty}$ and $\{b_m\}_{m=1}^{\infty}$ are also called Fourier coefficients and can be written in terms of $\{c_m\}_{m=-\infty}^{\infty}$ by the formulas: $a_0 = c_0$, $a_m = c_{-m} + c_m$, and $b_m = i^{-1}(c_{-m} - c_m)$, for $m > 0$. From these formulas, it follows easily (problem 2) that

$$a_0 = \frac{1}{2\pi} \int_{-\pi}^{\pi} f(t)\, dt \tag{18}$$

$$a_m = \frac{1}{\pi} \int_{-\pi}^{\pi} f(t) \cos mt\, dt, \quad m > 0 \tag{19}$$

$$b_m = \frac{1}{\pi} \int_{-\pi}^{\pi} f(t) \sin mt\, dt, \quad m > 0. \tag{20}$$

Example 1   Suppose that $f(x) = (\pi - |x|)^2$ on $[-\pi, \pi]$. Since $f(\pi) = 0 = f(-\pi)$, the function f can be extended to be a continuous function of period 2π on $\mathbb{R}$. Since $f(t)$ is even, all the coefficients b_m are equal to zero (problem 4). We calculate

$$a_0 = \frac{1}{2\pi} \int_{-\pi}^{\pi} (\pi - |t|)^2\, dt = \frac{1}{\pi} \int_0^{\pi} (\pi - t)^2\, dt = \frac{\pi^2}{3}$$

and by integration by parts,

$$
\begin{aligned}
a_m &= \frac{2}{\pi} \int_0^{\pi} (\pi - t)^2 \cos mt\, dt \\
&= \frac{2}{\pi m} \left\{ (\pi - t)^2 \sin mt \right\}_{t=0}^{t=\pi} + \frac{2}{\pi m} \int_0^{\pi} 2(\pi - t) \sin mt\, dt \\
&= -\frac{4}{\pi m^2} \left\{ (\pi - t) \cos mt \right\}_{t=0}^{t=\pi} \\
&= \frac{4}{m^2}.
\end{aligned}
$$

Thus, the Fourier series of f is

$$f(x) \sim \frac{\pi^2}{3} + \sum_{m=1}^{\infty} \frac{4}{m^2} \cos mx. \tag{21}$$

Because of the m^2 in the denominator, the series certainly converges, but we don't know whether it converges to $f(x)$, which is why we write the $\sim$ in (21). The graphs of f and the partial sums $S_1(x)$, $S_2(x)$, and $S_3(x)$ (see Figure 9.2.1) provide some numerical evidence that the series (21) converges to $f(x)$.

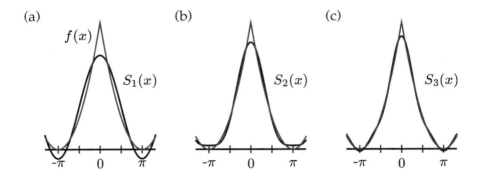

Figure 9.2.1

Example 2   Let f be the function $f(x) = -1$ on $[-\pi, 0]$ and $f(x) = 1$ on $(0, \pi)$. f is piecewise continuous on $[-\pi, \pi)$ with a jump discontinuity at $x = 0$. Thus the 2π periodic extension of f will have jump discontinuities at the points $x = 2n\pi$ for integral n. In addition, $f(\pi^-) = 1$ and $f(\pi^+) = -1$, so the periodic extension also has jump discontinuities at the points $x = (2n+1)\pi$ for integral n. See Figure 9.2.2(a). It is not hard to calculate the Fourier coefficients. Since f is an odd function and $\cos mx$ is even, $a_m = 0$ for all m. Furthermore,

$$b_m = -\frac{1}{\pi} \int_{-\pi}^{0} \sin mx \, dx + \frac{1}{\pi} \int_{0}^{\pi} \sin mx \, dx$$

$$= \begin{cases} \frac{4}{\pi m} & m = 1, 3, 5, \ldots \\ 0 & m = 2, 4, 6, \ldots \end{cases}$$

Thus, the Fourier series of f is

$$f(x) \sim \sum_{m=0}^{\infty} \frac{4}{\pi(2m+1)} \sin(2n+1)x. \qquad (22)$$

Notice that it is not at all evident that this series converges for any $x \neq 0$ since the coefficients decay only like m^{-1}. In fact, how could a sum of infinitely differentiable functions like $\sin mx$ and $\cos mx$ converge to a function like f which has jumps? Nevertheless, Figure 9.2.2, which graphs the partial sums, $S_1(x)$, $S_3(x)$, and $S_9(x)$, provides strong visual evidence that the partial sums of the series get close to f in some sense.

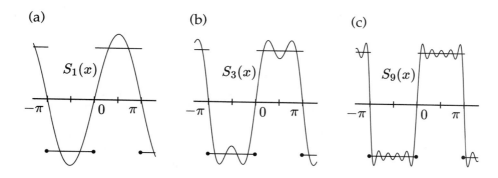

(a)   (b)   (c)

$S_1(x)$ $S_3(x)$ $S_9(x)$

Figure 9.2.2

Example 3 We return briefly to consider initial conditions for the heat equation considered in Section 9.1. Suppose that f is a continuous function defined on the interval $[0, L]$ such that $f(0) = 0 = f(L)$. Then $g(x) = f(\frac{xL}{\pi})$ is a continuos function on the interval $[0, \pi]$. By defining $g(x) = -g(-x)$, we can extend g to be a continuous odd function on $[-\pi, \pi]$, and since $g(0) = 0$, no jump discontinuity is introduced at $x = 0$. Furthermore, because $g(-\pi) = 0 = g(\pi)$, g extends to a unique 2π periodic function on $\mathbb{R}$. Since g is odd, the Fourier coefficients a_m are all zero and thus the Fourier series of g has only sine terms. Therefore, if the Fourier series of g converges to $g(x)$ at each x, then

$$f(x) = g\left(\frac{x\pi}{L}\right) = \sum_{m=1}^{\infty} b_m \sin \frac{m\pi x}{L},$$

so $f(x)$ can indeed be represented by formula (11). Thus the question posed in Section 9.1 is a special case of the Fourier series question posed in this section. If f is continuously differentiable and satisfies $f'(0) = 0 = f'(L)$, then the even extension of g to $[-\pi, \pi]$ can be extended to be a continuously differentiable 2π periodic function on $\mathbb{R}$ (problem 5). This extension is used in the case where the ends of the rod are insulated (problem 7 of Section 9.1) and results in a cosine series for the original function f.

Problems

1. Let $\{f_n\}$ be a sequence of functions on $\mathbb{R}$ which satisfy $f_n(x+2\pi) = f_n(x)$ for all n and x. Suppose that $f_n \to f$ pointwise. Prove that $f(x + 2\pi) = f(x)$ for all x.

2. Use (17) to prove formulas (18), (19), and (20).

3. Suppose that the Fourier series of f converges to $f(x)$ everywhere. Find a condition on the Fourier coefficients $\{c_m\}$ which holds if and only if f is real-valued.

4. Let f be a piecewise continuous function on $[-\pi, \pi]$. Show that $a_m = 0$ for all m if f is an odd function and $b_m = 0$ for all m if f is an even function.

5. Suppose that f is a continuously differentiable function on the interval $[0, L]$ that satisfies $f'(0) = 0 = f'(L)$. Show that $g(x) = f(\frac{xL}{\pi})$ defined on $[0, \pi]$ has a unique even, continuously differentiable, 2π periodic extension to $\mathbb{R}$.

6. Assume that the Fourier series for $f(x) = (\pi - |x|)^2$, which we computed in Example 1, converges to $f(x)$ for each x in $[-\pi, \pi]$. Prove that

$$\sum_{n=1}^{\infty} \frac{1}{n^2} = \frac{\pi^2}{6}.$$

7. Compute the Fourier coefficients of the periodic extension of the function $f(x) = x$ on $[-\pi, \pi)$. How does the periodic extension of the function $g(x) = x$ on $(-\pi, \pi]$ differ from the extension of f? Do the Fourier coefficients differ?

8. Show that the Fourier series of the function $f(x) = |x|$ on the interval $[-\pi, \pi]$ is

$$f(x) \sim \frac{\pi}{2} - \frac{4}{\pi} \sum_{n=1}^{\infty} \frac{1}{(2n-1)^2} \cos(2n-1)x.$$

9. Suppose that the Fourier coefficients of a piecewise continuous function f satisfy $\sum |a_m| < \infty$ and $\sum |b_m| < \infty$ and that the Fourier series converges to $f(x)$ for all x. Prove that f must be continuous.

10. Suppose that the Fourier coefficients of a piecewise continuous function f satisfy $\sum m^k |a_m| < \infty$ and $\sum m^k |b_m| < \infty$ for some fixed positive integer k and that the Fourier series converges to $f(x)$ for all x. Prove that f is k times continuously differentiable.

11. Let f be a continuous function of period 2π defined on $\mathbb{R}$.

   (a) Prove that $\int_a^{a+2\pi} f(x)\, dx$ is independent of a.

(b) Let f_a be the function $f_a(x) \equiv f(x - a)$. What is the relationship between the Fourier coefficients of f and the Fourier coefficients of f_a?

(c) Let f_k be the function $f_k(x) \equiv e^{ikx} f(x)$, where k is an integer. What is the relationship between the Fourier coefficients of f and the Fourier coefficients of f_k?

9.3 Pointwise Convergence

In this section we give conditions on f near a point x that guarantee that the Fourier series of f converges to $f(x)$ at x. The examples and problems in the last section suggest that the proof will be quite long and difficult. If the Fourier coefficients satisfy $\sum |a_m| < \infty$ and $\sum |b_m| < \infty$, then the Fourier series of f converges uniformly, so f would be continuous (problem 9 of Section 9.2). Thus, if f is not continuous, even at one point, we will not have these conditions on the coefficients. On the other hand, if we don't know that $\sum |a_m| < \infty$ and $\sum |b_m| < \infty$, how will we ever show that the Fourier series converges, let alone converges to $f(x)$? Example 2 of Section 9.2 indicates that there is a subtle phenomenon which we must analyze. In that case the function f has a jump discontinuity at $x = 0$. Nevertheless, the partial sums look like they are converging to the values of f to the right and left of $x = 0$. At $x = 0$ itself, the series converges to zero, which is the average of the value of f to the right ($+1$) and the value of f to the left (-1) of $x = 0$.

We begin by proving a lemma which generalizes the following explicit calculation. Suppose that f is a constant function, $f(x) = C$, on an interval $[a, b]$. Then,

$$\int_a^b f(t) e^{int} \, dt \;\; = \;\; \int_a^b C e^{int} \, dt$$

$$= \;\; C \left\{ \frac{e^{inb}}{in} - \frac{e^{ina}}{in} \right\}$$

$$\longrightarrow \;\; 0$$

as $n \to \infty$. A similar direct calculation (problem 1) shows that the same is true if f is piecewise constant.

Lemma 9.3.1 (The Riemann-Lebesgue Lemma)   Let f be piecewise continuous on the finite interval $[a, b]$. Then,

$$\lim_{n \to \infty} \int_a^b f(t) e^{\pm int} dt = 0. \tag{23}$$

Proof.   Let $a = a_0 < a_1 < a_2 < \ldots < a_N = b$ be a partition of the interval $[a, b]$ so that f is continuous on each subinterval (a_{j-1}, a_j). Since f is piecewise continuous, there is a unique continuous extension f_j of f on $[a_{j-1}, a_j]$, and by Corollary 3.5.4,

$$\int_a^b f(t) e^{\pm int} dt = \sum_{j=1}^N \int_{a_{j-1}}^{a_j} f_j(t) e^{\pm int} dt. \tag{24}$$

If each term in the finite sum on the right of (24) converges to zero as $n \to \infty$, then the left-hand side converges to zero. Thus, it is sufficient to prove the result in the case where f is continuous, which we henceforth assume. If f is constant, then the explicit calculation that we made above shows that (23) holds. So the idea of the proof is to approximate f by a piecewise constant function. Let $\varepsilon > 0$ be given. By Theorem 3.2.5, f is uniformly continuous on $[a, b]$, so we can choose a $\delta > 0$ so that

$$|f(x) - f(y)| \leq \frac{\varepsilon}{b - a} \qquad \text{if } |x - y| \leq \delta. \tag{25}$$

Choose a partition $a = p_0 < p_1 < p_2 < \ldots < p_N = b$ of $[a, b]$ so that $|p_i - p_{i-1}| \leq \delta$ for all i, and define g to be the function whose value on $[p_{i-1}, p_i)$ is $f(p_i)$. Then, by (25), $|f(t) - g(t)| \leq \frac{\varepsilon}{b-a}$ for all $t \in [a, b]$. Thus,

$$\left| \int_a^b (f(t) - g(t)) e^{\pm int} dt \right| \leq \int_a^b |f(t) - g(t)| dt \leq \varepsilon.$$

Therefore,

$$\left| \int_a^b f(t) e^{\pm int} dt \right| = \left| \int_a^b (f(t) - g(t)) e^{\pm int} dt + \int_a^b g(t) e^{\pm int} dt \right| \tag{26}$$

$$\leq \int_a^b |f(t) - g(t)| dt + \left| \sum_{j=1}^N f(p_j) \int_{p_{j-1}}^{p_j} e^{\pm int} dt \right| \tag{27}$$

$$\leq \varepsilon + \sum_{j=1}^N |f(p_j)| \left| \int_{p_{j-1}}^{p_j} e^{\pm int} dt \right|. \tag{28}$$

Each of the integrals in the finite sum goes to zero as $n \to \infty$, so

$$\limsup_{n\to\infty} \left| \int_a^b f(t)e^{\pm int}dt \right| \le \varepsilon.$$

Since $\varepsilon > 0$ was arbitrary,

$$\limsup_{n\to\infty} \left| \int_a^b f(t)e^{\pm int}dt \right| \le 0.$$

However, since $\left| \int_a^b f(t)e^{\pm int}dt \right| \ge 0$ for all n,

$$\liminf_{n\to\infty} \left| \int_a^b f(t)e^{\pm int}dt \right| \ge 0.$$

Therefore, by Corollary 6.1.2, $\lim_{n\to\infty} \left| \int_a^b f(t)e^{\pm int}dt \right|$ exists and equals zero, which implies that $\lim_{n\to\infty} \int_a^b f(t)e^{\pm int}dt = 0$.   ❏

The functions that we are considering are piecewise continuous, so the left-hand limit, $f(x^-)$, and the right-hand limit, $f(x^+)$, exist at every x. We say that the difference quotient of f has left-hand and right-hand limits at x if both of the limits,

$$\lim_{s \nearrow 0} \frac{f(x+s) - f(x^-)}{s}, \qquad \lim_{s \searrow 0} \frac{f(x+s) - f(x^+)}{s}$$

exist and are not infinite. Note that the value of f at x may not itself be involved in these difference quotients since it need not equal $f(x^-)$ or $f(x^+)$. Of course, if f is continuous at x, then $f(x^-) = f(x) = f(x^+)$. The following theorem gives sufficient conditions for the pointwise convergence of Fourier series.

❏ **Theorem 9.3.2** Suppose that f is piecewise continuous on $[-\pi, \pi]$ and periodic of period 2π on $\mathbb{R}$. Let $\{c_m\}$ be the Fourier coefficients of f and define $S_n(x) = \sum_{m=-n}^{n} c_m e^{imx}$. Then, at every point x where the left-hand and right-hand limits of the difference quotient of f exist,

$$\lim_{n\to\infty} S_n(x) = \frac{f(x^+) + f(x^-)}{2}. \tag{29}$$

Proof. By the definitions of S_n and c_m,

$$S_n(x) \;=\; \sum_{m=-n}^{n} \left(\frac{1}{2\pi} \int_{-\pi}^{\pi} f(t) e^{-imt}\, dt \right) e^{imx} \tag{30}$$

$$=\; \frac{1}{2\pi} \int_{-\pi}^{\pi} f(t) \sum_{m=-n}^{n} e^{im(x-t)}\, dt. \tag{31}$$

To find a simple expression for $D_n(x) \equiv \sum_{m=-n}^{n} e^{imx}$, we note that

$$(e^{ix} - 1) D_n(x) \;=\; e^{i(n+1)x} - e^{-inx}. \tag{32}$$

Solving (32) for $D_n(x)$ and multiplying the numerator and denominator by $e^{-ix/2}$, we find that

$$D_n(x) \;=\; \frac{\sin\left(n + \frac{1}{2}\right)x}{\sin \frac{1}{2}x}.$$

Substituting in (31) and making the change of variables $s = t - x$, we get

$$S_n(x) \;=\; \frac{1}{2\pi} \int_{-\pi}^{\pi} f(t) D_n(x - t)\, dt$$

$$=\; \frac{1}{2\pi} \int_{-\pi-x}^{\pi-x} f(x + s) D_n(-s)\, ds$$

$$=\; \frac{1}{2\pi} \int_{-\pi}^{\pi} f(x + s) D_n(s)\, ds.$$

In the last step we used the fact that D_n is even and that all integrals of $f(x + s) D_n(s)$ on intervals of length 2π are the same since both $f(x + s)$ and $D_n(s)$ are periodic of period 2π in s. Because $\frac{1}{2\pi} \int_{-\pi}^{\pi} D_n(s)\, ds = 1$ and $D_n(s)$ is even,

$$S_n(x) - \frac{f(x^+) + f(x^-)}{2} \;=\; \frac{1}{2\pi} \int_{-\pi}^{0} (f(x + s) - f(x^-)) D_n(s)\, ds$$

$$+ \frac{1}{2\pi} \int_{0}^{\pi} (f(x + s) - f(x^+)) D_n(s)\, ds$$

$$=\; \frac{1}{2\pi} \int_{-\pi}^{0} g(s) \sin\left(n + \tfrac{1}{2}\right)s\, ds \;+\; \frac{1}{2\pi} \int_{0}^{\pi} h(s) \sin\left(n + \tfrac{1}{2}\right)s\, ds,$$

where

$$g(s) = \frac{f(x + s) - f(x^-)}{\sin \frac{1}{2}s}, \qquad h(s) = \frac{f(x + s) - f(x^+)}{\sin \frac{1}{2}s}.$$

Since $\sin \frac{1}{2}s$ does not vanish on $[-\pi, 0)$, and f is piecewise continuous, g is piecewise continuous on $[-\pi, 0)$. Furthermore, if we write

$$g(s) = \frac{f(x+s) - f(x^-)}{s} \; \frac{s}{\sin \frac{1}{2}s},$$

we see that g has a left-hand limit at $s = 0$ because of the assumption that the difference quotient for f has a left-hand limit at x. Therefore, g is piecewise continuous on $[-\pi, 0]$. If we expand $\sin (n + \frac{1}{2})s$, we have

$$\frac{1}{\pi} \int_{-\pi}^{0} g(s) \sin (n + \tfrac{1}{2})s \, ds \;=\; \frac{1}{\pi} \int_{-\pi}^{0} (g(s) \cos \tfrac{s}{2}) \sin ns \, ds$$

$$+ \; \frac{1}{\pi} \int_{-\pi}^{0} (g(s) \sin \tfrac{s}{2}) \cos ns \, ds.$$

Now we write $\sin ns$ and $\cos ns$ in terms of e^{+ins} and e^{-ins}, and we use the Riemann-Lebesgue lemma to conclude that

$$\frac{1}{\pi} \int_{-\pi}^{0} g(s) \sin (n + \tfrac{1}{2})s \, ds \longrightarrow 0 \qquad \text{as } n \to \infty.$$

A similar proof, using the hypothesis that the difference quotient of f has a right-hand limit at x, shows that

$$\frac{1}{\pi} \int_{0}^{\pi} h(s) \sin (n + \tfrac{1}{2})s \, ds \longrightarrow 0 \qquad \text{as } n \to \infty,$$

which completes the proof. ❑

Corollary 9.3.3 If f is 2π periodic and continuously differentiable, then the Fourier series of f converges to $f(x)$ at every x.

Proof. If f is continuously differentiable then the right-hand and left-hand limits of the difference quotient exist at every $x \in \mathbb{R}$. Furthermore, since f is continuous, $f(x^+) = f(x) = f(x^-)$ for all x. Thus, by Theorem 9.3.2, $\lim_{n\to\infty} S_n(x) = f(x)$ for all x. ❑

Example 1 The function $f(x) = (\pi - |x|)^2$ introduced in Example 1 of Section 9.2 is continuous on $[-\pi, \pi]$, and since $f(\pi) = 0 = f(-\pi)$, the periodic extension to $\mathbb{R}$ is continuous. Note that f is the polynomial $(\pi - x)^2$ on the interval $(0, \pi)$ and is the polynomial $(\pi + x)^2$ on the interval $(-\pi, 0)$. Thus, the 2π periodic extension of f is continuously differentiable except possibly at the points $-\pi + 2\pi n$ and $2\pi n$, where n is an integer. At the points $-\pi + 2\pi n$, the left-hand and right-hand difference quotients both have limit zero, so the 2π periodic extension is continuously differentiable at these points. At $x = 0$ the derivative of f does not exist (see Figure 9.2.1). However,

$$\lim_{s \searrow 0} \frac{f(0 + s) - f(0^+)}{s} = \lim_{s \searrow 0} \frac{(\pi - s)^2 - \pi^2}{s} = -2\pi$$

and

$$\lim_{s \nearrow 0} \frac{f(0 + s) - f(0^-)}{s} = \lim_{s \nearrow 0} \frac{(\pi + s)^2 - \pi^2}{s} = 2\pi,$$

so both the right-hand and left-hand limits of the difference quotient exist at $x = 0$, although they are not equal. Thus, the right-hand and left-hand limits of the difference quotient of f exist at all x. Therefore, according to Theorem 9.3.2, the Fourier series of f converges to $f(x)$ at all x. That is,

$$f(x) = \frac{\pi^2}{3} + \sum_{n=1}^{\infty} \frac{4}{n^2} \cos nx.$$

Example 2   The step function, f, defined in Example 2 of Section 9.2, is continuous except for integral multiples of π. The left-hand and right-hand limits of the the difference quotient of f exist at every point $x \in \mathbb{R}$ and equal zero. Therefore, Theorem 9.3.2 guarantees that the Fourier series of f will converge at all x. Where f is continuous, that is, except at integral multiples of π, the Fourier series converges to $f(x)$. At each jump discontinuity, the average of $f(x^+)$ and $f(x^-)$ is 0, so the Fourier series will converge to zero at the points $x = n\pi$. This is exactly the behavior seen in Figure 9.2.2. Note that all the partial sums, $S_n(x)$, equal zero at $x = 0$, so they converge to zero as $n \to \infty$. The overshoot of the partial sums near the jump discontinuities is called the Gibbs phenomenon after the physicist J. W. Gibbs (1839 – 1903).

Let g be the function which equals -1 on the open interval $(-\pi, 0)$ and which equals 1 on the open interval $(0, \pi)$. Then, depending how we define g at 0 and at one or the other of the points $\pm\pi$, we get different 2π periodic extensions of g to the whole line $\mathbb{R}$. All of these different extensions (the function f above is one of them) have the same Fourier series since the choice of values at the points 0 and $\pm\pi$ does not affect the integrals for the Fourier coefficients.

One reason that Theorem 9.3.2 is hard to prove is that the hypotheses and conclusion are "local." If the difference quotient for f has right-hand

and left-hand limits at a point x (which says that f is reasonably regular near x), then the Fourier series of f converges at x. This is true no matter what the behavior of f is away from x. Thus the proof must show that the behavior on the rest of the interval doesn't matter.

A theorem similar to Theorem 9.3.2 was proven in 1829 by A. Dirichlet (1805 – 1859). Note that it gives sufficient conditions for convergence of a Fourier series but does not answer the question of whether the Fourier series of every continuous function converges. Most contemporaries of Dirichlet thought the answer was yes, but they were proved wrong by Du-Bois Reymond, who found in 1876 a continuous function whose Fourier series diverged at one point. The example, whose construction is too long to reproduce here, led many analysts to believe that there must be continuous functions whose Fourier series diverge everywhere. The answer, in fact, lies in between. The Fourier series of a continuous function can diverge on at most a set of "measure zero." Conversely, given a set of measure zero, there is a continuous function whose Fourier series diverges there. Sets of measure zero are defined in problem 10. Questions about the pointwise convergence of Fourier series played an important role in analysis for a century and a half following Fourier's *La Théorie Analytique de Chaleur* [17] and were not satisfactorally settled until the 1960s. For a beautiful historical discussion, see [29].

Problems

1. A function f is called **piecewise constant** on $[a, b]$ if there are finitely many disjoint subintervals (a_i, b_i) such that $[a, b] = \cup[a_i, b_i]$ and f is constant on each (a_i, b_i). Prove by direct calculation that if f is piecewise constant, then

$$\lim_{n \to \infty} \int_a^b f(t)e^{\pm int} \, dt = 0.$$

2. Let f be the 2π periodic extension of the function

$$f(x) = \begin{cases} 2 + 3x & -\pi < x < 0 \\ x^2 & 0 \le x < \pi \\ 1 & x = \pi. \end{cases}$$

   (a) For each $x \in \mathbb{R}$, find the right-hand and left-hand limits of the difference quotient of f.

   (b) For each $x \in \mathbb{R}$, to what number does the Fourier series of f converge?

3. Which of the following functions have difference quotients at $x = 0$ that have finite right-hand limits?

   (a) $f(x) = |x|$.

   (b) $f(x) = \sqrt{|x|}$.

   (c) $f(x) = \frac{\sin x}{x}$.

   (d) $f(x) = \sin \frac{1}{x}$.

   (e) $f(x) = x \sin \frac{1}{x}$.

   (f) $f(x) = x^2 \sin \frac{1}{x}$.

4. Let f be the 2π periodic extension of the function

$$f(x) \;=\; \begin{cases} 0 & -\pi < x < 0 \\ 5x & 0 \le x < \pi \\ 5\pi/2 & x = \pi. \end{cases}$$

   Prove that the Fourier series of f converges to $f(x)$ for all x.

5. (a) Compute the Fourier coefficients of the periodic extension of the function $f(x) = x$ on $[-\pi, \pi)$ (problem 7 of Section 9.2).

   (b) Compute the graphs of S_1, S_3, and S_{10} and compare them to the graph of f.

   (c) Where does the Fourier series of f converge? To what does it converge?

   (d) Show that the solution of the heat equation discussed in problem 5 of Section 9.1 satisfies the initial condition at all $x \in [0, \pi]$ except $x = \pi$.

6. Let f be the function

$$f(x) = \begin{cases} \pi + x & -\pi \le x \le 0 \\ \pi - x & 0 \le x \le \pi. \end{cases}$$

   Where does the Fourier series of f converge? To what function does it converge?

7. Let f be a twice continuously differentiable 2π periodic function. Prove that the Fourier series of f converges to f uniformly. Hint: use integration by parts to estimate the coefficients. Note that the same conclusion holds with a weaker hypothesis (Theorem 9.4.4).

8. Explain carefully why the coefficients $\{a_n\}$ of the solution $u(t, x)$ in problem 7 of Section 9.1 can be chosen so that the initial condition $u(0, x) = f(x)$ holds for all x if $f'(0) = 0 = f'(L)$.

9. Let f be a 2π periodic function that is infinitely often continuously differentiable. Explain why the derivatives of f are periodic with period 2π. Prove that the Fourier series for f may be differentiated term by term as often as we like and that we thereby obtain the Fourier series for the derivatives of f.

10. A subset $E \subseteq \mathbb{R}$ is said to have **measure zero** if, given any $\varepsilon > 0$, there is a countable family of intervals $\{I_n\}_{n=1}^\infty$ such that $E \subseteq \cup I_n$ and $\sum length(I_n) \leq \varepsilon$.

    (a) Show that any finite set of points has measure zero.

    (b) Show that the set $\{\frac{1}{n}\}$ has measure zero.

    (c) Show that any countable set has measure zero. Remark: There are uncountable sets of measure zero.

9.4 Mean-square Convergence

In the last section, we saw that the question of the pointwise convergence of Fourier series is quite delicate. We shall see in this section that mean-square convergence is fairly easy to establish. We begin from a more abstract point of view which will allow us to make connections with other problems in classical and modern analysis. If f and g are complex-valued piecewise continuous functions on an interval $[a, b]$, we define the **inner product**, (f, g), of f and g by

$$(f, g) \equiv \int_a^b f(x)\overline{g(x)}\, dx,$$

where $\overline{g(x)}$ denotes the complex conjugate of $g(x)$. The integration of complex-valued functions is discussed in Section 8.2. Elementary properties of the Riemann integral and complex numbers allow one to show easily that (problem 1)

(a) $(f, f) \geq 0$ and $(f, f) = 0$ if and only if $f = 0$.

(b) $(\alpha_1 f_1 + \alpha_2 f_2, g) = \alpha_1 (f_1, g) + \alpha_2 (f_2, g)$ for all α_1 and α_2 in $\mathbb{C}$.

(c) $\overline{(f, g)} = (g, f)$.

From (b) and (c), it follows that the inner product is conjugate linear in the second factor; that is, $(f, \alpha_1 g_1 + \alpha_2 g_2) = \overline{\alpha}_1 (f, g_1) + \overline{\alpha}_2 (f, g_2)$. The

L^2 norm, defined in Section 5.3, can be expressed in terms of the inner product by

$$\|f\|_2 = (f, f)^{\frac{1}{2}}.$$

If $\|f - f_n\|_2 \to 0$, then the sequence of functions $\{f_n\}$ is said to **converge to f in the L^2 norm**, or in **the mean-square sense**. The reader is asked to show in problem 3 that pointwise convergence for all x does not imply mean-square convergence, nor does mean-square convergence imply pointwise convergence.

Proposition 9.4.1 (The Cauchy-Schwarz Inequality)   For all piecewise continuous functions f and g,

$$|(f, g)| \quad \leq \quad \|f\|_2 \|g\|_2. \tag{33}$$

Proof.   The proof uses only the three simple properties of the inner product. First suppose that $\|g\|_2 = 1$.

$$
\begin{aligned}
0 \quad \leq \quad & \|f - (f, g)g\|_2^2 \\
= \quad & (f - (f, g)g, f - (f, g)g) \\
= \quad & (f, f) - (f, g)\overline{(f, g)} - (f, g)\overline{(f, g)} + (f, g)\overline{(f, g)}(g, g) \\
= \quad & \|f\|_2^2 - |(f, g)|^2,
\end{aligned}
$$

from which (33) follows in the case $\|g\|_2 = 1$. If g is the zero function, then (33) certainly holds, and if g is not, then $\|g\|_2 > 0$. Applying what we have just proven to the functions f and $h = g/\|g\|_2$, we find $|(f, h)| \leq \|f\|_2$ since $\|h\|_2 = 1$. Thus,

$$\frac{|(f, g)|}{\|g\|_2} = |(f, h)| \leq \|f\|_2,$$

which proves (33) in the general case.   ❏

Proposition 9.4.2   If f and g are piecewise continuous on $[a, b]$, then

(a) $\|f\|_2 \geq 0$ and $\|f\|_2 = 0$ if and only if $f = 0$.

(b) $\|\alpha f\|_2 = |\alpha| \, \|f\|_2$ for all $\alpha \in \mathbb{C}$.

(c) $\|f + g\|_2 \leq \|f\|_2 + \|g\|_2$.

Proof. Properties (a) and (b) follow immediately from similar properties of the inner product. By the Cauchy-Schwarz inequality,

$$
\begin{aligned}
\|f + g\|_2^2 &= (f, f) + (f, g) + (g, f) + (g, g) \\
&\leq \|f\|_2^2 + 2\|f\|_2\|g\|_2 + \|g\|_2^2 \\
&= (\|f\|_2 + \|g\|_2)^2,
\end{aligned}
$$

so (c) follows by taking the square root of both sides. ❏

Thus $\|\cdot\|_2$ has the three properties of a norm (see Section 5.8), which is why we have been referring to it as the L^2 norm. Since this norm comes from an inner product, we can introduce a notion of orthogonality.

Definition. Two piecewise continuous functions f and g are are said to be **orthogonal** if $(f, g) = 0$. A set (finite or infinite) of piecewise continuous functions, $\{\varphi_n\}_{n=1}^N$, is said to be an **orthonormal family** if $\|\phi_n\|_2 = 1$ for all n and $(\varphi_n, \varphi_m) = 0$ for all $m \neq n$.

It may seem strange to refer to two *functions* as orthogonal, but there is a good reason for this terminology. Recall that the "dot product" of two vectors in $\mathbb{R}^n$, $x = (x_1, x_2, \ldots, x_n)$ and $y = (y_1, y_2, \ldots, y_n)$, is defined by

$$
x \cdot y \equiv \sum_{j=1}^n x_j y_j.
$$

Notice that the dot product is a function from $\mathbb{R}^n \times \mathbb{R}^n$ to $\mathbb{R}$ that satisfies

(a) $x \cdot x \geq 0$ and $x \cdot x = 0$ if and only if $x = 0$.

(b) $(\alpha_1 x + \alpha_2 y) \cdot z = \alpha_1 x \cdot z + \alpha_2 y \cdot z$.

(c) $x \cdot y = y \cdot x$.

These three properties are completely analogous to the three properties of the L^2 inner product introduced at the beginning of this section, except for the extra complex conjugation. We say that two vectors in $\mathbb{R}^n$, x and y, are orthogonal if $x \cdot y = 0$; so we use the same terminology in the case of the L^2 inner product. We shall see that the concept of orthogonality for functions plays as fundamental a role as orthogonality in $\mathbb{R}^n$. This suggests that there is an underlying idea here which is worth studying futher. In project 4 we introduce the notion of inner product space and show that the Cauchy-Schwarz inequality holds in general. In project 5, we give examples of complete inner product spaces, called Hilbert spaces.

Of course, the orthonormal family that we have in mind is

$$\left\{ \frac{e^{inx}}{\sqrt{2\pi}} \right\}_{n=-\infty}^{\infty} \tag{34}$$

on the interval $[-\pi, \pi]$. The fact that the functions e^{inx} are orthogonal to each other enabled us to derive formula (17) for the Fourier coefficients. Dividing by $\sqrt{2\pi}$ ensures that each has L^2 norm equal to 1. Note that it is only for convenience in the definition that we have indexed the sequence of functions $\{\varphi_n(x)\}$ from $n = 1$ to $n = N$, where possibly $N = \infty$. The index can run over some other set (as in the above example), and the order of the functions in the sequence plays no role in the definition.

Suppose that f is a piecewise continuous function and $\{\varphi_n\}_{n=1}^{N}$ is a finite orthonormal family on the interval $[a, b]$. We want to find the linear combination of the functions φ_n that gives the best approximation to f in the mean-square sense. That is, we want to choose coefficients $\{c_n\}$ so that the norm of the difference $\|f - \sum c_n \varphi_n(x)\|_2$ is as small as possible. First, we calculate

$$
\begin{aligned}
\left\| f - \sum_{n=1}^{N} c_n \varphi_n \right\|_2^2 &= \left(f - \sum_{n=1}^{N} c_n \varphi_n, f - \sum_{n=1}^{N} c_n \varphi_n \right) \\
&= (f, f) - \sum_{n=1}^{N} c_n (\varphi_n, f) - \sum_{n=1}^{N} \bar{c}_n (f, \varphi_n) + \sum_{n=1}^{N} |c_n|^2 \\
&= \|f\|_2^2 - \sum_{n=1}^{N} |(f, \varphi_n)|^2 + \sum_{n=1}^{N} |c_n - (f, \varphi_n)|^2,
\end{aligned}
$$

where we completed the square in the last step. Since the first two terms on the right do not depend on the sequence $\{c_n\}$ and the third term is nonnegative, we see that $\|f - \sum_{n=1}^{N} c_n \varphi_n\|_2$ is smallest if we choose $c_n = (f, \varphi_n)$. In this case

$$\left\| f - \sum_{n=1}^{N} c_n \varphi_n \right\|_2^2 = \|f\|_2^2 - \sum_{n=1}^{N} |c_n|^2. \tag{35}$$

The numbers $c_n = (f, \varphi_n)$ are called the **generalized Fourier coefficients** of f with respect to the orthonormal family $\{\varphi_n\}_{n=1}^{N}$. Notice that since the left-hand side of (35) is nonnegative, we have

$$\| \sum_{n=1}^{N} c_n \varphi_n \|_2^2 = \sum_{n=1}^{N} |c_n|^2 \leq \|f\|_2^2. \qquad (36)$$

If $\{\varphi_n(x)\}$ is an infinite orthonormal family, then (36) holds for each finite N. Since the right-hand side doesn't depend on N, this shows that $\sum_{n=1}^{\infty} |c_n|^2$ is finite and

$$\sum_{n=1}^{\infty} |c_n|^2 \leq \|f\|_2^2,$$

which is known as **Bessel's inequality**. We summarize what we have proven in a theorem.

❏ **Theorem 9.4.3** Let f be a piecewise continuous function and let $\{\varphi_n\}$ be an orthonormal family of piecewise continuous functions on a finite interval $[a, b]$. Then, choosing $c_n = (f, \varphi_n)$ minimizes $\|f - \sum_{n=1}^{N} c_n \varphi_n\|_2$, and the sequence $\{c_n\}$ satisfies Bessel's inequality.

We can now use these concepts to analyze Fourier series. Since

$$\int_{-\pi}^{\pi} \frac{e^{inx}}{\sqrt{2\pi}} \frac{\overline{e^{imx}}}{\sqrt{2\pi}} \, dx = \frac{1}{2\pi} \int_{-\pi}^{\pi} e^{i(n-m)x} \, dx = \begin{cases} 1 & \text{if } n = m \\ 0 & \text{otherwise,} \end{cases}$$

the set of functions in (34) is an orthonormal family on the interval $[-\pi, \pi]$. The **Fourier coefficients** of a function f are defined by the inner products

$$c_m = (f, (2\pi)^{-1/2} e^{imx}) = \frac{1}{\sqrt{2\pi}} \int_{-\pi}^{\pi} f(x) e^{-imx} \, dx, \qquad (37)$$

and the **Fourier series** of f is

$$f(x) \sim \sum_{m=-\infty}^{\infty} c_m \frac{e^{imx}}{\sqrt{2\pi}}.$$

Note that the Fourier series is the same as (16), but the definition of c_m differs by a factor of $\sqrt{2\pi}$ from (17) since we have put the factor $\sqrt{2\pi}$ under e^{inx} so that $e^{inx}/\sqrt{2\pi}$ has L^2 norm equal to 1. If we define the partial sum, $S_n(f)$, of the Fourier series of f by

$$S_n(f)(x) = \sum_{m=-n}^{n} c_m \frac{e^{imx}}{\sqrt{2\pi}},$$

then the analogue of (36) is

$$\|S_n(f)\|_2^2 \leq \sum_{m=-n}^{n} |c_m|^2 \leq \|f\|_2^2, \tag{38}$$

and **Bessel's inequality** states

$$\sum_{m=-\infty}^{\infty} |c_m|^2 \leq \|f\|_2^2. \tag{39}$$

We can immediately use Bessel's inequality to improve Corollary 9.3.3.

❏ **Theorem 9.4.4**   Let f be a continuously differentiable function of period 2π. Then, the Fourier series of f converges to f uniformly.

Proof.   We already know from Corollary 9.3.3 that the Fourier series of f converges to f pointwise. By the Weierstrass M-test, we need only show that $\sum_{-\infty}^{\infty} |c_m| < \infty$ to conclude that the series converges uniformly. If we denote by $\{c_m'\}$ the Fourier coefficients of f', then integration by parts in (17) shows that $imc_m = c_m'$ for all $m \neq 0$. For each positive integer n, let $\mathcal{G}_n$ denote the set of integers $\mathcal{G}_n \equiv \{m \mid -n \leq m \leq n; \, m \neq 0\}$. Then, the discrete Cauchy-Schwarz inequality (problem 10 in Section 5.8) implies that

$$\sum_{m \in \mathcal{G}_n} |c_n| = \sum_{m \in \mathcal{G}_n} \frac{|c_m'|}{|m|}$$

$$\leq \left(\sum_{m \in \mathcal{G}_n} \frac{1}{m^2} \right)^{\frac{1}{2}} \left(\sum_{m \in \mathcal{G}_n} |c_m'|^2 \right)^{\frac{1}{2}}.$$

Since f' is a continuous function, Bessel's inequality for f' implies that $\sum |c_m'|^2 < \infty$, and we also know that $\sum \frac{1}{m^2} < \infty$. Thus, the right-hand side has a finite limit as $n \to \infty$. Therefore,

$$\sum |c_m| < \infty,$$

from which it follows that the Fourier series converges uniformly.   ❏

The following lemma plays a crucial role in the main theorem.

Lemma 9.4.5   Every periodic continuous function can be uniformly approximated by a periodic continuously differentiable function.

Proof. Let $j(x)$ be a continuously differentiable function on $\mathbb{R}$ such that $j(x) \geq 0$, $j(x) = 0$ outside the interval $[-1, 1]$, and $\int_{-\infty}^{\infty} j(x)\, dx = 1$. See problem 4 for the construction of such a function. For each $\delta > 0$, set $j_\delta(x) \equiv \delta^{-1} j(x/\delta)$. Then $j_\delta(x) \geq 0$, $j_\delta(x) = 0$ outside the interval $[-\delta, \delta]$, and $\int_{-\infty}^{\infty} j_\delta(x)\, dx = 1$. Suppose that $f(x)$ is a periodic continuous function, and define

$$g_\delta(x) \equiv \int_{-\infty}^{\infty} j_\delta(x - y) f(y)\, dy.$$

For each fixed x, the integrand $j_\delta(x - y) f(y)$ is a continuous function of y which is zero outside of the interval $[x - \delta, x + \delta]$, so the integral makes sense. In fact, for x in any fixed interval $[a, b]$, the only values of y that contribute to the integral are those in the interval $[a - \delta, b + \delta]$, so

$$g_\delta(x) = \int_{a-\delta}^{b+\delta} j_\delta(x - y) f(y)\, dy.$$

Since $j_\delta(x-y) f(y)$ is continuously differentiable in x on $[a, b] \times [a-\delta, b+\delta]$, Theorem 5.2.4 guarantees that g_δ is continuously differentiable on $[a, b]$. However, $[a, b]$ was arbitrary, so g_δ is continuously differentiable on $\mathbb{R}$. Furthermore, by making the change of variables $\tau = x - y$, we can write

$$g_\delta(x) = \int_{-\infty}^{\infty} j_\delta(\tau) f(x - \tau)\, d\tau,$$

from which it follows that g_δ has the same period as f. Finally,

$$
\begin{aligned}
|g_\delta(x) - f(x)| &= \left| \int_{-\infty}^{\infty} j_\delta(\tau)(f(x - \tau) - f(x))\, d\tau \right| \\
&= \left| \int_{-\delta}^{\delta} j_\delta(\tau)(f(x - \tau) - f(x))\, d\tau \right| \\
&\leq \sup_{-\delta \leq \tau \leq \delta} |f(x - \tau) - f(x)| \int_{-\infty}^{\infty} j_\delta(\tau)\, d\tau \\
&\leq \sup_{-\delta \leq \tau \leq \delta} |f(x - \tau) - f(x)|.
\end{aligned}
$$

Since f is periodic and continuous, f is uniformly continuous. Therefore, the term on the right goes uniformly to zero as $\delta \to 0$. Thus, $g_\delta \to f$ uniformly as $\delta \to 0$. ❏

❏ **Theorem 9.4.6**   Let f be a continuous function of period 2π and let $S_n(f) = \sum_{m=-n}^{n} c_m e^{-imx}$ be the partial sum of the Fourier series of f. Then,

$$\|f - S_n(f)\|_2 \;\rightarrow\; 0 \qquad \text{as } n \rightarrow \infty$$

and

$$\sum_{m=-\infty}^{\infty} |c_m|^2 \;=\; \int_{-\pi}^{\pi} |f(x)|^2 \, dx \qquad \text{(Parseval's relation)}.$$

Proof.   Let $\varepsilon > 0$ be given. By the lemma, we can choose a continuously differentiable function g which has period 2π and satisfies $\|f - g\|_2 \le \varepsilon/3$. By the triangle inequality [part (c) of Proposition 9.4.2],

$$\|f - S_n(f)\|_2 \;\le\; \|f - g\|_2 \,+\, \|g - S_n(g)\|_2 \,+\, \|S_n(g) - S_n(f)\|_2. \quad (40)$$

Since the formula for the Fourier coefficients, (17), depends linearly on the function under the integral, $S_n(g) - S_n(f) = S_n(g - f)$. Furthermore, by (38), $\|S_n(g - f)\|_2 \le \|g - f\|_2$, so,

$$\|f - S_n(f)\|_2 \;\le\; \frac{\varepsilon}{3} \,+\, \|g - S_n(g)\|_2 \,+\, \frac{\varepsilon}{3} \quad (41)$$

for all n. To estimate the middle term, notice that

$$\|g - S_n(g)\|_2^2 \;=\; \int_{-\pi}^{\pi} |g(x) - S_n(g)(x)|^2 \, dx \;\le\; 2\pi \|g - S_n(g)\|_\infty^2.$$

Since g is continuously differentiable, Theorem 9.4.4 guarantees that $\|g - S_n(g)\|_\infty \rightarrow 0$ as $n \rightarrow \infty$. Thus, $\|g - S_n(g)\|_2 \rightarrow 0$ as $n \rightarrow \infty$. Therefore, we can choose an N so that $\|g - S_n(g)\|_2 \le \varepsilon/3$ for $n \ge N$. Using this estimate in (41) gives

$$\|f - S_n(f)\|_2 \;\le\; \varepsilon \qquad \text{for } n \ge N,$$

which proves that $\|f - S_n(f)\|_2 \rightarrow 0$ as $n \rightarrow \infty$. Furthermore, formula (35) gives

$$\|f - S_n(f)\|_2 \;=\; \|f\|^2 \,-\, \sum_{m=-n}^{n} |c_m|^2.$$

The left-hand side goes to zero as $n \rightarrow \infty$, which implies that $\lim_{n \rightarrow \infty} \sum_{m=-n}^{n} |c_m|^2 = \|f\|^2$. This proves Parseval's relation.   ❏

It is worthwhile to notice the interesting structure of the proof and the important role of Lemma 9.4.5. We wanted to show that the left side of (40) is small for large n. The first term on the right is made small by using the lemma to choose a continuously differentiable g that is close to f. The second term is small for large n because the Fourier series of a continuously differentiable function (g) converges uniformly and therefore in mean-square sense. The third term is shown to be small by using Bessel's inequality. This suggests a way of proving that the conclusions of Theorem 9.4.6 hold for more general classes of functions. If f is a function which can be approximated in mean-square sense by a continuous function g, then the first term on the right of (40) can be made small. The second term can then be made small since we have just shown that the Fourier series of a continuous functions converges in the mean-square sense to the function. If Bessel's inequality holds for $f - g$ then the third term will also be small. This idea can be used to show that the Fourier series of a piecewise continuous functions converge in the mean-square sense to the function (problem 6) and that the same is true for functions with even nastier singularities (project 3). If we can find the most general class of functions which can be approximated in mean-square sense by continuous functions, for which $\int_a^b |f(x)|^2\, dx$ makes sense and for which Bessel's inequality holds, we should be able to prove that the conclusion of Theorem 9.4.6 holds for that class of functions. To carry out this program, one needs the Lebesgue integral.

Problems

1. Verify the three properties, (a), (b), and (c), of the inner product.

2. Let $\{f_n\}$ be a sequence of continuous functions on $[a, b]$ that converges to a function f uniformly. Prove that $f_n \to f$ in the mean-square sense.

3. Let $[a, b]$ be a finite interval.

    (a) Construct a sequence of continuous functions on $[a, b]$, $\{f_n\}$, so that $f_n \to 0$ pointwise but $\|f_n\|_2 \to \infty$. Hint: choose f_n to be a function which is tall on a small set and zero elsewhere.

    (b) Construct a sequence of continuous functions on $[a, b]$, $\{f_n\}$, so that $\|f_n\|_2 \to 0$ but $\{f_n(x)\}$ does not converge to 0 for any $x \in [a, b]$. Hint: find f_n with narrow graphs that march back and forth across $[a, b]$.

4. (a) Use pieces of polynomials to construct a continuously differentiable function $j(x)$ on $\mathbb{R}$ that is nonnegative, vanishes outside the interval $[-1, 1]$ and satisfies $\int_{-\infty}^{\infty} j(x)\, dx = 1$.

(b) Let $f(x) = 0$ for $x \geq 0$ and $f(x) = e^{-1/x^2}$ for $x < 0$. Prove that f is a C^∞ function on $\mathbb{R}$. Hint: see Example 4 in Section 6.4.

(c) Use the function f and its translates to construct a C^∞ function j that has the properties in (a).

5. For the step function of Example 2 in Section 9.2, compute approximately $\|f - S_{2n+1}\|_2$ for $n = 0, n = 1$, and $n = 4$.

6. (a) Let f be a piecewise continuous function on $[\pi, \pi]$. Prove that there is a sequence of continuous functions f_n on $[\pi, \pi]$ so that $f_n \to f$ in mean-square sense. Hint: connect the pieces.

(b) Use the idea of the proof of Theorem 9.4.6 to show that the Fourier series of a piecewise continuous function f converges to f in the mean-square sense.

7. Let f and g be continuous functions of period 2π with Fourier coefficients $\{c_n\}$ and $\{d_n\}$, respectively. Prove that

$$\int_{-\pi}^{\pi} f(x)\overline{g(x)}\, dx = \sum_{n=-\infty}^{\infty} c_n \overline{d_n}.$$

8. Use Parseval's relation and the function in Example 2 of Section 9.2 to prove that

$$\frac{\pi^2}{8} = \sum_{n=0}^{\infty} \left(\frac{1}{2n+1}\right)^2.$$

9. Let f be a continuously differentiable function on $[-\pi, \pi]$ such that $\int_{-\pi}^{\pi} f(x)\, dx = 0$. Prove that

$$\int_{-\pi}^{\pi} |f'(x)|^2\, dx \ \geq \ \int_{-\pi}^{\pi} |f(x)|^2\, dx. \tag{42}$$

Prove that strict equality holds in (42) if and only if $f(x) = a\cos x + b\sin x$ for some constants a and b. Hint: use Parseval's relation.

10. Suppose that f is a twice continuously differentiable function which is periodic of period 2π. Prove that

$$\int_{-\pi}^{\pi} |f'(x)|^2\, dx \ \leq \ \left(\int_{-\pi}^{\pi} |f(x)|^2\, dx\right)^{\frac{1}{2}} \left(\int_{-\pi}^{\pi} |f''(x)|^2\, dx\right)^{\frac{1}{2}}.$$

Inequalities such as this are called **Sobolev inequalities**.

Projects

1. Consider heat flow in a bar of length π with $\kappa = 1$ and insulated ends as in Problem 7 of Section 9.1. Let the initial temperatures be $f(x) = 0$ on $[0, \pi/2]$ and $f(x) = 10$ on $[\pi/2, \pi]$.

   (a) Compute the coefficients in the expansion $f(x) = \sum_{n=0}^{\infty} a_n \cos nx$.

   (b) We will approximate the solution, $u(t, x)$, of the heat equation by the first eight terms of its expansion $v(t, x) = \sum_{n=0}^{7} a_n e^{-\lambda_n t} \cos nx$. Graph v at times $t = 0, \frac{1}{4}, \frac{1}{2}, \frac{3}{4}$, and 1.

   (c) What properties of the solution discussed in Section 9.1 (and problem 7 of that section) can you observe in the graphs?

   (d) How close is v to the true solution at $t = 1$?

   (e) Compare v and $5 - e^{-t} \cos nx$ at $t = 1$. Why are they so close?

   (f) Investigate the influence of κ by graphing the approximate solutions at the times $t = 0, \frac{1}{4}, \frac{1}{2}, \frac{3}{4}$, and 1 in the cases $\kappa = 10$ and $\kappa = 1/10$.

2. The technique used to solve the heat equation in Section 9.1 can also be used to solve boundary value problems for other partial differential equations. In the simplest model for a vibrating string of length L that has fixed ends, the unknown function, $u(t, x)$, which represents the vertical displacment from equilibrium of the string at position x at time t, satisfies the **wave equation**

$$u_{tt}(t, x) - c^2 u_{xx}(t, x) = 0, \qquad 0 \le x \le L \tag{43}$$

and the boundary conditions

$$u(t, 0) = 0 = u(t, L). \tag{44}$$

The constant $c = T/\rho$ where T is the tension in the string and ρ is the density. We specify the initial displacements and the initial velocity of the string at each x.

$$u(0, x) = f(x) \tag{45}$$
$$u_t(0, x) = g(x) \tag{46}$$

For simplicity, we will assume that f is three times continuously differentiable, g is twice continuously differentiable, and $g(0) = f(0) = 0 = f(L) = g(L)$.

   (a) Follow the separation of variables method of Section 9.1 to show that for each positive integer n and any choice of the constants a_n and b_n the function

$$u_n(t, x) = \{a_n \sin \frac{n\pi ct}{L} + b_n \cos \frac{n\pi ct}{L}\} \sin \frac{n\pi x}{L}$$

   satisfies (43) and (44).

(b) Using Theorem 9.3.2 and the idea in Example 3 of Section 9.2, show that the constants $\{a_n\}$ and $\{b_n\}$ can be chosen so that the series

$$u(t, x) \quad = \quad \sum_{n=1}^{\infty} u_n(t, x) \tag{47}$$

converges and (45) and (46) hold.

(c) Prove that $u(t, x)$ is twice continuously differentiable in t and x and satisfies (43) and (44). Justify any term-by-term differentiations.

(d) The energy, E, of the solution, u, is given by the following expression

$$E(t) \quad = \quad \frac{1}{2} \int_0^L \rho(u_t(t, x))^2 + T(u_x(t, x))^2 \, dx.$$

Show that E is independent of t. Justify any differentiating under the integral sign.

(e) Suppose that v is another twice continuously differentiable function which satisfies (43) – (46). Show that $w = u - v$ is a solution of (43) and that its energy is zero. Use this to prove that $u(t, x) = v(t, x)$ for all $0 \leq x \leq L, 0 \leq t$, so the solution is unique.

(f) We saw in problem 10 of Section 9.2 that the faster the Fourier coefficients go to zero as $n \to \infty$, the more differentiable the function defined by the Fourier series is. This idea was used in Theorem 9.1.1 to show that the solution of the heat equation is always infinitely differentiable for $t > 0$. Do you think that the same is true for the wave equation?

3. Let $f(x) = |x|^{-\frac{1}{4}}$ on the interval $[-\pi, \pi]$.

(a) Explain why the improper Riemann integrals (f, e^{-inx}) and $\|f\|_2^2$ exist.

(b) If g is continuous, explain why Bessel's inequality will hold for $f - g$.

(c) Show that f can be approximated as closely as we like in the mean-square sense by a continuous function.

(d) Prove that the Fourier series of f converges to f in the mean-square sense.

(e) Prove that the Fourier series of $h(x) = \sin\frac{1}{x}$ converges to h in the mean-square sense on $[-\pi, \pi]$.

(f) Formulate theorems more general than Theorem 9.3.6 that guarantee mean-square convergence of Fourier series.

4. A vector space V is called an **inner product space** if there is a function, $(\cdot, \cdot)$, from $V \times V$ to $\mathbb{C}$ which satisfies the three properties, (a), (b), and (c), of the L^2 inner product listed at the beginning of Section 9.4. For $v \in V$, we define $\|v\| \equiv (v, v)^{\frac{1}{2}}$.

(a) Prove that $|(v, w)| \leq \|v\|\|w\|$ for all $v \in V$ and $w \in V$.

(b) Prove that $\| \cdot \|$ is a norm on V.

(c) Suppose that $v_n \to v$; that is, $\|v_n - v\| \to 0$. Prove that for every $w \in V$, $(v_n, w) \to (v, w)$.

(d) Vectors v and w are said to be **orthogonal** if $(v, w) = 0$. Let $\{\phi_n\}$ be an orthonormal family vectors in V; that is $\|\phi_n\| = 0$ for all n and $(\phi_n, \phi_m) = 0$ if $n \neq m$. Prove that for all $v \in V$,

$$\sum |(v, \phi_n)|^2 \leq \|v\|^2.$$

(e) The family $\{\phi_n\}$ is said to be an **orthonormal basis** for V if every vector $v \in V$ can be written $v = \sum c_n \phi_n$ for some $c_n \in \mathbb{C}$, where the series converges to v in the norm $\| \cdot \|$. If $\{\phi_n\}$ is an orthonormal basis, show that $c_n = (v, \phi_n)$ and $\|v\|^2 = \sum |(v, \phi_n)|^2$.

5. An inner product space is called a **Hilbert space** if it is complete, that is, if every Cauchy sequence in V has a limit in V. These spaces are named after David Hilbert (1862 – 1943).

(a) Let $\mathbb{C}^n$ denote the set of n-tuples of complex numbers with the usual vector addition and scalar multiplication. For any two such vectors, $z = (z_1, z_2, \ldots, z_n)$ and $w = (w_1, w_2, \ldots, w_n)$, we define the inner product by

$$(z, w) \equiv \sum_{j=1}^{n} z_j \overline{w}_j.$$

Prove that $\mathbb{C}^n$ is a Hilbert space.

(b) Let ℓ_2 denote the set of sequences $\{a_n\}_{n=1}^{\infty}$ such that $\sum |a_n|^2 < \infty$. For two sequences $a = \{a_n\}$ and $b = \{b_n\}$ in ℓ_2, we define the inner product of a and b by

$$(a, b) \equiv \sum_{n=1}^{\infty} a_n \overline{b}_n.$$

(c) Use the discrete Cauchy-Schwarz inequality to show that $\sum |a_n \overline{b}_n| < \infty$, so the definition of inner product makes sense.

(d) Verify that $(\cdot, \cdot)$ is an inner product on ℓ_2.

(e) What is an orthonormal basis for ℓ_2?

(f) Prove that ℓ_2 is a Hilbert space. Hint: this is difficult; follow the outline of the proof of Theorem 5.8.1.

CHAPTER 10

Probability Theory

In this chapter we show how many of the analytical tools which we have developed, such as sequences, series, limit theorems, and metric spaces, are used in probability theory. In Section 10.1 we introduce discrete random variables, using the Bernoulli, binomial, and Poisson random variables as examples. In Section 10.2 we show how simple probabilistic ideas and the concept of metric are used in coding theory. Continuous random variables are discussed in Section 10.3. Finally, in Section 10.4 we develop more advanced applications of metric space concepts to probability theory. Chebyshev's inequality and the weak law of large numbers are covered in the projects.

10.1   Discrete Random Variables

We begin by reviewing some of the basic concepts of probability theory using the experiment of rolling two dice as an example. Probability theory involves the analysis of real-valued functions defined on sets of possible outcomes of experiments. The set of possible outcomes is called a **sample space** and the functions are called **random variables**.

Suppose that we roll two dice, one green and one red. Let m and n be the values showing on the green die and the red die, respectively, after we roll. The set of possible outcomes, S, consists of pairs of integers (m, n) where m and n are between 1 and 6. The following are functions defined on S and take values in $\mathbb{R}$:

$$X(m, n) = m + n, \quad Y(m, n) = m, \quad Z(m, n) = n.$$

Given an outcome (m, n), the value of X is the sum of the dice. The values of Y and Z are the numbers on the green die and the red die, respectively. Because X, Y and Z are $\mathbb{R}$-valued functions defined on the set of outcomes of the experiment, they are random variables. Y and Z

take values between 1 and 6 and X takes values 2 through 12; that is, the range of X is the set of integers between 2 and 12.

Definition. A random variable whose values lie in a finite or countable subset of $\mathbb{R}$ is called **discrete**.

If X is a discrete random variable, we are very interested in the probabilities that it takes on different values. We denote the probability that the value of X is r by

$$P\{X = r\}.$$

If we make assumptions about the experiment, we may be able to calculate these probabilities.

There are 36 distinct possible outcomes when we roll the two dice. Assume that each outcome is equally likely. Let's calculate the probabilities that X takes various values. Since there is only one outcome, $(1,1)$, so that the sum of the dice is 2, we have $P\{X = 2\} = \frac{1}{36}$. On the other hand, there are two different outcomes, $(1,2)$ and $(2,1)$, so that the sum is 3, and therefore $P\{X = 3\} = \frac{2}{36}$. By counting the different ways that we could have $m + n = r$ for any given r, we can compute $P\{X = r\}$ for any r. For example, $P\{X = 7\} = \frac{6}{36}$ and 7 is the most likely sum. Notice that $P\{X = 1\} = 0$ and $P\{X = \pi\} = 0$ since 1 and π are not values that X can take.

If X is a discrete random variable which can take the values $\{a_n\}$, we require that

$$\sum_n P\{X = a_n\} \; = \; 1,$$

which says simply that X must take one of these values. If A is any subset of $\mathbb{R}$, then we define the probability that the value of X is in A, denoted $P\{X \in A\}$, by

$$P\{X \in A\} \; \equiv \; \sum_{a_n \in A} P\{X = a_n\},$$

where the sum is over all n such that $a_n \in A$. This makes sense because we are saying that the probability that the value of X lies in A is the sum of the probabilities that X takes on each of the different numbers in A. The function whose value at a_n is $P\{X = a_n\}$ is called the **mass density** of the discrete random variable X.

If we denote the complement of A in $\mathbb{R}$ by A^c, we note that

$$P\{X \in A\} + P\{X \in A^c\} \; = \; \sum_{a_n \in A} P\{X = a_n\} + \sum_{a_n \in A^c} P\{X = a_n\} \; = \; 1.$$

Thus, we always know that $P\{X \in A^c\} = 1 - P\{X \in A\}$. We remark that we often write $P\{a \leq X \leq b\}$ instead of $P\{X \in [a,b]\}$.

Let X be the random variable in the dice experiment and let A be the closed interval $A = [-1, 3]$. Then $P\{X \in A\} = \frac{1}{36} + \frac{2}{36} = \frac{3}{36}$ since 2 and 3 are the only possible values of X in the interval $[-1, 3]$. Now, suppose that $A = [3, \infty)$. Then,

$$P\{3 \leq X \leq \infty\} = 1 - P\{-\infty \leq X < 3\} = 1 - P\{X = 2\} = 1 - \tfrac{1}{36}.$$

Two discrete random variables, X and Y, are said to be **independent** if for any two subsets A and B of $\mathbb{R}$,

$$P\{X \in A \text{ and } Y \in B\} = P\{X \in A\}P\{Y \in B\}. \tag{1}$$

Consider the random variables Y and Z in the dice experiment. Let A be the set $\{2, 3\}$ and let B be the set $\{4\}$. Then, $P\{Y \in A\} = \frac{1}{3}$ and $P\{Z \in B\} = \frac{1}{6}$. On the other hand, the only outcomes where both $Y \in A$ and $Z \in B$ are true are $(2, 4)$ and $(3, 4)$. Thus, $P\{X \in A \text{ and } Y \in B\} = \frac{2}{36}$, so

$$P\{X \in A \text{ and } Y \in B\} = P\{Y \in A\}P\{Z \in B\} \tag{2}$$

for these two particular sets, A and B. With a little more work, one can show that (2) is true for all choices of A and B, so the random variables Y and Z are independent. On the other hand, consider the random variables X and Y with the same two sets, $A = \{2, 3\}$ and $B = \{4\}$. Then, $P\{X \in A\} = \frac{1}{36} + \frac{2}{36} = \frac{1}{12}$ and $P\{Y \in B\} = \frac{1}{6}$. However, there are no outcomes so that X has the value 2 or 3 and Y has the value 4. Thus,

$$P\{X \in A \text{ and } Y \in B\} = 0 \neq \tfrac{1}{12} \cdot \tfrac{1}{6} = P\{X \in A\}P\{Y \in B\},$$

so the random variables X and Y are not independent. This makes sense because the likelihood that X will take any particular value will "depend" on what the value of Y is.

The three special kinds of discrete random variables arise frequently.

Example 1 (Bernoulli)   A discrete random variable which takes only the two values 0 and 1 is called a **Bernoulli** random variable. Bernoulli random variables typically arise in experiments with two outcomes. For example, if one flips a coin and assigns X the value 1 if the result is heads and takes the value 0 if the result is tails, then X is a Bernoulli random variable. In the discussion of Markov chains in Section 2.3, the state of the phone at the nth check is a Bernoulli random variable be-cause the phone is either free or busy. A Bernoulli random variable is completely specified by giving the probability $P\{X = 1\} = p$ since then $P\{X = 0\} = 1 - p$. Then X is said to be Bernoulli with parameter p.

Example 2 (Binomial)   Let $n > 0$ be a positive integer and suppose that $0 < p < 1$. A random variable X which takes the values $0, 1, \ldots, n$ with probabilities

$$P\{X = k\} \quad = \quad \frac{n!}{(n-k)!k!} p^k (1-p)^{n-k} \tag{3}$$

is called a **binomial** random variable (with parameters n and p). Using the binomial theorem, we can check that the probabilities add up to 1:

$$\sum_{k=0}^{n} P\{X = k\} \quad = \quad \sum_{k=0}^{n} \frac{n!}{(n-k)!k!} p^k (1-p)^{n-k}$$
$$= \quad (p + (1-p))^n$$
$$= \quad 1.$$

Binomial random variables typically arise as the sums of independent Bernoulli random variables. Suppose, for example, that we have a coin that comes up heads with probability p. Let X_1 be the random variable whose value is 1 if the first flip is a head and whose value is 0 if the first flip is a tail. Similarly, let X_2 be the random variable whose value is 1 if the second flip is a head and whose value is 0 if the second flip is a tail. Suppose that we assume that the first and second flips are independent. The probability that they are both heads is

$$P\{X_1 = 1 \ \text{ and } \ X_2 = 1\} \quad = \quad P\{X_1 = 1\}P\{X_2 = 1\} \quad = \quad p^2.$$

Similarly, the probability that the first flip is a head and the second flip is a tail is $p(1 - p)$. Now suppose we flip the coin n times and each flip is independent of the other flips. Let X_k be the random variable that is 1 if the kth flip is heads and 0 if the kth flip is tails. Define the random variable $X \equiv X_1 + X_2 + \cdots + X_n$. Then, the value of X is just the number of heads in n flips of the coin, so the possible values for X are the integers 0 through n. To compute $P\{X = k\}$, notice that the probability that any particular configuration of k heads will occur is $p^k(1-p)^{n-k}$ because the flips are independent. Since there are $\frac{n!}{(n-k)!k!}$ different choices for the positions of the k heads in n flips, we see that (3) holds; that is, X is a binomial random variable.

Example 3 (Poisson)   Suppose $\lambda > 0$, and let X be a random variable that takes the values $0, 1, 2, \ldots$ with probabilities

$$P\{X = k\} = e^{-\lambda}\frac{\lambda^k}{k!}.$$

X is called a **Poisson** random variable with parameter λ. Notice that

$$\sum_{k=0}^{\infty} P\{X = k\} = e^{-\lambda}\sum_{k=0}^{\infty}\frac{\lambda^k}{k!} = e^{-\lambda}e^{\lambda} = 1$$

as required. A random variable is often assumed to be a Poisson random variable if it is the number of occurrences of a rare event, such as the number of radioactive decays in a given period of time or the number of misprints on a page.

Notice that there is no mention of "sample space" in the definitions of Bernoulli, binomial, and Poisson random variables. The definitions simply specify the mass density of the random variables, that is, the range and the probability of each value in the range. Thus, two experiments and sample spaces S_1 and S_2 may be entirely different, but if the random variables X_1 on S_1 and X_2 on S_2 have probabilities

$$P\{X_1 = k\} = e^{-\lambda}\frac{\lambda^k}{k!} = P\{X_2 = k\},$$

then they are both Poisson random variables with parameter λ.

Definition.   Let X be a discrete random variable which takes values $\{a_n\}$. If $\sum_n |a_n|P\{X = a_n\}$ is finite, then the **expected value** of X, denoted $E(X)$, is defined by

$$E(X) \equiv \sum_n a_n P\{X = a_n\}$$

and X is said to have **finite expectation**.

The expected value, which is also called the **mean**, is the weighted average of the possible values of X, with the weight of each value given by its probability of occurrence. Note that once we know the mass density of a discrete random variable, we can compute its mean without knowing what the underlying experiment is or the meaning of X. A simple example of a random variable which does not have finite expectation is given in problem 6. The means of the three standard random variables defined above are easy to calculate.

Proposition 10.1.1

(a) If X is Bernoulli with parameter p, then $E(X) = p$.

(b) If X is binomial with parameters n and p, then $E(X) = np$.

(c) If X is Poisson with parameter λ, then $E(X) = \lambda$.

Proof. If X is Bernoulli with parameter p, then

$$E(X) \; = \; 0 \cdot P\{X = 0\} + 1 \cdot P\{X = 1\} \; = \; p.$$

If X is binomial with parameters n and p, then

$$
\begin{aligned}
E(X) \;&=\; \sum_{k=0}^{n} k \frac{n!}{(n-k)!k!} p^k (1-p)^{n-k} \\[2mm]
&=\; np \sum_{k=1}^{n} \frac{(n-1)!}{(n-k)!(k-1)!} p^{k-1} (1-p)^{(n-1)-(k-1)} \\[2mm]
&=\; np \sum_{k=0}^{n-1} \frac{(n-1)!}{((n-1)-k)!(k)!} p^k (1-p)^{(n-1)-k} \\[2mm]
&=\; np.
\end{aligned}
$$

Finally, if X is Poisson with parameter λ, then

$$E(X) = \sum_{k=0}^{\infty} k \frac{\lambda^k}{k!} e^{-\lambda} = \lambda e^{-\lambda} \sum_{k=1}^{\infty} \frac{\lambda^{(k-1)}}{(k-1)!} = \lambda e^{-\lambda} \sum_{k=0}^{\infty} \frac{\lambda^k}{k!} = \lambda,$$

which completes the proof.   ❑

The following example shows how power series arise naturally in probability theory.

Example 4 Consider an experiment in which we flip a fair coin until we get heads. Let X be the number of flips which we make. Then, $P\{X = 1\} = \frac{1}{2}$. If the experiment ends after two flips, then we got tails on the first flip and heads on the second. This happens with probability $(\frac{1}{2})^2$.

Similarly, if the experiment ends after k flips, we got $k - 1$ tails in a row followed by a head, so $P\{X = k\} = (\frac{1}{2})^k$. We wish to compute $E(X) = \sum_{k=1}^{\infty} k(\frac{1}{2})^k$. Since

$$2E(X) = \sum_{k=1}^{\infty} k(\tfrac{1}{2})^{k-1} = \sum_{k=0}^{\infty}(k+1)(\tfrac{1}{2})^k,$$

we calculate the power series

$$\sum_{k=0}^{\infty}(k+1)x^k = \frac{d}{dx}\left(\sum_{k=0}^{\infty} x^k\right)$$

$$= \frac{d}{dx}\left(\frac{1}{1-x}\right)$$

$$= \frac{1}{(1-x)^2}$$

for $|x| < 1$. In the first step we used Theorem 6.3.3 and the fact that the radius of convergence of the geometric series is 1. Substituting $x = \frac{1}{2}$, we see that $2E(X) = (1 - \frac{1}{2})^{-2}$, so $E(X) = 2$.

Let S be a sample space and let X be a discrete random variable on S that takes values $\{a_n\}$. If ψ is any real-valued function on $\mathbb{R}$, then the composition $\psi \circ X$ is a discrete random variable on S that takes the values $\{\psi(a_n)\}$. The following theorem shows how the expectation of $\psi \circ X$ can be computed.

❏ **Theorem 10.1.2** Let X be a random variable with range $\{a_n\}$ and let ψ be a real-valued function on $\mathbb{R}$. If the series $\sum \psi(a_n)P\{X = a_n\}$ converges absolutely, then $\psi \circ X$ has finite expectation and

$$E(\psi \circ X) = \sum \psi(a_n)P\{X = a_n\}. \tag{4}$$

Proof. The series $\sum \psi(a_n)P\{X = a_n\}$ converges absolutely by hypothesis, so by Theorem 6.2.5, we may rearrange it in any way that we like and it will have the same sum. If we denote the values of $\psi \circ X$ by $\{b_j\}$, then

$$\sum_n \psi(a_n)P\{X = a_n\} = \sum_j \sum_{\{n \,|\, \psi(a_n)=b_j\}} b_j P\{X = a_n\}$$

$$= \sum_j b_j \sum_{\{n \,|\, \psi(a_n)=b_j\}} P\{X = a_n\}$$

$$= \sum_j b_j P\{\psi \circ X = b_j\}$$

$$= E(\psi \circ X).$$

If we follow exactly the same steps but replace $\psi(a_n)$ by $|\psi(a_n)|$ and b_j by $|b_j|$, we see that $\sum_j b_j P\{\psi \circ X = b_j\}$ converges absolutely. ❑

If X and Y are random variables on a sample space S, then it makes sense to add or multiply X and Y since they are functions on S with values in $\mathbb{R}$. If X and Y have finite expectation, then $E(X+Y) = E(X)+E(Y)$ (problem 9). However, there is in general no simple relationship between $E(XY)$ and $E(X)$ and $E(Y)$, except in the case where X and Y are independent.

❑ **Theorem 10.1.3**   Suppose that X and Y are random variables on the same sample space and that X and Y are independent. Suppose X and Y have finite expectation. Then XY has finite expectation and

$$E(XY) \;=\; E(X)E(Y). \tag{5}$$

Proof.   Let the values of X be $\{a_i\}$ and the values of Y be $\{b_j\}$. We begin by considering the double sum

$$\sum_{i,j} |a_i b_j| P\{X = a_i \text{ and } Y = b_j\}.$$

All the terms are positive in this double sum. Thus, by Theorem 6.2.6, if we show that it converges in any rearrangement, then it converges in all rearrangements and the sum is always the same. Since X and Y are independent,

$$\sum_{i,j} |a_i b_j| P\{X = a_i \text{ and } Y = b_j\} \;=\; \sum_{i,j} |a_i||b_j| P\{X = a_i\} P\{Y = b_j\}$$

$$= \sum_i |a_i| \left(\sum_j |b_j| P\{Y = b_j\} \right) P\{X = a_i\}$$

$$= \left(\sum_j |b_j| P\{Y = b_j\} \right) \left(\sum_i |a_i| P\{X = a_i\} \right)$$

$$< \infty.$$

In the last step we used the hypothesis that X and Y have finite expectations. This proves that XY has finite expectation. Equation (5) is proved by following the same steps with $|a_i b_j|$ replaced by $a_i b_j$. ❑

In the projects we define the variation and standard deviation of a discrete random variable and outline the proofs of Chebyshev's inequality and the weak law of large numbers.

Problems

1. Consider the experiment where two dice are rolled, and let the random variable X be the sum of the faces.

    (a) Compute $P\{X = r\}$ for all r.

    (b) Compute the mean of X.

    (c) Compute the standard deviation of X.

2. Consider an experiment where we flip two fair coins simultaneously. Let X denote the number of heads plus the square of the number of tails.

    (a) Compute $P\{X = r\}$ for all r.

    (b) Compute the mean of X.

    (c) Compute the standard deviation of X.

3. Let X be a Bernoulli random variable with $P\{X = 1\} = p$. Compute the standard deviation of X.

4. In a bin of oranges each orange is good with probability .8 and bad with probability .2, independent of the other oranges. You select five oranges at random. What is the probability that

    (a) all five are good?

    (b) all five are bad?

    (c) exactly three are good?

    (d) at least three are good?

    (e) fewer than two are good?

5. The expected number of misprints on a page in a manuscript is five. If we assume that the number of misprints is a Poisson random variable, what is the probability that a particular page will have:

    (a) no misprints?

    (b) less than or equal to three misprints?

(c) four or more misprints?

6. Let X be a random variable taking values in $\mathbb{N}$ such that $P\{X = k\} = \frac{6}{\pi^2 k^2}$.

   (a) Prove that $\sum_{k=1}^{\infty} P\{X = k\} = 1$. Hint: see problem 6 in Section 9.2.

   (b) Prove that X does not have finite expectation.

7. Let $0 < p < 1$ and consider an experiment in which we flip a coin which comes up heads with probability p until we get heads. Let X be the number of flips which we make. Compute $E(X)$.

8. Let X be the random variable which is the sum of the faces in the experiment of rolling two dice. Compute $E(X)$ and $E(X^2)$.

9. Let X and Y be discrete random variables on the same sample space and suppose that both X and Y have finite expectation. Prove that for any c and d, the random variable $cX + dY$ has finite expectation and

$$E(cX + dY) \;=\; cE(X) \;+\; dE(Y).$$

10. Consider the experiment of rolling two dice, one red and one green. Compute the expectations of the following random variables:

    (a) the sum of the faces?

    (b) three times the green face minus the red face?

    (c) the product of the faces?

11. Suppose that $Y(\lambda)$ is a Poisson random variable with parameter λ. Let A be any subset of $\mathbb{R}$ and define $P_A(\lambda) \equiv P\{Y(\lambda) \in A\}$. Prove that P_A is a continuous function of λ on $(0, \infty)$.

10.2   Coding Theory

Probability theory and metrics arise in a natural way in coding theory. To explain why, we start with a simple example.

Example 1   Suppose that we wish to transmit a binary message which has two digits, that is, our message is 00, 01, 10, or 11. Suppose that each digit has a probability p of being transmitted correctly. Assuming that the digits are transmitted independently, the probability that the message received will be the same as the one transmitted is p^2. Intuitively, we should be able to improve the reliability of transmission by adding redundancy to the message. For example, suppose that we simply repeat each message three times. Instead of sending 00, we send 000000 and instead of sending 01 we send 010101, and so forth. These four binary strings

$$000000 \quad 010101 \quad 101010 \quad 111111$$

will be our **code words**. Let S be the set of all strings of 0's and 1's of length 6, and let ρ be the discrete metric on S. That is, if $\{x_i\}_{i=1}^6$ and $\{y_i\}_{i=1}^6$ are binary strings of length 6, then

$$\rho(\{x_i\}, \{y_i\}) = \sum_{i=1}^6 \delta(x_i, y_i)$$

where $\delta(x, x) = 0$ and $\delta(x, y) = 1$ if $x \neq y$. Thus the distance between two strings is just the number of places in which they differ.

Here is the idea. Notice that each of the four code words which we want to send is a distance ≥ 3 from each of the other code words. Suppose that we send the code word C_s and that it is received as a binary string C_r with one error in it. Suppose that C_o is a code word different from C_s. Then, by the triangle inequality for metrics,

$$3 \leq \rho(C_s, C_o) \leq \rho(C_s, C_r) + \rho(C_r, C_o).$$

Since $\rho(C_s, C_r) = 1$ by assumption, we see that $\rho(C_r, C_o) \geq 2$; that is, the received string is a distance ≥ 2 from all other code words. Thus, if we receive a string which has one error in transmission, we can correct the error by replacing the received string by the unique code word which has distance 1 from it. Using this scheme, we can correctly decode a received string if it is in fact correct or if it has exactly one error. The probability of correct transmission is p^6, and the probability for transmission with exactly one error is $6p^5(1 - p)$. Thus, we will correctly decode the string with probability $p^6 + 6p^5(1 - p)$. If, for example, $p = .9$, then

$$p^2 = .81 \quad and \quad p^6 + 6p^5(1 - p) = .886,$$

so by adding redundancy we have achieved an improvement in the reliability of our transmission channel. On the other hand, if $p = .7$, then

$$p^2 = .49 \quad and \quad p^6 + 6p^5(1 - p) = .42,$$

so the reliability has been decreased. This raises several natural questions. For which p does a repetition of three times improve the reliability? Suppose that we repeat each two-digit message five times. Then the four code words would be a distance ≥ 5 apart and we would be able to correctly decode received strings with no errors, one error, or two errors. For what p will this code improve reliability? Can we achieve any desired level of reliability less than 1? For which p? These questions are investigated further in problems 1, 2, and 9. For obvious reasons, these codes are called repetition codes.

Figure 10.2.1

Example 1 shows the kind of question that motivated the development of coding theory. We are given the characteristics of an information transmission channel, in this case the probability p. We know what kind of information we wish to transmit; in the example the information comes as a pair of binary digits. The problem is to design an encoding device and a decoding device to improve the reliability of the channel. Usually there are other constraints as well. Suppose that there are N binary code words of length n. Then

$$R = \frac{\log_2 N}{n}$$

is called the **information rate** of the channel. In Example 1, if we transmit the four words 00, 01, 10, and 11 directly, then $R = 1$ since $N = 4$ and $n = 2$. If we use the repetition code that repeats each word three times, then $R = \frac{1}{3}$ since $N = 4$ and $n = 6$. It can be shown (problem 9) that if $p > \frac{3}{4}$ and we allow enough repetitions, then we can achieve any desired reliability strictly less than 1, but at the cost of a very low information rate. The challenge is to design codes which have both high reliability and high information rates. To see that this is an interesting question with deep connections to linear algebra we will describe the Hamming (7,4) code, named after one of the founders of the subject, R. Hamming (1915 –).

Suppose that we wish to encode signals consisting of four binary digits. We put the four digits in the 3rd, 5th, 6th, and 7th positions of a seven digit binary vector $(x_1, x_2, x_3, x_4, x_5, x_6, x_7)$. The entries in the 1st, 2nd, and 4th positions are defined by

$$
\begin{aligned}
x_1 &= x_3 + x_5 + x_7 && (6) \\
x_2 &= x_3 + x_6 + x_7 && (7) \\
x_4 &= x_5 + x_6 + x_7 && (8)
\end{aligned}
$$

where we use binary arithmetic, so $1+1=0$. There are 2^4 choices for the 3rd, 5th, 6th, and 7th positions, and the other positions are then determined. Thus, there are 16 code words, each of length 7. Since $x_i + x_i = 0$ for each i, we can write the equations which define x_1, x_2, and x_4 as

$$
\begin{aligned}
x_1 + x_3 + x_5 + x_7 &= 0 \\
x_2 + x_3 + x_6 + x_7 &= 0 \\
x_4 + x_5 + x_6 + x_7 &= 0
\end{aligned}
$$

Thus, among the 128 binary vectors of length 7, the 16 code words are the binary vectors $v = (x_1, x_2, x_3, x_4, x_5, x_6, x_7)$ such that $Hv = 0$ where H is the matrix

$$
H \equiv \begin{pmatrix}
1 & 0 & 1 & 0 & 1 & 0 & 1 \\
0 & 1 & 1 & 0 & 0 & 1 & 1 \\
0 & 0 & 0 & 1 & 1 & 1 & 1
\end{pmatrix}.
$$

The set of binary strings of length n is the Cartesian product of $\mathbb{Z}_2$ with itself n times, so it is denoted by $\mathbb{Z}_2^n$. The set $\mathbb{Z}_2^n$ is a vector space because we can add binary vectors by adding their components (using binary arithmetic) and we can use $\mathbb{Z}_2$ as the field of scalars (problem 7). These abstract concepts are not used in our calculations below, just the fact that H is a linear transformation from $\mathbb{Z}_2^7$ to $\mathbb{Z}_2^3$.

Since H is a linear transformation, the set of vectors v such that $Hv = 0$ is a vector subspace of $\mathbb{Z}_2^7$. That is, if C_1 and C_2 are code words, then $C_1 - C_2$ is a code word. This makes it easy to compute the distance between code words. As in Example 1, we define a metric on the set of seven-digit binary strings by

$$
\rho(\{x_i\}, \{y_i\}) = \sum_{i=1}^{7} \delta(x_i, y_i).
$$

Then, $\rho(C_1, C_2) = \rho(C_1 - C_2, 0)$, and $\rho(C_1 - C_2, 0)$ is simply the number of 1's in the code word $C_1 - C_2$. Suppose that $C_1 \neq C_2$, and let $(x_1, x_2, x_3, x_4, x_5, x_6, x_7) \equiv C_1 - C_2$. We shall show that $C_1 - C_2$ has at least three 1's. If there are three or more 1's in the 3rd, 5th, 6th, and 7th positions, there is nothing more to prove. From equations (6) $-$ (8), it follows easily that if exactly two out of x_3, x_5, x_6, x_7 equal 1, then exactly one of x_1, x_2, x_4 equals 1. And, if exactly one out of x_3, x_5, x_6, x_7 equals 1, then exactly two of x_1, x_2, x_4 equal 1. Thus,

$$\rho(C_i, C_j) \geq 3, \qquad \text{for all } C_i \neq C_j.$$

If C is a code word we let $B_1(C)$ denote the set of words within a distance 1 of C. Now, let e_i denote the string which is all 0's except for a 1 in the ith position. Then He_i is just the ith column of H; in particular,

$$He_i \neq \begin{pmatrix} 0 \\ 0 \\ 0 \end{pmatrix}.$$

Thus, if C is a code word, the seven distinct strings $C + e_i$, $i = 1, 2, ..., 7$, which are a distance 1 from C, are not code words since

$$H(C + e_i) \; = \; HC \; + \; He_i \; = \; He_i \; \neq \; \begin{pmatrix} 0 \\ 0 \\ 0 \end{pmatrix}.$$

Thus, $B_1(C)$ contains one code word and seven distinct binary strings which are not code words. Let C_i and C_j be distinct code words and suppose $x \in B_1(C_i)$ and $y \in B_1(C_j)$. Then, by the triangle inequality for metrics,

$$3 \; \leq \; \rho(C_i, C_j) \; \leq \; \rho(C_i, x) \; + \; \rho(x, y) \; + \; \rho(y, C_j).$$

Since $\rho(C_i, x) \leq 1$ and $\rho(C_j, y) \leq 1$, we must have $\rho(x, y) \geq 1$ which implies that $x \neq y$. Thus, the sets $B_1(C_i)$ are disjoint. Therefore, their union contains $16 \times 8 = 128$ distinct binary vectors, that is, all the binary vectors of length 7.

We choose the decoding algorithm which assigns to each received signal the unique code word within a distance 1. With this encoding and decoding scheme, a code word transmitted with zero errors or with one error is decoded correctly. Thus, the probability of correct transmission is $p^7 + 7p^6(1 - p)$ as compared to p^4 for the direct transmission of the four original digits. If $p = .9$, for example, then $p^4 = .66$ and $p^7 + 7p^6(1 - p) = .85$. The information rate is $\frac{4}{7}$, so we have achieved a dramatic improvement in reliability with only a modest reduction in the information rate.

The Hamming matrix has the nice property that the ith column is just the binary representation of i reading from bottom to top. Suppose that the four digits that we wish to transmit are encoded in a code word C and that a single error occurs in transmission. Then the transmitted word can be written as $C + e_i$ for some i. If we apply H to the received word, we obtain $H(C+e_i) = He_i$, which is the ith column of H. Since the ith column is just the binary representation of i, we can see immediately, by applying H to the received word, in which digit the error occurs. For example, suppose that the sender wishes to transmit the signal 1101. Thus, $x_3 = 1, x_5 = 1, x_6 = 0$, and $x_7 = 1$. Using (6) – (8), the sender determines that $x_1 = 1, x_2 = 0$, and $x_4 = 0$ and transmits the signal

$$1010101. \tag{9}$$

Suppose that we receive the signal

$$1000101 \tag{10}$$

because of an error in the transmission line. Applying H to the column vector $(1, 0, 0, 0, 1, 0, 1)$, we obtain the vector

$$\begin{pmatrix} 1 \\ 1 \\ 0 \end{pmatrix}.$$

Since this is not the zero vector we know that there is an error in the signal, and if there is only one error, it is in the third position since 011 is the binary representation of 3. Therefore, we can correct (10) to obtain the code word (9), thus recovering the signal that the sender wished to transmit.

For more information about the history and mathematical development of coding theory, see [42] or [32].

Problems

1. (a) Generate the graph of the function $f(p) = p^6 + 6p^5(1 - p) - p^2$ and use it to show that there is a p_o so that the triple repetition code in Example 1 improves reliability if $p > p_o$ and hurts reliability if $p < p_o$.

    (b) Use Newton's method to get a good estimate of p_o.

2. Suppose that we wish to transmit a binary message which has two digits, 00, 01, 10, or 11. We use a repetition code which repeats each message five times.

   (a) How many code words are there? How many possible transmitted signals are there?

   (b) Explain why all the code words are a distance ≥ 5 apart. Explain why it follows that we can correctly decode any transmitted signal with ≤ 2 errors.

   (c) Suppose that the transmission channel sends an individual digit correctly with probability p. What is the probability that the transmitted message will have ≤ 2 errors?

   (d) If $p = .95$, compare the reliability of this coding scheme with the reliability of sending the two digits directly.

   (e) If $p = .7$, compare the reliability of this coding scheme with the reliability of sending the two digits directly.

   (f) Find a p_o so that this code improves reliability if $p > p_o$ and hurts reliability if $p < p_o$.

   (g) What is the information rate of this channel?

3. Suppose that we encode two binary digits as the first two digits of a three-digit binary string and choose the third digit so that the sum of all the digits is zero. Explain why this code can detect single errors but cannot correct them. What is the information rate of this code?

4. Prove that the Hamming $(7, 4)$ code has information rate $R = \frac{4}{7}$.

5. Suppose that you are receiving signals which employ the Hamming $(7, 4)$ code. Decode the following received signals:

   (a) 0010111.

   (b) 0110111.

   (c) 0111101.

   (d) 0111100.

6. Let $\mathcal{S}$ be the set of all possible words of length n that can be transmitted by an information channel, and let $\mathcal{C} \subseteq \mathcal{S}$ be the set of code words. Let ρ be the discrete metric on $\mathcal{S}$. The set $\mathcal{C}$ is called a **perfect code** if there is an integer m so that the union of the sets of radius m centered at the code words $C \in \mathcal{C}$ equals $\mathcal{S}$ and, furthermore, each pair of code words is a distance at least $2m + 1$ apart. Which of the following codes is perfect?

   (a) The repetition code of problem 1.

   (b) The parity check code of problem 3.

   (c) The $(7, 4)$ Hamming code.

7. Let $\mathbb{Z}_2^n$ be the set of n-tuples of 0's and 1's. We define addition by

$$(x_1, x_2, \ldots, x_n) + (y_1, y_2, \ldots, y_n) \equiv (x_1 + y_1, x_2 + y_2, \ldots, x_n + y_n),$$

where the plus signs on the right mean binary addition. For $z \in \mathbb{Z}_2$ we define scalar multiplication by

$$z \cdot (x_1, x_2, \ldots, x_n) \equiv (z \cdot x_1, z \cdot x_2, \ldots, z \cdot x_n),$$

where on the right $z \cdot x_i$ means multiplication in $\mathbb{Z}_2$. Use the fact that $\mathbb{Z}_2$ is a field (problem 7 of Section 1.1) to show that $\mathbb{Z}_2^n$ satisfies the definition of vector space (with $\mathbb{R}$ replaced by $\mathbb{Z}_2$) given in Section 5.8.

8. Suppose that we wish to transmit a single binary digit on a channel which has a probability p of correct transmission for each digit sent. For a given odd integer n we encode and decode as follows. We transmit the digit n times. If there are more 1's than 0's, we decode the signal as a 1. If there are more 0's than 1's, we decode the signal as a 0.

   (a) What is the probability that the decoded signal is the signal that was encoded?

   (b) Suppose that $p > \frac{1}{2}$. Prove that if we choose n large enough, the probability of correct transmission can be made larger than any $\beta < 1$. Hint: use the weak law of large numbers; see project 3.

9. Suppose that we wish to transmit pairs of binary digits as in Example 1 on a channel which has a probability p of correct transmission for each digit sent. Let n be a given odd integer. We encode each pair of digits by repeating it n times.

   (a) What are the four code words?

   (b) Using the discrete metric introduced in Example 1, explain why each pair of code words is at least a distance n apart.

   (c) Explain why a message with $\leq (n - 1)/2$ errors can be correctly decoded.

   (d) What is the probability that the decoded signal is the signal that was encoded?

   (e) Suppose that $p > \frac{3}{4}$. Prove that if we choose n large enough, the probability of correct transmission can be made larger than any $\beta < 1$. Hint: use the weak law of large numbers; see project 3.

10.3   Continuous Random Variables

In many situations, one wants to analyze random variables whose values do *not* lie in a countable set, for example, the height of a randomly chosen man, or the amount of time that we have to wait before some chance event occurs. One can force such problems to be discrete by measuring height or time in discrete units (for example, inches or seconds) and rounding measurements up or down, but usually this is cumbersome and unnatural. Often, one assumes that the probability that the value of X lies in a particular set of real numbers is given by integrating a "density function" over the set. Suppose that f is a nonnegative function such that

$$\int_{-\infty}^{\infty} f(x)\,dx \;=\; 1. \tag{11}$$

By writing (11), we are assuming implicitly that the improper Riemann integral exists. We say that f is the **density function** for a random variable X if

$$P\{a \le X \le b\} \;=\; \int_{a}^{b} f(x)\,dx$$

for all a and b, including $a = -\infty$ and $b = \infty$. Condition (11) simply guarantees that the total probability is 1.

Example 1   Consider the function

$$f(x) \;=\; \begin{cases} 0, & \text{if } x \le 1 \\ \frac{2}{x^3}, & \text{if } x > 1. \end{cases}$$

Since,

$$\lim_{c \searrow 1}\; \lim_{d \to \infty} \int_{c}^{d} \frac{2}{x^3}\,dx \;=\; \lim_{c \searrow 1}\; \lim_{d \to \infty} \left(\frac{1}{c^2} - \frac{1}{d^2} \right) \;=\; 1,$$

we see that (11) is satisfied. If X is a random variable with density f, then

$$P\{\tfrac{1}{2} \le X \le 2\} \;=\; \int_{1}^{2} \tfrac{2}{x^3}\,dx \;=\; \tfrac{3}{4}$$

and

$$P\{-2 \le X \le 1\} \;=\; \int_{-2}^{1} 0\,dx \;=\; 0.$$

If a random variable X has a density f, then the probability that the value of X lies between a and b is just the area under f between a and b.

See Figure 10.3.1. Notice that

$$P\{X = a\} = \int_a^a f(x)\,dx = 0,$$

so if a random variable has a density function, then the probability of taking on any particular value is zero. If

$$\int_{-\infty}^{\infty} |x| f(x)\,dx < \infty, \qquad (12)$$

then we say that X has **finite expectation** and define

$$E(X) = \int_{-\infty}^{\infty} x f(x)\,dx. \qquad (13)$$

$E(X)$ is called the **expected value** of X or the **mean**. We often write $\mu \equiv E(X)$ for short. Note that by problem 12 of Section 3.6, the improper integral (13) exists if the improper integral (12) exists. The **variance** of X, σ^2, is defined by

$$\sigma^2 \equiv \int_{\infty}^{\infty} (x - \mu)^2 f(x)\,dx$$

if the improper integral exists. These definitions are analogous to the definitions of mean and variance in the case where X is a discrete random variable (see Section 10.1 and project 1). By using Riemann sums, the analogy can be made precise (problem 13). As in the discrete case, σ is called the **standard deviation**.

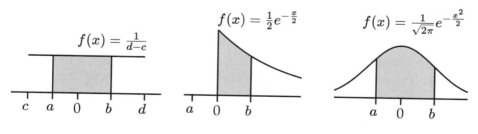

Figure 10.3.1

Three special kinds of density functions arise frequently.

Example 2 (Uniform)   Let $f(x) = \frac{1}{d-c}$ on a finite interval $[c, d]$ and $f(x) = 0$ for x outside the interval. A random variable whose density is f is said to be **uniformly distributed** on the interval $[c, d]$.

Example 3 (Exponential)   Suppose that $\lambda > 0$ and define f by

$$f(x) \;=\; \begin{cases} \lambda e^{-\lambda x}, & x \geq 0 \\ 0, & x < 0. \end{cases}$$

A random variable with density f is said to be **exponentially distributed** with parameter λ. Often waiting times, that is, the amount of time until some event occurs, are assumed to be exponentially distributed. The mean and standard deviation of a random variable which is exponentially distributed with parameter λ are both equal to $1/\lambda$ (problem 7).

Example 4 (Normal)   Let $\mu \in \mathbb{R}$ and $\sigma > 0$, and define f by

$$f(x) \;\equiv\; \frac{1}{\sigma\sqrt{2\pi}}\, e^{\frac{-(x-\mu)^2}{2\sigma^2}}.$$

A random variable with density f is said to be **normally distributed** with mean μ and standard deviation σ. If one computes the mean and standard deviation by using the definitions above, one obtains μ and σ, respectively, which is why this terminology is used. If $\mu = 0$ and $\sigma = 1$, then X is called a **standard normal** random variable.

Note that the probability $P\{a \leq X \leq b\}$ can easily be evaluated in the case of a uniform or exponentially distributed random variable because we can use the Fundamental Theorem of Calculus to evaluate the integral explicitly. Since there is no elementary function whose derivative is the normal density, $P\{a \leq X \leq b\}$ must be evaluated by numerical methods such as those discussed in Section 3.4. There are tables which give the approximate values of $P\{a \leq X \leq b\}$ for a large number of different a's and b's. Fortunately, the integrals $P\{a \leq X \leq b\}$ in the case of general μ and σ can be related to the integrals in the case $\mu = 0$ and $\sigma = 1$ by a simple change of variables (problem 10). Thus, we only need tables of values for the case of a standard normal random variable.

The fact that many random variables are normal can be explained by a deep theorem in probability theory called the central limit theorem, which says that a random variable is approximately normal if it is the sum of many independent, "similar" random variables. For example, imagine that a man's total height is the sum of small "boosts". Each boost is present or absent depending on whether a particular nucleotide in his DNA is present or absent in a particular position. If the nucleotides in different positions are independent, then height should be approximately normally distributed. This is a gross oversimplification of course, but it gives the idea of how such an argument might be constructed.

There is a natural way to unify our treatment of discrete random variables and random variables with densities. If X is a random variable, we define

$$F(x) \equiv P\{-\infty < X \leq x\}.$$

That is, $F(x)$ is just the probability that the value of X lies in the interval $(-\infty, x]$. F is called the **cumulative distribution function** of X. As x gets larger $P\{-\infty < X \leq x\}$ cannot decrease so F is a monotone increasing function. Since the value of X is some real number, we must have

$$\lim_{x \to \infty} F(x) = P\{-\infty < X < \infty\} = 1 \tag{14}$$

and

$$\lim_{x \to -\infty} F(x) = P\{\emptyset\} = 0. \tag{15}$$

Furthermore, since F is monotone increasing, the limits from the left and right, $F(x^-)$ and $F(x^+)$, exist for each x. Thus, the only possible discontinuities of F are jump discontinuities (problem 7 in Section 3.5). Any real-valued function on $\mathbb{R}$ that is monotone increasing and satisfies (14) and (15) is called a **cumulative distribution function** even if no random variable is specified.

Suppose that X is a discrete random variable which takes on only finitely many values, for example, a Bernoulli or binomial random variable. Let $a_1, a_2, \ldots, a_N$ denote the possible values listed in increasing order. Then

$$F(x) = P\{-\infty < X \leq x\} = \sum_{a_n \leq x} P\{X = a_n\}. \tag{16}$$

Thus F is zero on the interval $(-\infty, a_1)$, equal to $P\{X = a_1\}$ on the interval $[a_1, a_2)$, equal to $P\{X = a_1\} + P\{X = a_2\}$ on the interval $[a_2, a_3)$, and so forth. See Figure 10.3.2. F is constant except for jump discontinuities at each a_n, and the size of the jump, $F(a_n^+) - F(a_n^-)$, at a_n is equal to $P\{X = a_n\}$.

In general, a discrete random variable X takes on countably many values $\{a_n\}$. In this case the right-hand side of (16) may be an infinite series if infinitely many of the numbers a_n are less than or equal to a particular x. The series always converges by the comparison test since it consists of a subset of the terms of the series $\sum P\{X = a_n\}$, which converges and sums to 1. If $\{a_n\}$ has no limit points (e.g a Poisson random variable), then by the Bolzano-Weierstrass theorem, there can be at most finitely many values a_n in any given finite interval. These points can be ordered, and again we get the simple picture in Figure 10.3.2.

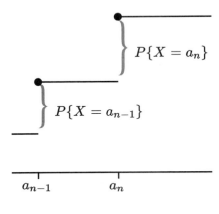

Figure 10.3.2

If $\{a_n\}$ has lots of limit points, there may not be a natural ordering of $\{a_n\}$. For example, suppose that the values $\{a_n\}$ are the rational numbers. Formula (16) is true but it is much harder to visualize the graph of F since, in any interval about a limit point of $\{a_n\}$, F will have infinitely many steps. As before, the possible values of X are just the points of discontinuity of F, and the probabilities are just the sizes of the jumps at these points. Thus, the mass density of a discrete random variable can be recovered from its cumulative distribution function. A cumulative distribution function which is constant except for finite or countably many jumps is called **discrete**.

If the random variable X has a density f, then

$$F(x) \equiv P\{-\infty < X \leq x\} = \int_{-\infty}^{x} f(t)\, dt.$$

The improper integral on the right exists because $\int_{c}^{x} f(t)\, dt$ is monotone increasing and bounded by 1 as $c \to -\infty$. If f is continuous, as in the case of a normal random variable, then by the Fundamental Theorem of Calculus (Theorem 4.2.5), F is continuously differentiable and $F'(x) = f(x)$. Thus the density f can be recovered from the cumulative distribution function F. If f is only piecewise continuous, as in the case of uniform or exponential random variables, then F is differentiable and $F'(x) = f(x)$ except at the points where f jumps. In both cases, the cumulative distribution function is continuous. A random variable X with a continuous cumulative distribution function is called a **continuous random variable**. Note that this does *not* refer to the continuity of X as a function from the sample space to $\mathbb{R}$. Of course, a cumulative distribution function may be neither discrete nor continuous. However, it can always be written as a convex combination of a discrete cumulative distribution function and a continuous cumulative distribution function (problem 12).

If F is the cumulative distribution function of a random variable X, then for any $a < b$,

$$
\begin{aligned}
P\{a < X \le b\} &= P\{-\infty < X \le b\} - P\{-\infty < X \le a\} \\
&= F(b) - F(a).
\end{aligned}
$$

Thus, if F is continuous,

$$
\lim_{a \to b} P\{a < X \le b\} = 0.
$$

Since $0 \le P\{X = b\} \le P\{a < X \le b\}$ for all $a < b$, we conclude that $P\{X = b\} = 0$. Therefore, continuous random variables take on specific values with probability zero. We saw this before in the special case when X has a density, for example, when X is uniform, exponential, or normal. This raises the natural question of what kinds of sets can have positive probability. The following example shows that this question is deeper than it looks.

Example 5 (the Cantor set and function)   We will describe the Cantor set, C, which is a subset of $[0, 1]$, by saying which points are *not* in C. First, we exclude the middle third $(\frac{1}{3}, \frac{2}{3})$ of the interval $[0, 1]$. We then exclude the middle thirds, namely, $(\frac{1}{9}, \frac{2}{9})$ and $(\frac{7}{9}, \frac{8}{9})$, of the two intervals that remain. Now there are four remaining intervals, and we exclude their middle thirds, and so forth. The Cantor set is the collection of points in $[0, 1]$ that remain after we have carried out this procedure infinitely often. A straightforward calculation with geometric series shows that the sum of the lengths of the excluded intervals equals 1, so in that sense C is a very small set. On the other hand, there is another characterization of C which allows one to show that C is uncountable (see below), so in that sense C is a very large set.

Decimal expansions were discussed in project 4 of Chapter 6. The word "decimal" is used, of course, because one is writing a given number as a sum of powers of 10. In similar fashion, one can show that every number x satisfying $0 < x < 1$ can be written

$$x \ = \ \sum_{n=1}^{\infty} \frac{a_n}{3^n},$$

where, for each n, $a_n = 0, 1$, or 2. This is called the **ternary** expansion of x. Given x, the sequence $\{a_n\}$ is uniquely determined except when x is of the form $q/3^n$ for some integers n and q, in which case there are exactly two expansions, one ending in a string of 0's and the other ending in a string of 2's. Conversely, if $\{a_n\}$ is any sequence of 0's, 1's, or 2's, the series converges to a real number x that is in the interval $[0, 1]$. The proofs of these statements are similar to those outlined in project 4 of Chapter 6.

It is not too difficult to see that the Cantor set is just the set of x in $[0, 1]$ whose ternary expansions have no 1's. For example, the first middle third we eliminated, $(\frac{1}{3}, \frac{2}{3})$, consists of numbers such that $a_1 = 1$ in their ternary expansions. Similarly, if $a_1 \neq 1$ but $a_2 = 1$, then x is in the interval $(\frac{1}{9}, \frac{2}{9})$ or the interval $(\frac{7}{9}, \frac{8}{9})$, depending on whether $a_1 = 0$ or $a_1 = 2$. If a number x has two ternary expansions, then it is in the Cantor set if one of the expansions has no 1's. This shows that there is a one-to-one correspondence between C and the set of all sequences of 0's and 2's. Since a straightforward modification of the proof of Theorem 1.3.6 shows that this set of sequences is uncountable, C must be uncountable too.

We shall now define a function g on $[0, 1]$. If the ternary expansion of x has no 1's, we set $N = \infty$ and otherwise let N be the index of the first place in the ternary expansion of x where a 1 occurs. Set $b_n = \frac{1}{2}a_n$ for $n < N$ and $b_N = 1$, and define

$$g(x) \ \equiv \ \sum_{n=1}^{\infty} \frac{b_n}{2^n}.$$

This function, g, is called the **Cantor function**. Notice that the value of g is $\frac{1}{2}$ for all x in $(\frac{1}{3}, \frac{2}{3})$ since in that case $a_1 = 1$. Similarly, the value of g is $\frac{1}{4}$ on $(\frac{1}{9}, \frac{2}{9})$ and the value of g is $\frac{3}{4}$ on $(\frac{7}{9}, \frac{8}{9})$. Continuing in this fashion, one can see that g is constant on each of the intervals in the complement of the Cantor set. Furthermore, g is monotone increasing and continuous. The monotonicity can be proved by checking cases, and the continuity holds because two numbers that are very close have ternary expansions which are identical for a large number of terms.

Define a function F by

$$F(x) \equiv \begin{cases} 0, & x < 0 \\ g(x), & 0 \leq x \leq 1 \\ 1, & x > 1. \end{cases}$$

Then F is monotone increasing, continuous, and satisfies (14) and (15). Therefore, it satisfies all the properties of a continuous cumulative distribution function. If there is a random variable X with cumulative distribution function F, what sets of values would have positive probability? Well, the intervals $(-\infty, 0)$ and $(1, \infty)$ have zero probability since F is constant on each. Similarly, each of the intervals in the complement of the Cantor set in $[0, 1]$ has zero probability since g is constant on each of those intervals. Thus, with probability 1, the value of X must lie in C. This is true even though the probability that X takes on any particular value in C is zero since F is continuous. Furthermore, away from the Cantor set F is differentiable but its derivative is zero, so X does not have a density function f.

This example indicates why more advanced applications of analysis to probability theory usually begin with a thorough study of sets and measures, functions that assign sizes to sets.

Problems

1. Suppose that after a new car is purchased, the number of years until the first major repair is a random variable X with density

$$f(x) = \begin{cases} xe^{-x} & if\ x \geq 0 \\ 0 & if\ x < 0. \end{cases}$$

   (a) Show that $\int_{-\infty}^{\infty} f(x)\, dx = 1$.

   (b) Compute $P\{0 \leq X \leq 2\}$.

   (c) Compute $P\{X > 2\}$.

   (d) Compute $P\{-5 \leq X \leq 2\}$.

2. Suppose that X is a random variable with a continuous density f that has finite variance.

   (a) Prove that X has finite expectation.

   (b) Prove that $\sigma^2 = E(X^2) - E(X)^2$.

3. Find the mean and variance of the random variable in problem 1.

4. Find the mean of the random variable in Example 1 and show that the variance is not finite.

5. (a) Suppose that the departure time of a bus is uniformly distributed between 1p.m. and 1:10p.m. If you arrive at 1:03 p.m., what is the probability that you will have missed the bus?

   (b) A point is chosen at random on the interval $[0, 4]$. What is the probability that it will be within $\frac{1}{8}$ of π? What is the probability that it will be within $\frac{1}{8}$ of an integer?

6. Suppose that X is uniformly distributed on the interval $[c, d]$. Find the mean and standard deviation of X. Draw the graph of the cumulative distribution function of X.

7. Prove that the mean and standard deviation of a random variable that is exponentially distributed with parameter λ are both equal to $1/\lambda$. Compute explicitly the cumulative distribution function and verify that it has the right properties.

8. Let $f_n(x)$ be the density of an exponentially distributed random variable X_n with parameter λ_n. Suppose that $\lambda_n \to \lambda > 0$.

   (a) Prove that $\{f_n\}$ converges uniformly to the density of an exponentially distributed random variable with parameter λ.

   (b) Let X be an exponentially distributed random variable with parameter λ. Prove that for all a and b (finite or infinite),

   $$\lim_{n \to \infty} P\{a \le X_n \le b\} = P\{a \le X \le b\}.$$

9. Let f be the density of a standard normal random variable. Prove that $\int_{-\infty}^{\infty} f(x)dx = 1$ by the following trick:

   $$\left(\int_{-\infty}^{\infty} f(x)\, dx \right)^2 = \left(\int_{-\infty}^{\infty} f(x)\, dx \right) \left(\int_{-\infty}^{\infty} f(y)\, dy \right)$$

   $$= \int_{-\infty}^{\infty} \int_{-\infty}^{\infty} f(x)f(y)\, dx\, dy.$$

   Now use polar coordinates in the plane to do the integral on the right explicitly. Note that this uses the fact that one can compute double integrals by iterating the integrals (see project 4 of Chapter 4), as well as a change of variables formula for multiple integrals.

10. Let Y be normally distributed with parameters μ and σ, and suppose that X is a standard normal random variable. Prove that

   $$P\{a \le Y \le b\} = P\{\frac{a - \mu}{\sigma} \le X \le \frac{b - \mu}{\sigma}\}.$$

   Hint: make a change of variables and use Theorem 4.4.5.

11. Generate the graphs of the cumulative distribution function of the following random variables:

    (a) A Bernoulli random variable with $p = \frac{1}{4}$.
    (b) A binomial random variable with $p = \frac{3}{4}$ and $n = 6$.
    (c) A Poisson random variable with $\lambda = \frac{1}{2}$.

12. Let F be a cumulative distribution function.

    (a) Prove that the set of points where F is discontinuous is countable. Hint: at how many points could the jump of F be $\geq \frac{1}{n}$?
    (b) For each point of discontinuity a_n, let $p_n \equiv F(a_n^+) - F(a_n^-)$, and define $F_d(x) \equiv \sum_{a_n \leq x} p_n$. Show that F_d is monotone increasing and constant except for countably many jumps.
    (c) Define $F_c(x) = F(x) - F_d(x)$. Prove that F_c is continuous and monotone increasing.
    (d) Prove that there is a discrete cumulative distribution function G_d, a continuous cumulative distribution function G_c, and a real number α satisfying $0 \leq \alpha \leq 1$, so that

    $$F(x) = \alpha G_d(x) + (1 - \alpha)G_c(x).$$

13. Let f be the density function for a random variable X on a sample space S. For simplicity we will assume that f is continuous and that $f(x) = 0$ outside of the finite interval $[a, b]$. Divide $[a, b]$ into N subintervals $[x_{i-1}, x_i]$, each of length $\delta = \frac{b-a}{N}$, and let $\bar{x}_i$ denote the midpoint of each interval. Define Y_N to be the random variable on S which takes the value $\bar{x}_i$ on the set $\{s \in S \mid X(s) \in [x_{i-1}, x_i)\}$.

    (a) Prove that Y_N is a discrete random variable with probabilities

    $$P\{Y_N = \bar{x}_i\} = \int_{x_{i-1}}^{x_i} f(x)dx.$$

    (b) Write formulas for the mean, μ_N, and standard deviation, σ_N, of Y_N. Use Corollary 3.3.2 to prove that $\mu_N \to \mu$ and $\sigma_N \to \sigma$ as $N \to \infty$, where μ and σ are the mean and standard deviation of X.
    (c) Prove that for all $c < d$,

    $$P\{c \leq X \leq d\} = \lim_{N \to \infty} P\{c \leq Y_N \leq d\}.$$

14. Let f be a continuous function on $\mathbb{R}^2$. So that we don't have to consider improper integrals, we shall assume that $f(x, y) = 0$ for all (x, y) outside of some square. Let X and Y be random variables on a sample space and suppose that

    $$P\{a \leq X \leq b \text{ and } c \leq Y \leq d\} = \int_c^d \int_a^b f(x, y)\,dx\,dy \qquad (17)$$

for all $a \leq b$ and $c \leq d$. The function f is called the **joint density** of X and Y, respectively.

(a) Explain why the functions f_X and f_Y on $\mathbb{R}$ defined by

$$f_X(x) \equiv \int_{-\infty}^{\infty} f(x, y)\, dy, \qquad f_Y(y) \equiv \int_{-\infty}^{\infty} f(x, y)\, dx,$$

are the densities of X and Y.

(b) For $(x, y) \in \mathbb{R}^2$, define

$$F(x, y) \equiv \int_{-\infty}^{y} \int_{-\infty}^{x} f(s, t)\, ds dt.$$

F is called the **joint cumulative distribution function** of X and Y. Using Theorem 4.2.5, Theorem 5.2.4, and Project 4 [part (e)] of Chapter 4, give a careful justification of the formula

$$\frac{\partial^2}{\partial x \partial y} F(x, y) = f(x, y).$$

(c) X and Y are said to be **independent** if

$$P\{a \leq X \leq b \text{ and } c \leq Y \leq d\} = P\{a \leq X \leq b\}P\{c \leq Y \leq d\}$$

for all $a \leq b$ and $c \leq d$. Prove that X and Y are independent if and only if $f(x, y) = f_X(x)f_Y(y)$.

10.4   The Variation Metric

In many applications of probability theory, it is extremely useful to have a way of measuring when two random variables are "close". For simplicity, we shall restrict our attention to random variables that take values in $\mathbb{N} \cup \{0\}$. To motivate the mathematical development, we begin by explaining heuristically why a binomial random variable is approximately a Poisson random variable if p is small, n is large, and np has moderate size. Throughout we shall denote by $Y(\lambda)$ a Poisson random variable with parameter λ. Recall that $Y(\lambda)$ has mass density

$$P\{Y(\lambda\} = k) = e^{-\lambda}\frac{\lambda^k}{k!}.$$

Suppose that X is a binomial random variable with parameters n and p, and let $\lambda = np$. Then, for $0 \le k \le n$,

$$P\{X = k\} = \frac{n!}{(n-k)!k!}\, p^k(1-p)^{n-k} \tag{18}$$

$$= \frac{n!}{(n-k)!k!}\left(\frac{\lambda}{n}\right)^k\left(1-\frac{\lambda}{n}\right)^{n-k} \tag{19}$$

$$= \frac{n(n-1)\ldots(n-k+1)}{n^k}\,\frac{(1-\lambda/n)^n}{(1-\lambda/n)^k}\,\frac{\lambda^k}{k!}. \tag{20}$$

Now, suppose that p is small and n is large. Then λ/n is small. Suppose that k is small compared to n. Then (see problem 1),

$$\left(1-\frac{\lambda}{n}\right)^n \approx e^{-\lambda} \tag{21}$$

$$\frac{n(n-1)\ldots(n-k+1)}{n^k} \approx 1 \tag{22}$$

$$\left(1-\frac{\lambda}{n}\right)^k \approx 1, \tag{23}$$

where we use $\approx$ to mean "approximately equal to." Substituting (21), (22), and (23) in (20), we obtain

$$P\{X = k\} \approx e^{-\lambda}\frac{\lambda^k}{k!} = P\{Y(\lambda) = k\}.$$

Though we have used the assumption that k is small compared to n, both the binomial and the Poisson probabilities are small for large k since p is small and are thus close to each other. This suggests that for any set $A \subseteq \mathbb{N} \cup \{0\}$,

$$P\{X \in A\} \approx P\{Y(\lambda) \in A\}. \tag{24}$$

We shall see later that this is true and derive a bound for the difference.

The approximation (24) is the reason that the number of occurrences of rare events is often assumed to have a Poisson distribution. Here is an example.

Example 1 Suppose that the average number of earthquakes each year is 2.5. A reasonable, simple assumption would be that the probability of a quake on any particular day is $p = 2.5/365$. Let X_i be the random variable which has the value 1 if there is a quake on the ith day of the year and equals zero otherwise. Then the value of $X \equiv \sum_{i=1}^{365} X_i$ is the number of quakes during the year. If we assume that the X_i are independent and set $p = P\{X_i = 1\}$, then X is a binomial random variable with $n = 365$ and $p = 2.5/365$. Since n is large, p is small, and $\lambda \equiv np = 2.5$ has moderate size, the mass density of X should be well approximated by the mass density of a Poisson random variable with the same mean, that is, by $Y(2.5)$. The probability that there will be no earthquakes in the year is given by X as

$$P\{X = 0\} = \left(\frac{362.5}{365}\right)^{365} = .081$$

and by Y as

$$P\{Y = 0\} = e^{-2.5} = .082 .$$

Similarly, the probability that there will be ≤ 2 quakes is given by X as

$$\left(\frac{362.5}{365}\right)^{365} + 365\left(\frac{362.5}{365}\right)^{364}\left(\frac{2.5}{365}\right) + \frac{(365)(364)}{2!}\left(\frac{362.5}{365}\right)^{363}\left(\frac{2.5}{365}\right)^2$$

$$= .5434$$

and by Y as

$$e^{-2.5} + e^{-2.5}(2.5) + e^{-2.5}\frac{(2.5)^2}{2!} = .5438.$$

In both cases, the approximation by the Poisson is very good. It is clear that the calculations with the Poisson random variable are simpler. What we need is an error bound.

We briefly describe two other situations in which the question of measuring the closeness of two mass densities is both natural and important. Mathematical details can be found in the references.

Example 2   Let's consider what it means to "randomize" the order of a deck of playing cards by repeated shuffling. There are 52! possible orderings of the cards. A shuffle is a well-defined probabilistic procedure which, when applied to any particular ordering, gives each of the 52! orderings with various probabilities (some of which could be zero). For example, a simple (inefficient) shuffle would be to choose a card at random from the deck and reinsert it at a randomly chosen place. Let's assume that we have chosen a specific method of shuffling. Since there are 52! possible orderings of the cards, we can label the orderings by the numbers $1, 2, \ldots, 52!$. Let the value of the random variable X_n be the label after n shuffles. Let U be the random variable which takes on each of the label values $1, 2, \ldots, 52!$ with probability $\frac{1}{52!}$; that is, U takes on each label value with equal probability. A reasonable mathematical interpretation of the question "How random is the deck after n shuffles?" is the question "How close is the mass density of X_n to the mass density of U?" since U assigns equal probabilities to each ordering. If one uses a "riffle shuffle", one can show that the densities are quite close if $n \geq 7$. For an excellent introduction to the mathematics of card shuffling, see [38], where the riffle and other shuffles are formally defined.

Example 3   In Example 5 of Section 5.6, we discussed the use of metrics in molecular biology. Even when we have determined how close two DNA sequences are, difficult questions in probability theory are involved in the interpretation of the results. Suppose that in two long sequences we find two short subsequences that match perfectly. How likely is this event if the letters in each sequence are chosen randomly (with or without independence in neighboring positions) from the DNA alphabet A, G, C, T. In order to make calculations, one must typically argue that, under appropriate assumptions, the distribution of the random variable of interest (in this case the length of the longest identical substrings) is approximately given by one of the standard, simple distributions of probability theory, for example, the Poisson distribution or the binomial distribution. Thus one wants to know when two discrete probability distributions are close. See [30] for an excellent discussion of the applications of metrics, combinatorics, and probability theory to molecular biology.

We are now ready to define the distance between random variables.

Definition.   Let X and Y be two random variables which take values in $\mathbb{N} \cup \{0\}$. Define

$$\rho_v(X, Y) \equiv \sup_{A \subseteq \mathbb{N}} |P\{X \in A\} - P\{Y \in A\}| \qquad (25)$$

$\rho_v(X, Y)$ is called the **variation** of X and Y.

This is a natural way to measure distance since if $\rho_v(X, Y)$ is small then the probabilities that the values of X and Y will be in any particular set A will be close to each other. Note, however, that $\rho_v(X, Y)$ does not compare the random variables themselves, only their mass densities. In particular, $\rho_v(X, Y) = 0$ if X and Y have the same mass densities. So X and Y need not even be random variables on the same sample space. These remarks are made explicit by the following theorem, which shows how to compute $\rho_v(X, Y)$ in terms of the mass density of X and the mass density of Y.

❑ **Theorem 10.4.1**   Let X and Y be random variables with values in $\mathbb{N} \cup \{0\}$ that have mass densities $p_n = P\{X = n\}$ and $q_n = P\{Y = n\}$, respectively. Then,

$$\rho_v(X, Y) \;=\; \frac{1}{2} \sum_{n=0}^{\infty} |p_n - q_n|. \tag{26}$$

Furthermore, for all such random variables X, Y, and Z,

(a) $\rho_v(X, Y) \geq 0$.

(b) $\rho_v(X, Y) = \rho_v(Y, X)$.

(c) $\rho_v(X, Y) \leq \rho_v(X, Z) + \rho_v(Z, Y)$.

Proof.   We shall prove (26), from which the properties (a), (b), and (c) follow quite easily (problem 2). Let $S^+ \equiv \{n \in \mathbb{N} \cup \{0\} \,|\, p_n \geq q_n\}$ and $S^- \equiv \{n \in \mathbb{N} \cup \{0\} \,|\, p_n < q_n\}$. For any $A \subseteq \mathbb{N} \cup \{0\}$,

$$P\{X \in A\} - P\{Y \in A\} \;=\; \sum_{n \in A} p_n - \sum_{n \in A} q_n \tag{27}$$

$$=\; \sum_{n \in A} (p_n - q_n) \tag{28}$$

$$=\; \sum_{n \in A \cap S^+} (p_n - q_n) \;+\; \sum_{n \in A \cap S^-} (p_n - q_n). \tag{29}$$

Since the first term in (29) is positive and the second is negative, it follows that

$$\sum_{n \in A \cap S^-} (p_n - q_n) \;\leq\; P\{X \in A\} - P\{Y \in A\} \;\leq\; \sum_{n \in A \cap S^+} (p_n - q_n),$$

which implies

$$\sum_{n \,\epsilon\, S^-} (p_n - q_n) \;\leq\; P\{X \,\epsilon\, A\} - P\{Y \,\epsilon\, A\} \;\leq\; \sum_{n \,\epsilon\, S^+} (p_n - q_n).$$

Subtracting $\sum_{\mathbb{N}} q_n = 1$ from $\sum_{\mathbb{N}} p_n = 1$, we find that

$$
\begin{aligned}
0 \;=\; \sum_{\mathbb{N} \cup \{0\}} (p_n - q_n) \;&=\; \sum_{S^+}(p_n - q_n) + \sum_{S^-}(p_n - q_n) \\
&=\; \sum_{S^+}(p_n - q_n) - \sum_{S^-}|p_n - q_n|,
\end{aligned}
$$

so

$$\sum_{S^+}(p_n - q_n) \;=\; \sum_{S^-}|p_n - q_n|. \tag{30}$$

It follows that

$$-\sum_{n \,\epsilon\, S^+}(p_n - q_n) \;\leq\; P\{X \,\epsilon\, A\} - P\{Y \,\epsilon\, A\} \;\leq\; \sum_{n \,\epsilon\, S^+}(p_n - q_n),$$

so

$$|P\{X \,\epsilon\, A\} - P\{Y \,\epsilon\, A\}| \;\leq\; \sum_{n \,\epsilon\, S^+}(p_n - q_n).$$

Since we can always choose $A = S^+$, this implies that

$$\sup_{A \subseteq \mathbb{N} \cup \{0\}} |P(X \,\epsilon\, A) - P(Y \,\epsilon\, A)| \;=\; \sum_{n \,\epsilon\, S^+}(p_n - q_n).$$

It follows from (30) that

$$\sum_{n \,\epsilon\, S^+}(p_n - q_n) \;=\; \frac{1}{2} \sum_{n \,\epsilon\, \mathbb{N} \cup \{0\}}|p_n - q_n|,$$

which proves (26).   ❏

Note that (a), (b), and (c) are just the properties of a metric except that $\rho(X,Y) = 0$ does not imply that $X = Y$. In fact, $\rho_v(\cdot, \cdot)$ *is* a metric on the set of sequences $\{p_n\}_{n=0}^{\infty}$ of nonnegative numbers such that $\sum p_n = 1$. That is, ρ_v is a metric on the set of mass densities, We follow common practice and write $\rho_v(X,Y)$ even though ρ_v depends only on the mass densities of X and Y, not on the random variables themselves. We now prove a theorem that relates the properties of $\rho_v(X,Y)$ to probabilistic properties of X and Y.

❏ **Theorem 10.4.2** Let X and Y be random variables with values in $\mathbb{N} \cup \{0\}$. Then,

    (a) Then $\rho_v(X, Y) \leq P\{X \neq Y\}$.

    (b) If Z is a random variable with values in $\mathbb{N} \cup \{0\}$ which is independent of X and Y, then

$$\rho_v(X + Z, Y + Z) \leq \rho_v(X, Y).$$

    (c) Let $\{X_i\}_{i=1}^N$ and $\{Y_i\}_{i=1}^N$ be families of mutually independent random variables that take values in $\mathbb{N} \cup \{0\}$. Then,

$$\rho_v\left(\sum X_i, \sum Y_i\right) \leq \sum_{i=1}^N \rho_v(X_i, Y_i).$$

Proof. To prove (a), we note that for every $A \subseteq \mathbb{N} \cup \{0\}$,

$$
\begin{aligned}
P\{X \in A\} \;&=\; P\{X \in A \text{ and } Y \in A\} + P\{X \in A \text{ and } Y \in \tilde{A}\} \\
&\leq\; P\{Y \in A\} + P\{X \neq Y\},
\end{aligned}
$$

so $P\{X \in A\} - P\{Y \in A\} \leq P\{X \neq Y\}$. Reversing the roles of X and Y gives $P\{Y \in A\} - P\{X \in A\} \leq P\{X \neq Y\}$, which proves (a).

Recall that two random variables X and Y are independent if for all sets $A \subseteq \mathbb{R}$ and $B \subseteq \mathbb{R}$,

$$P\{X \in A \text{ and } Y \in B\} \;=\; P\{X \in A\}P\{Y \in B\}.$$

For $A \subseteq \mathbb{N} \cup \{0\}$, define $A - \{n\} \equiv \{m \in \mathbb{N} \cup \{0\} \mid m = j - n \text{ for some } j \in A\}$. Then, to prove (b), we compute

$$P\{X + Z \in A\} \;=\; \sum_{n=0}^{\infty} {}^{\prime}\, P\{X \in A - \{n\} \text{ and } Z = n\} \tag{31}$$

$$=\; \sum_{n=0}^{\infty} P\{X \in A - \{n\}\}P\{Z = n\} \tag{32}$$

$$\leq\; \sum_{n=0}^{\infty} [P\{Y \in A - \{n\}\} + \rho_v(X, Y)]P\{Z = n\} \tag{33}$$

$$=\; \sum_{n=0}^{\infty} P\{Y \in A - \{n\}\}P\{Z = n\} + \rho_v(X, Y) \tag{34}$$

$$=\; \sum_{n=0}^{\infty} P\{Y \in A - \{n\} \text{ and } Z = n\} + \rho_v(X, Y) \tag{35}$$

$$=\; P\{Y + Z \in A\} + \rho_v(X, Y). \tag{36}$$

In step (32) we used the hypothesis that Z is independent of X and in step (35) that Z is independent of Y. In step (34) we used the fact that $\sum_{n=0}^{\infty} P\{Z = n\} = 1$. Combining this inequality with the inequality obtained by reversing the roles of X and Y, we obtain

$$|P\{X + Z \in A\} - P\{Y + Z \in A\}| \le \rho_v(X, Y),$$

from which (b) follows.

To prove (c), let $N = 2$. Then, by the triangle inequality and the result of part (b),

$$\begin{aligned}
\rho_v(X_1 + X_2, Y_1 + Y_2) &\le \rho_v(X_1 + X_2, X_2 + Y_1) + \rho_v(X_2 + Y_1, Y_1 + Y_2) \\
&\le \rho_v(X_1, Y_1) + \rho_v(X_2, Y_2).
\end{aligned}$$

Continuing in this manner, we see that part (c) follows by an induction argument (problem 8). ❑

The proofs of the next three results all depend on the following two ideas. First, since $\rho_v(X, Y)$ depends only on the mass densities of X and Y, its value does not change if Y is replaced by another random variable $\widetilde{Y}$ with the same density. Second, suppose that Y is a Poisson random variable with mean λ. Then, for every choice of $\mu_i > 0$ so that $\mu_1 + \mu_2 + \ldots + \mu_n = \lambda$, there is a sample space S and mutually independent Poisson random variables $Y(\mu_i)$ so that $\widetilde{Y} \equiv \sum Y(\mu_i)$ is a Poisson random variable with mean λ. Thus $\rho_v(X, Y) = \rho_v(X, \widetilde{Y})$. See problems 12, 13, and 14.

Corollary 10.4.3 Let $Y(\mu_1)$ and $Y(\mu_2)$ be Poisson random variables with means $\mu_1 \le \mu_2$. Then,

$$\rho_v(Y(\mu_1), Y(\mu_2)) \le \mu_2 - \mu_1.$$

Proof. Let $\widetilde{Y}(\mu_1)$ and $\widetilde{Y}(\mu_2 - \mu_1)$ be independent Poisson random variables with means μ_1 and $\mu_2 - \mu_1$ and define $\widetilde{Y}(\mu_2) \equiv \widetilde{Y}(\mu_1) + \widetilde{Y}(\mu_2 - \mu_1)$. Then $\widetilde{Y}(\mu_2)$ is Poisson with mean μ_2. Therefore,

$$\begin{aligned}
\rho_v(Y(\mu_1), Y(\mu_2)) &= \rho_v(\widetilde{Y}(\mu_1), \widetilde{Y}(\mu_2)) \\
&= \rho_v(\widetilde{Y}(\mu_1), \widetilde{Y}(\mu_1) + \widetilde{Y}(\mu_2 - \mu_1)) \\
&\le P\{\widetilde{Y}(\mu_1) \ne \widetilde{Y}(\mu_1) + \widetilde{Y}(\mu_2 - \mu_1)\} \\
&= P\{\widetilde{Y}(\mu_2 - \mu_1) \ne 0\}
\end{aligned}$$

$$= \ 1 - e^{-(\mu_2 - \mu_1)}$$

$$\leq \ \mu_2 - \mu_1.$$

In the third step we used part (a) of Theorem 10.4.2. The last step follows from the Mean Value Theorem.   ❑

Thus we have an upper bound for how close the mass densities of two Poisson random variables are if the means are close.

Example 4   Suppose that a geologist is asked to predict how likely it is that there will be ≤ 2 earthquakes in a certain region next year. She assumes that the number of earthquakes is given by a Poisson random variable. However, she needs to determine the right mean to use, and she does so by a statistical analysis of historical data. In such situations it is not always clear how much of the historical data to use since older data may be more unreliable. If she uses data since 1950 she gets a mean of 2.5 (as in Example 1), and if she uses data back to 1900 she gets a mean of 2.4. How much difference does it make which mean she uses? According to Corollary 10.4.3,

$$|P\{Y(2.5) \ \epsilon \ A\} \ - \ P\{Y(2.4) \ \epsilon \ A\}| \ \leq \ 2.5 \ - \ 2.4 \ = \ .1$$

for any set A. If $A = \{0, 1, 2\}$, then we calculated in Example 1 that

$$P\{Y(2.5) \ \epsilon \ \{0, 1, 2\}\} \ = \ e^{-2.5} + (2.5)e^{-2.5} + \frac{(2.5)^2}{2!}e^{-2.5} \ = \ .5438,$$

and similarly,

$$P\{Y(2.4) \ \epsilon \ \{0, 1, 2\}\} \ = \ e^{-2.4} + (2.4)e^{-2.4} + \frac{(2.4)^2}{2!}e^{-2.4} \ = \ .5697.$$

Thus we see that the two predictions indeed differ by less than 0.1.

❑ **Theorem 10.4.4 (Le Cam's Inequality)**   Let $X_1, X_2, ..., X_n$ be independent Bernoulli random variables with probabilities $p_1, p_2, ..., p_n$. Suppose that $Y(\sum_{i=1}^n p_i)$ is a Poisson random variable with mean $\sum_{i=1}^n p_i$. Then,

$$\rho_v(\sum_{i=1}^n X_i, Y(\sum_{i=1}^n p_i)) \ \leq \ \sum_{i=1}^n p_i^2. \tag{37}$$

Proof. We first prove the result in the case $n = 1$. Let X be a Bernoulli random variable with $p = P\{X = 1\}$. Then,

$$
\begin{aligned}
2\rho_v(X, Y(p)) &= \sum_{n=0}^{\infty} |P\{X = n\} - P\{Y(p) = n\}| \\
&= \sum_{n=0}^{1} |P\{X = n\} - P\{Y(p) = n\}| + \sum_{n=2}^{\infty} |P\{Y(p) = n\}| \\
&= |(1 - p) - e^{-p}| + |p - pe^{-p}| + \sum_{n=2}^{\infty} |P\{Y(p) = n\}| \\
&= (e^{-p} - pe^{-p} + 2p - 1) + (1 - e^{-p} - pe^{-p}) \\
&= 2p(1 - e^{-p}).
\end{aligned}
$$

Since $(1 - e^{-p}) \leq p$ by the Mean Value Theorem, we conclude that

$$
\rho_v(X, Y(p)) \;\leq\; p^2. \tag{38}
$$

For the general case, we replace $Y\left(\sum_{i=1}^n p_i\right)$ by the sum $\sum_{i=1}^n \tilde{Y}(p_i)$ of independent Poisson random variables $\tilde{Y}(p_i)$ with corresponding means p_i. And we replace $\tilde{X}_1, \tilde{X}_2, ..., \tilde{X}_n$, by independent Bernoulli random variables with probabilities $p_1, p_2, ..., p_n$, so that the families $\{\tilde{X}_i\}$ and $\{\tilde{Y}_i\}$ are mutually independent. This can be done by a product construction similar to that described in problem 13. Thus, by part (c) of Theorem 10.4.2 and (38),

$$
\begin{aligned}
\rho_v\left(\sum_{i=1}^n X_i, Y\left(\sum_{i=1}^n p_i\right)\right) &= \rho_v\left(\sum_{i=1}^n \tilde{X}_i, \sum_{i=1}^n \tilde{Y}(p_i)\right) \\
&\leq \sum_{i=1}^n \rho_v(\tilde{X}_i, \tilde{Y}_i(p_i)) \\
&\leq \sum_{i=1}^n p_i^2. \qquad \square
\end{aligned}
$$

Corollary 10.4.5   Let $\{X_i\}_{i=1}^n$ be a sequence of independent Bernoulli random variables each with mean p, and let $X = \sum_{i=1}^n X_i$. Let $Y(np)$ be a Poisson random variable with mean np. Then,

$$
\rho_v(X, Y(np)) \;\leq\; np^2.
$$

Proof. This is the special case of (37) when $p_i = p$ for all i. $\square$

Example 1 (revisited)   In Example 1, $p = \frac{2.5}{365}$ and

$$np^2 = 365 \left(\frac{2.5}{365}\right)^2 = .017.$$

Thus, whatever the set $A \subseteq \mathbb{N} \cup \{0\}$, we should have

$$|P\{X \in A\} - P\{Y(np) \in A\}| \leq .017.$$

This was indeed true for the two cases, $A_1 = \{0\}$ and $A_2 = \{0, 1, 2\}$, that we compared explicitly in Example 1.

Problems

1. For fixed k, show that the expressions on the left sides of (21), (22), and (23) converge to the respective right sides as $n \to \infty$.

2. Complete the proof of Theorem 10.4.1 by verifying that ρ_v satisfies the properties (a), (b), and (c).

3. Suppose that on the average 1 out of every 75 items coming off an assembly line is defective. Assume that each item coming off has probability $\frac{1}{75}$ of being defective. Use a binomial random variable and a Poisson random variable to calculate the probability that the next batch of 75 items will have

   (a) no defective items.

   (b) two or more defective items.

4. Every day during January you buy a lottery ticket. Each ticket has a chance of winning equal to $\frac{1}{100}$. Use a binomial random variable and a Poisson random variable to calculate the probability that you will win

   (a) exactly once.

   (b) more than once.

   (c) not at all.

5. Assume that each of the four letters in the DNA alphabet occurs independently in each position with probability $\frac{1}{4}$. Suppose that we have a DNA strand of length 8. Use a binomial random variable and a Poisson random variable to calculate the probability that the strand contains

   (a) no C's.

   (b) exactly two C's.

   (c) less than or equal to two C's.

6. Let X be a random variable which takes the values $0, 1$, and 2 with probabilities $p_0 = \frac{1}{4}$, $p_1 = \frac{1}{2}$, and $p_2 = \frac{1}{4}$, and let Y be a random variable which takes the values $0, 1, 2$, and 3, each with probability $\frac{1}{4}$. Compute $\rho_v(X, Y)$.

7. For each integer $n \geq 0$, let X_n be a random variable which takes the values $0, 1, 2, \ldots, n$, each with probability $\frac{1}{n+1}$.

   (a) Compute $\rho_v(X_n, X_m)$.

   (b) Compute $\lim_{n \to \infty} \rho_v(X_n, X_m)$.

8. Provide the details of the induction argument for part (c) of Theorem 10.4.2.

9. Compute the error bound for the difference between the binomial and the Poisson random variables in the situation described in problem 3.

10. Compute the error bound for the difference between the binomial and the Poisson random variables in the situation described in problem 4.

11. Compute the error bound for the difference between the binomial and the Poisson random variables in the situation described in problem 5. What is the point of this problem?

12. Let $Y(\mu_1)$ and $Y(\mu_2)$ be independent Poisson random variables with means μ_1 and μ_2. Prove that $Y(\mu_1) + Y(\mu_2)$ is a Poisson random variable with mean $\mu_1 + \mu_2$. Hint: since $Y(\mu_1)$ and $Y(\mu_2)$ are independent,

$$P\{Y(\mu_1) + Y(\mu_2) = n\} = \sum_{k=0}^{n} P\{Y(\mu_1) = k \text{ and } Y(\mu_2) = n - k\}$$

$$= \sum_{k=0}^{n} P\{Y(\mu_1) = k\} P\{Y(\mu_2) = n - k\}.$$

13. Suppose that λ, μ_1, and μ_2 are positive numbers satisfying $\lambda = \mu_1 + \mu_2$. Let $S \equiv \mathbb{N} \cup \{0\} \times \mathbb{N} \cup \{0\}$, and define random variables X_1 and X_2 on S by

$$X_1(m, n) = m, \qquad X_2(m, n) = n.$$

Assume that for each m and n,

$$P\{X_1 = m \text{ and } X_2 = n\} = e^{-\mu_1} \frac{\mu_1^m}{m!} e^{-\mu_2} \frac{\mu_2^n}{n!}.$$

   (a) Prove that X_1 is a Poisson random variable with mean μ_1 and X_2 is a Poisson random variable with mean μ_2.

   (b) Prove that X_1 and X_2 are independent.

   (c) Prove that $X_1 + X_2$ is a Poisson random variable with mean $\lambda = \mu_1 + \mu_2$.

14. Let $Y(\lambda)$ be a Poisson random variable with mean λ. Let $\mu_i > 0$ for $i = 1, 2, \ldots, n$ and suppose that $\lambda = \mu_1 + \mu_2 + \ldots + \mu_n$. Show how to create a sample space S and random variables $X_1, X_2, \ldots, X_n$ so that

    (a) X_i is a Poisson random variable with mean μ_i.

    (b) the random variables X_i are mutually independent.

    (c) the mass density of $X_1 + X_2 + \ldots + X_n$ is the same as the mass density of $Y(\lambda)$.

15. Let $\{X_i\}_{i=1}^n$ be a sequence of independent Bernoulli random variables each with mean $p(n)$, which depends on n. Define $X(n) = \sum X_i$. Suppose that as $n \to \infty$, we have $p(n) \to 0$ and $np(n) \to \beta$. Prove that

$$\rho_v(X(n), Y(\beta)) \;\leq\; np(n)^2 + |\beta - np(n)|.$$

Hint: use $Y(np(n))$ and the triangle inequality.

Projects

1. The purpose of this project is to introduce the variance and develop some of its properties. Let X be a discrete random variable with values $\{a_n\}$ and suppose that X^2 has finite expectation. That is, the series $\sum a_n^2 P\{X = a_n\}$ converges absolutely. Since $|a_n| \leq a_n^2 + 1$, it follows that X has finite expectation. We denote $E(X)$ by μ. Since $(X - \mu)^2 = X^2 - 2\mu X + \mu^2$, it follows from problem 9 in Section 10.1 that $(X - \mu)^2$ has finite expectation too. We define

$$\sigma^2 \equiv E((X - \mu)^2).$$

and call σ^2 the **variance** of X. It is sometimes denoted $\mathrm{Var}(X)$. It's square root, σ, is called the **standard deviation**.

    (a) Suppose that X takes the values 1.5, 2, and 2.5 with probabilities $\frac{1}{4}$, $\frac{1}{2}$, and $\frac{1}{4}$, respectively. Compute μ and σ^2.

    (b) Suppose that X takes the values 0, 2, and 4 with probabilities $\frac{1}{4}$, $\frac{1}{2}$, and $\frac{1}{4}$, respectively. Compute μ and σ^2. What do you conclude about the meaning of σ^2?

    (c) Compute the variance of a Bernoulli random variable with parameter p.

    (d) Show that $\sigma^2 = E(X^2) - E(X)^2$. Hint: expand $(X - \mu)^2$ and use problem 9 in Section 10.1.

    (e) Let X be the sum of the faces in the experiment of rolling two dice. What is σ^2?

(f) Let $X_1, X_2, \ldots, X_n$ be independent random variables each having the same range and probabilities. Suppose $\sigma^2 \equiv \text{Var}(X_i)$ is finite and set $\mu = E(X_i)$. Define $S_n \equiv X_1 + X_2 + \ldots + X_n$. Prove that $\text{Var}(S_n) = n\sigma^2$. Hint: note that $E(S_n) = n\mu$ and expand:

$$(S_n - n\mu)^2 = \left(\sum_{j=1}^{n} (X_j - \mu) \right)^2$$

$$= \sum_{j=1}^{n} (X_j - \mu)^2 + \sum_{i \neq j}^{n} (X_i - \mu)(X_j - \mu);$$

then use problem 6 of Section 10.1 and Theorem 10.1.3.

2. The purpose of this project is to sketch the proofs and uses of Markov's inequality and Chebyshev's inequality. Throughout we assume that X is a discrete random variable which takes values $\{a_n\}$.

(a) Suppose that X has finite expectation. Prove that, for each $t > 0$,

$$P\{|X| \geq t\} \leq \frac{E(|X|)}{t}.$$

This is known as **Markov's inequality**. Hint: by Theorem 10.1.2, $E(|X|) = \sum |a_n| P\{X = a_n\}$; now eliminate the terms in which $|a_n| < t$.

(b) Markov's inequality is useful when we have information only about the mean of a random variable. Suppose that the number of defective toasters produced by a factory each day is a random variable with mean 40. What can you say about the probability that more than 60 defective toasters will be produced on a given day?

(c) Suppose that X has finite variance σ^2. Let $\mu = E(X)$. Prove that for each $\delta > 0$,

$$P\{|X - \mu| \geq \delta\} \leq \frac{\sigma^2}{\delta^2}.$$

This is known as **Chebyshev's inequality**. Hint: notice that $|X - \mu| \geq \delta$ if and only if $(X - \mu)^2/\delta^2 \geq 1$.

(d) Chebyshev's inequality can be used to provide information when we have information only about the mean and variance of a random variable. Suppose that the manager of the toaster factory in part (b) has the additional information that the standard deviation of the number of defective toasters is 4. What can you say about the probability that the number of defective toasters will be between 30 and 50 on a given day?

(e) How would the manager of the toaster factory get this information about the mean and the standard deviation?

(f) Chebyshev's inequality can also be used to provide an easy estimate when one knows the density of X but the density is complicated. Suppose that the manager of an alarm clock factory knows that 3% of the clocks produced are defective. He is negotiating a special order of 10,000 clocks, and to sweeten the deal he's going to agree to give a full refund if more than m clocks are defective. How should he choose m so that the chance of giving a refund is less than 5%? Hint: you need to use the fact that the variance of a binomial random variable with parameters p and n is $np(1 - p)$.

3. The purpose of this project is to introduce the weak law of large numbers. Let X_i be a sequence of independent discrete random variables all having the same density and finite variance. Let $\mu = E(X_i)$ and for each n define $S_n = X_1 + X_2 + \ldots + X_n$. We want to prove that for all $\delta > 0$,

$$\lim_{n \to \infty} P\left\{\left|\frac{S_n}{n} - \mu\right| \geq \delta\right\} = 0. \tag{39}$$

This is known as **the weak law of large numbers**.

(a) Prove that $E(S_n/n) = \mu$.

(b) Let $\sigma^2 = \mathrm{Var}(X_i)$. Prove that $\mathrm{Var}(S_n/n) = \sigma^2/n$. Hint: use part (f) of Project 1.

(c) Use Chebyshev's inequality to prove (39).

(d) Suppose that we flip a fair coin repeatedly and that $X_i = 1$ if the ith flip is a head and $X_i = 0$ if the ith flip is a tail. Explain carefully what the weak law of large numbers means in this case. Why does this make sense?

(e) Suppose that we flip a coin, which comes up heads with probability $p > \frac{1}{2}$, repeatedly. Let $X_i = 1$ if the ith flip is a head and $X_i = 0$ if the ith flip is a tail. Let $S_n = X_1 + X_2 + \ldots + X_n$. Prove that

$$\lim_{n \to \infty} P\left\{\frac{S_n}{n} \leq \frac{1}{2}\right\} = 0.$$

Why does this make sense?

4. In this project we outline an example, due to Charles Peskin, which illustrates why some biological systems are stochastic rather than deterministic. Suppose that it is the job of a group of 100 neurons to transmit a real number p satisfying $0 < p < 1$. One can think of p as the scaled intensity of some variable. Each neuron in our group either fires (produces a 1) or doesn't fire (produces a 0).

The deterministic scheme. The number p has a binary expansion $p = .n_1 n_2 n_3 \ldots$ where each n_i is either 0 or 1. We order the cells from 1 to 100.

The kth cell senses the kth binary digit of p and fires if and only if $n_k = 1$. Thus, the output of our group of 100 neurons that is read at a higher level is a string of 0's and 1's giving the first 100 binary digits of p.

The stochastic scheme. Each neuron fires independently with probability p. The output that is sensed at a higher level is the average output of the group of neurons, that is, the number firing divided by the total number of neurons.

(a) Compare and contrast the accuracy of the two schemes. Hint: for the stochastic scheme, use Chebyshev's inequality.

(b) Compare and contrast the simplicity of the two schemes.

(c) Compare and contrast the stability of the two schemes. Hint: suppose that a particular neuron dies.

(d) In each scheme how difficult would it be to improve accuracy and stability if there were selective pressure to do so?

Bibliography

[1] Bartle, R., and D. Sherbert, *Introduction to Real Analysis,* 2nd ed., John Wiley & Sons, Inc., New York, 1992.

[2] Birkhoff, G., and G.-C. Rota, *Ordinary Differential Equations,* 4th ed., John Wiley & Sons, Inc., New York, 1989.

[3] Boltazzini, U., *The Higher Calculus: A History of Real and Complex Analysis from Euler to Weierstrass,* Springer-Verlag, New York, 1986.

[4] Braun, M., *Differential Equations and Their Applications,* 4th ed., Springer-Verlag, New York, 1993.

[5] Burden, R., and J. Faires, *Numerical Analysis,* 5th ed., PWS-Kent, Boston, 1993.

[6] Churchill, R., and J. Brown, *Complex Variables and Applications,* 4th ed., McGraw-Hill, Inc., New York, 1984.

[7] Coddington, E., and N. Levinson, *Theory of Ordinary Differential Equations,* McGraw-Hill, Inc., New York, 1955.

[8] Devaney, R., *A First Course in Chaotic Dynamical Systems,* Addison-Wesley, Reading, Mass., 1992.

[9] Devaney, R., *An Introduction to Chaotic Dynamical Systems,* 2nd ed., Addison-Wesley, Reading, Mass., 1989.

[10] Edelstein-Keshet, L., *Mathematical Models in Biology,* Random House, New York, 1988.

[11] Edgar, G., *Measure, Topology, and Fractal Geometry,* Springer-Verlag, New York, 1990.

[12] Edwards, C., *Advanced Calculus of Several Variables,* Academic Press, New York, 1973.

[13] Edwards, C., *The Historical Development of the Calculus*, Springer-Verlag, New York, 1979.

[14] Ewing, G., *Calculus of Variations with Applications*, W. W. Norton & Co., Inc., New York, 1969.

[15] Fauvel, J., and J. Gray, *History of Mathematics: A Reader*, Macmillan Education Ltd., London, 1987.

[16] Feller, W., *An Introduction to Probability Theory and Its Applications*, Vols. I, II, John Wiley & Sons, New York, 1970, 1971.

[17] Fourier J., *La Théorie Analytique de Chaleur*, Didot, Paris, 1822.

[18] Gelfand, I., and S. Fomin, *Calculus of Variations*, Prentice Hall, Englewood Cliffs, NJ, 1963.

[19] Goldberg, R., *Methods of Real Analysis*, 2nd ed., John Wiley & Sons, Inc., New York, 1976.

[20] Grattan-Guinness, I., *The Development of the Foundations of Mathematical Analysis from Euler to Riemann*, MIT Press, Cambridge, Mass., 1970.

[21] Hirsch, M., and S. Smale, *Differential Equations, Dynamical Systems, and Linear Algebra*, Academic Press, New York, 1974.

[22] Hochstadt, H., *Integral Equations*, John Wiley & Sons, Inc., New York, 1973.

[23] Hoffman, M., and J. Marsden, *Basic Complex Analysis*, W. H. Freeman and Company, New York, 1987.

[24] Hoffman, M., and J. Marsden, *Elementary Classical Analysis*, W. H. Freeman and Company, New York, 1987.

[25] Hoppensteadt, F., and C. Peskin, *Mathematics in Medicine and the Life Sciences*, Springer-Verlag, New York, 1991.

[26] John, F., *Partial Differential Equations*, 4th ed., John Wiley & Sons, Inc., New York, 1982.

[27] Kincaid, D., and W. Cheney, *Numerical Analysis: Mathematics of Scientific Computing*, Brooks/Cole, Pacific Grove, 1991.

[28] Kline, M. , *Mathematical Thought from Ancient to Modern Times*, Oxford University Press, New York, 1972.

[29] Körner, T., *Fourier Analysis*, Cambridge University Press, Cambridge, England, 1988.

[30] Lander, E., and M. S. Waterman, *Calculating the Secrets of Life*, National Academy Press, Washington, D.C., 1995.

[31] Lawler, G., *Introduction to Stochastic Processes*, Chapman & Hall, 1995.

[32] McEliece, R., *The Theory of Information and Coding*, Addison-Wesley, Reading, Mass., 1977.

[33] Rosen, K., *Elementary Number Theory and Its Applications*, Addison-Wesley, Reading, Mass., 1993.

[34] Ross, K., *Elementary Analysis: The Theory of Calculus*, Springer-Verlag, New York, 1980.

[35] Ross, S., *A First Course in Probability Theory*, 4th ed., MacMillan, New York, 1994.

[36] Rudin, W., *Principles of Mathematical Analysis*, 3rd ed., McGraw-Hill Inc., New York, 1976.

[37] Scharlau, W., and H. Opolka, *From Fermat to Minkowski: Lectures on the Theory of Numbers and Its Historical Development*, Springer-Verlag, New York, 1985.

[38] Snell, J. L., *Topics in Contemporary Probability and Applications*, CRC Press, Boca Raton, Fla., 1995.

[39] Steele, J. , "Le Cam's Inequality and Poisson Approximation," *Mathematical Monthly*, **91**(1994), pp. 116 – 123.

[40] Stillwell, J., *The Geometry of Surfaces*, Springer-Verlag, New York, 1992.

[41] Strauss, W. , *Partial Differential Equations, An Introduction*, John Wiley & Sons, Inc., New York, 1992.

[42] van Lint, J., *Introduction to Coding Theory*, Springer-Verlag, New York, 1982.

Symbol Index

$\mathbb{R}$	real numbers	1
$\mathbb{R}^2$	Euclidean plane	7, 40
$\mathbb{R}^n$	n dimensional Euclidean space	157, 211
R	radius of convergence	245
$Ran(f)$	range of f	8
S^c	complement of S	7
$\backslash S$	complement of S	7
sup	supremum	53, 81
$T^{(n)}(x, x_o)$	n^{th} Taylor polynomial	135
$U_P(f)$	upper sum	88
$\mathbb{Z}$	integers	7
$\mathbb{Z}_2$	integers modulo 2	5
ϵ	is contained in (a set)	6
$\notin$	is not contained in (a set)	7
$\pi(n)$	number of primes $\leq n$	265
ρ	metric	196
ρ_2	Euclidean metric on $\mathbb{R}^n$	196
ρ_v	variational metric	389
$\emptyset$	empty set	7
$\int_a^b f(x)dx$	Riemann integral	89
$\int_C f(z)dz$	integral on a contour C in $\mathbb{C}$	308
$\leq, <, >, \geq$	order relations	2
$\neq$	not equal to	
$\cup$	union	7
$\cap$	intersection	7
$\equiv$	is equivalent to	3
$\xrightarrow{f}$	function f	8
$\rightarrow$	converges to (numbers)	29
$\rightarrow$	converges to (functions)	168
$\times$	Cartesian product	7
$|\cdot|$	absolute value	3, 254
$\|\cdot\|$	norm	212
$\|\cdot\|_p$	L^p norm	179
$\|\cdot\|_\infty$	sup norm	175
$\sim$	Fourier series of a function	333
$\approx$	approximately equal to	134
$\blacksquare$	beginning of theorem, end of proof	3

Index